APPLIED SCIENCE IN THE
CASTING OF METALS

APPLIED SCIENCE IN THE CASTING OF METALS

Edited by

K. STRAUSS
Foseco International Ltd., Birmingham

With many thanks for your contribution

K. Strauss.

PERGAMON PRESS

Oxford · New York · Toronto
Sydney · Braunschweig

Pergamon Press Ltd., Headington Hill Hall, Oxford
Pergamon Press Inc., Maxwell House, Fairview Park, Elmsford, New York 10523
Pergamon of Canada Ltd., 207 Queen's Quay West, Toronto 1
Pergamon Press (Aust.) Pty. Ltd., 19a Boundary Street, Rushcutters Bay,
N.S.W. 2011, Australia
Vieweg & Sohn GmbH, Burgplatz 1, Braunschweig

First edition 1970

Library of Congress Catalog Card No. 78-104787

PRINTED IN GREAT BRITAIN BY NEILL & CO. LTD., EDINBURGH

08 015711 4

CONTENTS

v

INTRODUCTION

"APPLIED science in the casting of metals" is a comprehensive title implying coverage of all the processes applicable to metallurgical operations. However, this book is limited mainly to those processes in the development of which this company has played a significant and pioneering part and a few explanatory words as to how this title has been selected may be desirable.

The Foseco organization has been engaged in the research and development of chemical products for the metallurgical industry since 1932.

A great deal of the fundamental work of investigation on which these products are based has been carried out in the Foseco laboratories and the resulting materials are available and widely used throughout the world. Most of the research work which preceded these products has never been published in books or technical journals.

The majority of this research work concerned the chemical treatment of molten metals and alloys such as the oxidizing flux degassing technique for alloys of copper and nickel, the removal of hydrogen from aluminium and its alloys with hexa-chloro-ethane and other degassing agents and the grain refinement of aluminium and magnesium alloys.

Since 1945, however, the basis of our activities has broadened and considerable time and effort has been devoted to the development and practical application of exothermic feeding aids for castings and ingots. Coatings for sand moulds and cores, as well as gravity and pressure dies, for all metals, have also been developed alongside sand additives for overcoming the incidence of sand defects in castings.

Exothermic feeding was in its infancy in 1945 and the role played by fluorides and other compounds in the heat generating reaction was not realized until much research work in our laboratories clarified the chemistry of this rather involved process. An extract from the report dealing with this investigation is published here for the first time.

Considerable work has also been done on the inoculation of grey cast irons and the production of high tensile and nodular graphite cast irons. In particular, progress has been made in solving the problem of fading by applying inoculation directly to the metal in the mould cavity.

These points will explain the choice of title and why, for instance, "Chemistry Applied to Metal Castings" would have been too restrictive.

Some information is included on the continuous casting of steel, vacuum degassing of steel and the spray process of steel manufacture. This is not because we claim any significant contribution to these developments but in order to give an up to date survey of the technology now applied in the steel industry. The short introductory chapter on steel-making by Professor W. Austin serves a similar purpose.

We have tried to avoid the use of proprietary names of products, but in some cases this was not possible, especially in the chapters dealing with exothermic and insulating materials. The reason for this is, of course, that most of the tests which we are now publishing were carried out with our own products and results with similar competitive compounds are very rarely available. The results are typical for these products only and that is why we use the names under which they are known in the trade.

As apart from one or two exceptions, each chapter has been written by a different author, there are obviously variations in style and presentation and also slight overlapping in the contents of some chapters (especially in those sections dealing with the feeding of castings). It was felt preferable to print each author's contribution intact, even if this meant a certain amount of repetition, thus avoiding constant references to previous or foregoing chapters.

All the chapters with the exception of that on corebinders, have been written by members of our organization, who have considerable knowledge in their own field. The chapter on corebinders was contributed by Messrs. E. Parkes, G. Westwood and R. Grigg of Fordath Ltd., a company with which the Foseco organization has very close ties.

Our thanks are due to all the authors who have contributed one or more chapters to this book. Mr. J. L. Francis has not only written several of the chapters, he has greatly assisted in the critical reading and editing of most of the manuscripts.

Mr. M. L. A. Kanssen and Mrs. L. Bourn of our Publicity Department have drawn or redrawn many of the diagrams and sketches. Mrs. R. Bassett of Foseco (F.S.) Ltd. and particularly Mrs. B. M. Muldoon have assisted in proof reading and the latter has also helped in compiling the index. To all of these our grateful thanks.

A. STEEL

CHAPTER 1

MANUFACTURE OF IRON AND STEEL

G. WESLEY AUSTIN

THE TREATMENT OF LIQUID IRON AND STEEL

The iron and steel industry is a traditional one with a history dating back from *ca.* 1500 B.C.; it developed rapidly from 1856, in which year Bessemer announced his pneumatic steel-making process based on molten pig iron.

Historically a malleable iron or steel was produced in the mushy state direct from iron ore and charcoal in a small shallow furnace. The product was hammered to consolidate the grains of iron or steel and to force out the slag. Such a process can be regarded as a direct process. Later a pig iron containing up to 6% of elements other than iron was smelted in a taller furnace, provided with an air blast. These furnaces were the forerunners of the modern blast furnaces. The product from these furnaces was used for iron castings, or made into wrought iron by a puddling process and used as such, or the bars were carburized and then melted in a crucible, to an homogeneous crucible steel.

After 1856 the pig iron was converted into steel in a Bessemer, or later a Thomas converter in which an air blast was blown through the molten iron to remove the unwanted elements. The invention and use of regenerative gas-fired furnaces followed, also that of electrically heated furnaces. These enabled pig iron and scrap in various proportions to be melted and worked into steel. The main production of steel, throughout the world, thus became mainly an indirect process via pig iron; this route remains the standard route today.

Iron constitutes about 5% of the accessible part of the earth's crust. Geological processes have resulted in concentration of the iron-forming ore deposits of various richnesses up to 70% iron. Much lower concentrations, down to 25%, have been and are still worked. The present world trend is to use rich deposits, transporting them in very large ships thousands of miles across oceans to the countries where they are smelted.

The ores are mainly oxides of iron and are acid in nature. A reducing agent is necessary to remove the oxygen and it is usually coke. Because the gangue is mainly acid and refractory, a basic flux, usually lime is used in smelting, forming with the acid gangue a complex fusible silicate, the slag.

The location and size of steel-producing plants in a country is altering. Formerly small smelters were built on or near the ore fields, or the supplies of coking coal, or the industrial centres. Now new plants are built mostly on the coast with port facilities for very large ships. The size of the individual plants is growing rapidly. There are now over thirty plants in the free world each with a production exceeding two million tons of steel per year.

The figures for the production of iron and steel are very large. They are of a different order from those of all other metals. They give in general an indication for each country, of the economic power of that country. Figures for the steel production of selected countries are given in Table 1. It can be seen from the table that the world production of steel is still increasing. The rise is more rapid in those countries which are not yet fully industrialized. The manpower employed in the industry represents a not unimportant proportion of the working population, being in 1966, 575,547 in the U.S.A., 312,513 in Japan, and 284,640 in the U.K.

TABLE 1. *Crude Steel Production—the World and Selected Countries*[1]

Country (000 tons)	1955	1960	1965	1966
Total C.E.C.A.	52,777	73,076	85,991	85,105
United Kingdom	20,017	24,694	27,483	24,704
United States	108,647	91,920	122,490	124,700
Japan	9,408	22,138	41,161	47,769
U.S.S.R.	45,271	65,292	91,000	96,900
World	270,000	333,500	446,100	459,500

Percentage share of some countries in world steel production:

	1955	1960	1965	1966
C.E.C.A.	19·5	22·1	19·3	18·6
United Kingdom	7·6	7·5	6·2	5·4
United States	40·2	27·8	27·4	27·1
Japan	3·5	6·7	9·2	10·4
U.S.S.R.	16·8	19·8	20·4	21·1
World	100·0	100·0	100·0	100·0

Molten pig iron is produced in the blast furnace. The furnace outline is shown diagrammatically in Fig. 1. The principal reaction, temperature and streams of reagents and gases are also shown. It is a strong vertical refractory-lined, almost cylindrical furnace. Ore, coke and flux are fed in at the top through an arrangement of gastight doors. The air blast is heated in stoves up to 1200°C, and may be enriched with oxygen. Hydrocarbons may be injected at a level above the air inlets.

The hot blast on entering the furnace burns the coke, also heated in its descent against the upward streaming hot gases, in a narrow zone to carbon dioxide: $C + O_2 \rightarrow CO_2$. This gas is almost immediately reduced by the excess incandescent coke present to carbon monoxide: $CO_2 + C \rightarrow 2CO$, the usual gas producer reaction. The lower iron oxides are reduced by carbon, by direct reduction: $FeO + C \rightarrow Fe + CO$. The oxides of manganese, phosphorus and silicon more stable than those of iron are also reduced by direct reduction by carbon: the proportions reduced are about two-thirds of the manganese, all the phosphorus, and the silicon depending on the temperature and the acidity of the slag. The sulphur in the iron depends largely on the total sulphur content of the charge, the basicity of the slag and the temperature.

The hot gases, mainly carbon monoxide and nitrogen, pass upwards, carrying out as far as carbon monoxide is concerned in counter-current, and at a temperature which is falling as the gases ascend, the partial reduction of the ore by indirect reduction: $Fe_2O_3 + CO \rightarrow 2FeO + CO_2$, and also: $Fe_2O_3 + 3CO \rightarrow Fe_2 + 3CO_2$. Overall the picture is that of the ore being gradually reduced

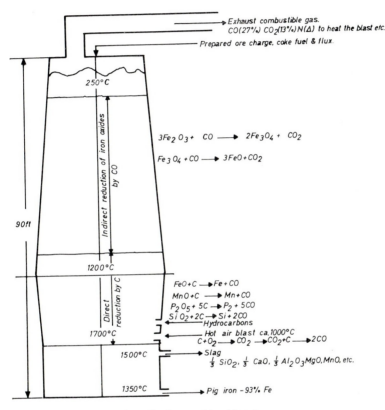

FIG. 1. The iron-making blast furnace.

with rising temperature in passing from the top of the furnace to the hearth. The coke consumption is reduced as the proportion of the indirect reduction increases: i.e. as the proportion of carbon dioxide in the waste gas increases.

The ore is prepared by crushing, sizing and sintering, to give a feed of uniform size and increased reducibility. Using rich ores, fully prepared, hydrocarbon injection, high-blast temperatures and very large furnaces (5000 tons/day) a coke consumption of half a ton of coke, per ton of pig iron made, may be obtained. The average good practice coke consumption is some 25% above this figure.

The slag composition is very roughly one-third silica, one-third lime and one-third alumina and magnesia. The more basic the slag, i.e. the higher the lime, the more firmly it holds sulphur and so the lower the sulphur in the iron. But this higher melting point slag necessitates a higher coke consumption, so the sulphur of the iron is often allowed to rise to one-quarter of 1 % and is removed from the iron by a special process outside the furnaces. The slag is used for many purposes, road-making, cement, slag wool, granulated fillers, etc.

The irons may be classified summarily into non-phosphoric (hematite) in which the phosphorus content is below 0·06%, and phosphoric, in which the phosphorus is higher, often considerably so, being up to 2%. The No. 1, or highest foundry grade grey iron, contains about 3% silicon and the carbon is mainly graphitic and it is comparatively soft: at $1\frac{1}{2}\%$ of silicon (the old forge iron), the carbon is about equally graphitic and combined: below 1 % silicon the carbon is mainly combined and the fracture is white. The latter iron is that used in the main for the basic steel-making process.

The molten iron may be cast into pigs in sand beds or in casting machines, or transported molten to the steelworks.

There are processes in use and under development using gaseous and oil fuels for the direct production of sponge or pelletized iron, but these are based mainly on local comparatively inexpensive supplies of gas and a suitable ore. Some extension of these plants and processes is likely.

Scrap iron may be regarded as a very rich iron ore and may be added to the process at any stage of the route from the blast furnace to the steel furnace. It may also be melted and carburized to a first-class pig in a hot-blast cupola and then converted to steel by almost any of the steel-making processes, but perhaps most profitably by an oxygen-blown process.

TREATMENT OF MOLTEN IRON BETWEEN BLAST FURNACE AND STEELWORKS

The molten iron for the steelworks is generally transported in refractory-lined insulated steel ladles. The capacity of these varies from 50 tons to several hundred tons. The very large ladles are usually torpedo shaped. The iron may be so conveyed over long distances sometimes over the state railway systems. The ladles of molten metal act as buffer stocks, or they may be emptied into a large refractory-lined storage vessel, a mixer, often fired, of up to 2000 tons capacity.

Some desulphurization is usually a necessity in the case of irons made under conditions of acid smelting and this may be carried out to some extent in the transport by special ladles before the iron is poured into the mixer. The formation of sulphides, insoluble in the molten iron, those of sodium, calcium, magnesium and manganese, is the principle used. The commonest reagent added to the ladle is soda, or soda + lime. This process, the Brassert process, is more effective under reducing conditions as is the case with pig iron, $3\frac{1}{2}\%$ carbon, but not always so with steel. The main reaction is as follows: $FeS + Na_2O + C \rightarrow Fe + Na_2S + CO$. Dispersion, agitation with intimate contact, is necessary for maximum effect. If soda ash is employed the CO_2 evolved increases the agitation. Blast-furnace slag should be excluded. Desulphurization below 0·04% sulphur is difficult. The process is often carried out in two stages for better desulphurization. It may also be carried out as a continuous process. Incidental difficulties may arise with fuming and by serious attack on refractories.

Lime and soda (9:1) may be injected into molten iron with a carrier gas. Excellent results have been obtained with this technique. Other reagents, notably calcium carbide or calcium silicate may be used. (Note the presence of the insoluble sulphide formed + the deoxidant.) Magnesium is used for the desulphurization of iron in the production of nodular iron. It is generally used as an alloy with nickel or in a pressurized ladle, to moderate the reactions.

The intimate contact between the reagent and metal provided by the carrier gas technique may be obtained by mechanical means as in the shaking ladle (Kalling and Eketorp) or by the centrifugal whisk which provides circulatory pumping and mixing (Oesberg). The sulphur can be reduced by these means to very low figures using lime and soda, or calcium carbide.

Desulphurization also takes place in the mixer where the iron sulphide/manganese reaction occurs: $FeS + Mn \rightarrow Fe + MnS$. The manganese sulphide separates in view of its low solubility in iron and passes into the mixer slag. Some 30% of the sulphur may be removed in the mixer. The reaction also occurs in the transport ladles.

Several reagents are thus available for the desulphurization of molten pig iron and there are many methods of applying them. The use of magnesium is expected to increase.

Some basic irons with a high silicon content are often desiliconized before use in basic steel making. The process may be carried out by air or oxygen blowing in an acid-lined converter, or oxygen may be lanced into the molten iron which is contained in a ladle or rotary furnace.

THE MANUFACTURE OF STEEL

The main production of steel in the world is by indirect processes based on pig iron, and by remelting the scrap arising in the steelworks, and also by melting industrial scrap. The world production of crude steel is given for four selected years in Table 1. The world output of pig iron was just over 70% of these values so the difference is largely provided by scrap.

Given the range of chemical composition of the irons produced in the blast furnace, and that of commoner steels, it is clear that a proportion of each element must be removed in its conversion to steel, certainly from basic pig iron. Table 2 displays these differences.

Since the unwanted presence of these elements in the pig iron is due to the powerful reducing conditions in the blast furnace, the opposite conditions are used for their removal, i.e. oxidizing conditions. The oxidizing agent employed is oxygen, as gas, or combined as oxygen in iron ore, but mainly as ferrous oxide in the slag. Scrap of about the composition desired in the finished steel may be

TABLE 2

	Pig irons		Non-alloy steels	
	Hematite	Basic	Mild	Medium
Carbon	3·5	3·5	0·15–0·25	0·45
Silicon	2·5	0·9	0·03	0·15–0·25
Manganese	+0·5	1·0	0·6	0·7
Sulphur	0·05	0·06	0·05	0·05
Phosphorus	0·05	1·5	0·05	0·05

melted directly in an electric furnace and oxidized or reduced as necessary. Processes employing gas/metal reactions are very much faster than those in which the oxygen is combined as iron oxide.

The heat of oxidation of the elements it is desired to remove, in the case of oxidation with oxygen gas, are not only sufficient to raise the temperature of the molten pig iron above that of steel ready for tapping, but also to melt a quantity of scrap, which usually constitutes up to one-third the total charge weight. Silicon can contribute a 300°C rise for 1% of the element present, phosphorus 200°C and manganese 100°C. Some of the oxide metal reactions are as follows:

$$FeO + C \rightarrow Fe + CO$$
$$Fe_3O_4 + 4C \rightarrow 3Fe + 4CO$$
$$Fe_2O_3 + 3C \rightarrow 2Fe + 3CO$$
$$2FeO + Si \rightarrow 2Fe + SiO_2$$
$$5FeO + 3CaO + 2P \rightarrow 5Fe + 3CaO\ P_2O_5$$

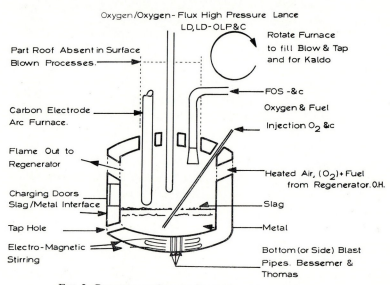

FIG. 2, Converter or furnace for steel-making reactions.

The carbon monoxide gas leaves the molten bath of metal as a boil and the other oxides enter the slag. The oxidation by means of iron ore normally absorbs heat, so external heating must be provided for processes in which this indirect oxidation is the main reaction.

The converter or furnace in which the steel-making reactions are carried out (Fig. 2, schematic) consists of a strong steel-plated structure, refractory lined and either stationary or capable of rotation about one or more axes. If the gas metal reactions are employed as in the Acid Bessemer, Thomas, LD, Kaldo processes, the vessel is cylindrical and the gas may be blown through the metal from the bottom or side, or as a high-speed jet from a lance arranged above the slag/metal surface. If oxide/metal reactions are employed the furnace is flatter with a large hearth and the reactions occur mainly at the slag/metal interface so this is as large as possible: the interface is agitated by the evolution of carbon monoxide. The bath may also be fed with oxygen from an immersed or surface lance. Heat for such hearth furnaces is provided by regenerative firing with preheated air, and gaseous or oil fuel or from electric energy.

The more important types of steel-making furnaces are illustrated schematically in Fig. 2. The approximate furnace capacity in modern installations may be taken as:

Open hearth furnaces	500 tons
Bottom blown converters	80 tons
LD, LD–AC, etc., and Kaldo	150–200 tons
Electric furnaces	100–200 tons

The relative cost of steel production by various processes under conditions in the United Kingdom, is given in the Benson Report, *The Steel Industry, Stage 1*, 1966. It was assumed that the charges were in the ratio 77/33% pig iron/scrap, and that the output of the plants would be $1\frac{1}{2}$ million tons per year, fixed capital remunerated at $12\frac{1}{2}\%$ and depreciation at 4% per annum or 90% of plant cost.

TABLE 3[2]

Process	Units of cost		
	Conversion cost/ per ton	Depn. and remuneration of fixed capital/ per ton	Total
LD low phos. iron	74·5	25·5	100·0
LD high phos. iron	79·7	26·8	106·5
Ajax high phos. iron	86·3	39·8	126·1
Fixed open hearth plus oxygen	95·2	43·1	138·3
Fixed open hearth	92·9	46·8	139·7
Kaldo low phos. iron	116·0	29·6	145·6
Tilting open hearth, high phos. iron	101·8	49·5	151·3
Kaldo high phos. iron	122·1	31·2	153·3

The refractory lining will be acid or basic depending upon whether the slag produced in the process is acid, with silica about 50%, or basic, with lime about 50%. The ferrous oxide will range from 10% to 30%. The type of slag will be dependent upon whether phosphorus is removed or not: if the former the slag must be rendered basic to hold the phosphoric oxide produced. If phosphorus is not to be removed, as in the acid processes, the oxidation of silicon, manganese and iron gives a fusible siliceous manganese iron oxide slag. Silicon is readily and first oxidized in all the processes, then manganese, carbon throughout the process, and phosphorus at the end of the air blown process (after blow), and early in the open hearth and top blown processes because of the early formation of a fluid basic oxidizing slag. A low temperature also favours the removal of phosphorus.

The composition of the bath is determined at frequent intervals on samples taken from the bath, particularly when the end point is approached. Rapid physical and chemical methods are employed. Automation, based on gravimetric and volumetric techniques and reaction rate, is becoming employed to a greater extent, temperature measurements are more refined and are often continuous.

Because the processes are based on oxidation, the concentration of oxygen in the melt increases as the process proceeds, particularly as carbon is eliminated. The solubility of oxygen in pure iron at its melting point is 0·25% (0·04% carbon is in equilibrium with 0·035% oxygen).

Large volumes of nitrogen are blown through the molten bath in the atmospheric air-blown process. Steels made by these processes contain a relatively high nitrogen content with consequent higher strength and lower ductility and toughness than steels of otherwise the same composition made by other processes. The original steep transition toughness curve typical for the low nitrogen content steels, falls towards higher temperature ranges with increasing nitrogen. Strain ageing is also more pronounced the higher the nitrogen content, the higher strength and lower elongation and toughness need, however, be no disadvantage for certain applications particularly where the section is small, and the temperature never very low. Steels are now made in which these phenomena are diminished by appropriate alloying and/or mechanical treatment.

Sulphur can be removed in each of the basic processes, but a hghly basic fluid slag is necessary. It is probably more economical to remove this element to or below the final level desired, by special treatments such as those described on molten pig iron. The favourable reducing conditions are then present. The use of these processes will probably increase. Sulphur is readily removed in electric furnace processes which employ a second strongly reducing carbide slag after the removal from the furnace of the first oxidizing slag.

The oxygen remaining in the molten iron as FeO after the various elements have been reduced to the required level is removed by deoxidation. Otherwise it would precipitate on solidification and destroy the cohesion of the steel. Deoxidants in increasing efficiency are manganese, silicon and aluminium, which on reaction with ferrous oxide give rise to their oxides which are insoluble in the molten steel. The deoxidation products, in view of their low density, mostly rise to the surface of the metal and collect in the slag. Such as remain in the steel are called the indigenous inclusions. The deoxidants may be added to the furnace or to the ladle or to both. The deoxidation is very critical and important.

The steel, when finished by deoxidation and addition of alloys, is tapped or run into the ladle. Some furnace slag is entrained with the metal stream and mixed in the general turbulence with the deoxidation products, which may not have completely separated from the metal. Such products, with eroded refractories, form the inclusions, found in the solid steel.

If deoxidation with carbon, and a resultant gaseous product, such as carbon monoxide, is carried out by lowering the absolute pressure, the oxide inclusion content is naturally lower. It may be reduced by as much as 50% under comparable conditions. This reduction may be achieved in vacuum degassing processes. These processes were, however, introduced mainly for the removal of hydrogen from molten steel.

The presence of hydrogen in contents of 6–7 p.p.m. in alloy steels, unless special cooling cycles were employed, often resulted in the formation in the solid steel of hairline cracks, small round fissures with a bright surface, more numerous nearer the ingot axis and more prevalent the larger the ingot. The cracks resulted from transformation stresses and the pressure due to the assembly of molecular hydrogen at preferred sites. The increasing production of basic electric furnace steel and the use of a second strongly reducing slag, for deoxidation and desulphurization, with consequent higher hydrogen content, focussed attention on the defect. The innovation to remove the hydrogen consisted of the employment of a vacuum chamber for degassing a stream of steel, expanding into droplets or creating new surfaces continuously in the chamber. The hydrogen is reduced to 2 p.p.m. or below by these processes.

Summarizing, it may be concluded that steels of a wide variety of mechanical, chemical and physical properties can be made by a range of processes and new ones are under active development. The properties desired in the steel depend primarily on its chemical composition. Great progress has been made in both physical and chemical methods of rapidly carrying out this control of composition. The major problems remaining are those of heterogeneity and the removal of inclusions (oxides, silicates, sulphides, furnace slag, refractories). Positive progess towards the solution of these problems is being made. Vacuum treatment, vacuum melting and electro-slag remelting and appropriate flux additions are contributions to these ends.

THE TREATMENT OF LIQUID STEEL

The making of steel up to the point at which it is tapped from the furnace into the casting ladle has been previously described in the foregoing. Separation, by gravity, of such furnace slag as flowed with the steel from the furnace, occurs in the ladle almost completely, and a heat barrier is set up by the slag between the molten steel and the atmosphere. Deoxidation products, entrained

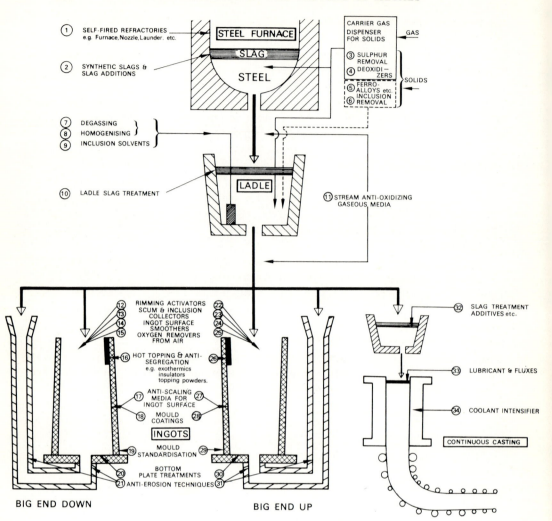

FIG. 3. Improved yield and quality techniques.

slag and eroded refractories continue to rise through the steel and are caught in the slag often forming a lower layer. The position at the ladle and therefrom onwards is shown diagrammatically in Fig. 3 "Improved yield and quality techniques". The points of application of such techniques are indicated on the diagram. As regards the yield of product from the ladle, it is surprising how low this can be. Table 4, "Steel conversion factors", is rearranged from Table No. 72, British Iron and Steel Federation, *Annual Statistics 1963*. The table shows that, excluding steel castings, which are usually made in non-permanent moulds, from 21 % to 45 % of the steel ingot weight cast is lost as regards its utilization as a final steelworks product for which it was intended. The weighted average loss for all non-alloy products is 26 %. Table 5 "Summary estimate of conversion losses", attempts to apportion these losses but the figures cannot be regarded as strictly quantitative.

The conversion losses listed in Table 5 are now dealt with in some detail. The continuous recording of ladle weight during teeming enables a closer control of teeming to be carried out with a reduction of short pours. Surface

TABLE 4. *Steel Conversion Factors*

Product	Multipliers	
	Ingot to product	Product to ingot
Non-alloy		
Castings	0·498	2·008
Tyres, wheels and axles	0·551	1·815
Forgings	0·553	1·808
Flat products, tubes, etc.		
Cold rolled strip	0·684	1·462
Tubes, pipes, fittings	0·688	1·453
Plates, 3 mm thick and over	0·708	1·412
Tin plate (incl. black plate)	0·719	1·390
Sheets, under 3 mm	0·748	1·337
Hot strip	0·755	1·324
Sections, etc.		
Bright bars	0·717	1·395
Heavy rails, sleepers, etc.	0·726	1·377
Colliery arches and light rails	0·733	1·364
Light sections and bars	0·757	1·321
Wire rods	0·772	1·295
Heavy sections and bars	0·793	1·261
Weighted average: non-alloy	0·736	1·359
Alloy	0·640	1·563
Average: all qualities	0·730	1·370

TABLE 5. *Summary Estimate of Conversion Losses*

	%	
	Average	Range
1. Defective and short poured ingots	1	1–2
2. Scaling losses on reheating	3	2½–4
3. Deseaming	1	½–3
4. Selective dressing: ingot turning, etc.	½	½–6
5. Bottom crop	2	1½–3
6. Top crop	9½	7–15
7. Rolling imperfections/rejections	1½	1–4
8. Adjustment to specified dimensions	7½	7–8
	26	21–45

defects in ingots and their products are defined in I.S.I. Special Report No. 44, 1951.[3] Sketches and photographs in the report clearly indicate the types of defect defined. Causes are given in the majority of cases and means of avoidance can often be deduced therefrom and a reduction in this conversion loss effected.

The ingot surface is of great importance. Techniques for its improvement by restraints on the metal stream itself, by treatment of the mould surface, and by the addition of fluxing, gathering and lubricating reagents to the rising meniscus, and the production of non-oxidizing atmospheres can now effect considerable improvements.

Factors such as mould design, section, corners and profile play a great part in the prevalence of ingot defects. Co-operation between steelmakers using moulds of different design for given ingot weight could, by statistical and computor investigation, throw much light on the desirable trends which might be followed in ingot mould design to reduce the incidence of ingot surface defects.

The scaling losses from the ingot and semi-product reheating processes represent quite a proportion of the total loss. It is possible to regard part of this loss as a benefit, for with the removal of the scale some ingot surface defects are undoubtedly removed. Generally, the scaling losses are regarded as almost unavoidable in view of the size and shape of the ingot, the temperatures necessary for mechanical working, and the furnace atmospheres.

Reduction of these losses can be effected along two main lines. First, by protecting the metal surface against oxidation by applied coatings, and secondly, by reducing the oxidizing nature of the heating furnace gases. The development of heat- and oxygen-resisting coatings has been in progress for many years, and although successful at temperatures lower than those employed for reheating for rolling, no entirely successful coating for pre-rolling treatment has yet been evolved. Research will doubtless continue. The other line of progress will no doubt be to heat the soaking furnace and reheating the furnace electrically, or by radiant tubes; the furnace atmosphere could be the nitrogen exhaust from tonnage oxygen plants. A careful and improved control of the normal combustion gases employing modern continuous gas analysers could also contribute substantially to this end.

Over and above the grosser ingot defects which may be severe enough to warrant rejection of the ingot, less marked and sub-surface defects may develop in the semi-finished and finished products. Surface treatments at an early stage may eliminate some defects of this type and also save material. Deseaming with oxygen gas, at the slab or bloom stage, or selective dressing with gases, or mechanical tools, working on the hot or the cold metal, or ingot turning, may be employed to prepare a better surface. Although metal losses arise from these treatments, continuous in-line processing is simplified. The very fact that such treatments are carried out on a large scale confirms the importance of better ingot surfaces and of devoting considerable resources to this problem.

The loss due to bottom crop results mainly from the irregular outline of the base of the ingot and its unfavourable crystalline structure, giving planes of weakness at the meeting of the columnar zones growing from the bottom and from the sides, including also in some cases a part of the weaker and inclusion richer sedimentary cone. Further, the action of the rolls is often to elongate the outer surface over the core, giving rise to "fish tail". Improvement may be effected by evolving locally a more favourable profile for the bottom part of the ingot. Appropriate bottom plate dressings, and maintenance of the desired profile and surface finish by mechanical treatment, or building up eroded areas with dressings, with weld metal or thermits, can reduce these losses.

The figures in Table 5 indicate that the major loss on conversion from ingot to product is that due to top crop. Deoxidized (killed) molten steel on solidification contracts about 4%. Further contraction of the solidified shell on to the inner portion reduces the contraction cavity, or pipe, to some 2–3% of the ingot. Hotter, more rapid pouring and increasing size increases this volume.

Ingot mould design plays an important part in determining the gross, macro- and the micro-structure of the ingot and the yield from it. Benedicks[4] has emphasized the importance of ingot shape, and particularly the effect of the location of the ingot thermal centre, which has a slower cooling rate than its surroundings. He compares fully the solidification for the two widely different cases of the thermal centre in the lower half of the axis, and in the upper half, i.e. big end down and big end up. He concluded that ingots of the second type are very desirable for the production of quality steels. He further concluded that effective heating of the top end would not only reduce pipe, but would probably avoid accumulations of slag in the ingot top. Refractory feeder heads, gas and electric heating have been used with some success to effect this.

Fenton[5] showed that the heat absorption of the hot top should be low which could be accomplished by using insulators as covering, or by heating the hot top. Perhaps the most effective means are to line the hot top with an exothermic material and to reduce the ratio of surface to volume. He also concluded that heat losses upwards should be reduced by insulating, or exothermic powders.

Schoeberl,[6] in a carefully conducted series of works experiments, showed that the ingot feeder head using firebrick tops varied from 12–20% depending on size, quality, etc. If combustible tiles were used, the feeder head could be reduced further by 6·5% and the billet yield increased by 6%. If exothermic hot tops were used on large mill, and small forging ingots the head could be reduced by 8–12% and the yield increased 7–10%. The advantages of exothermic hot tops for the size of ingots experimented with are clear from these experiments.

Again, Stroganov[7] has shown in a paper on the thermal balance for the feeder head region in ingots, that most of the heat lost by the feeder head region of the ingot is used to heat the lining and the casing of the hot top. Increased efficiency will thus result from the use of a lining with low specific heat and of minimum weight. Losses through the meniscus are reduced by covering it with exothermic or refractory powders.

New refractory light weight insulating materials are now available, which are also of low thermal conductivity and heat capacity, and the use of these has confirmed these views, particularly for larger ingots.

During solidification of killed steels, the steel considered so far, very little gas is evolved, but gas is evolved during the solidification of rimming steel. The bulk of the steel made in the world (possibly four-fifths) is rimming steel. The major constituent of the gas evolved from rimming steel is carbon monoxide, formed by the reaction between iron oxide and carbon as the temperature falls and solidification occurs. Much of the carbon monoxide rises and escapes from the ingot, carrying with it small particles of molten steel which burn in the air, giving an appearance of "effervescence". But many bubbles are trapped by the advancing solidifying face and so counteract the liquid to solid shrinkage. The ingot surface level is altered favourably. The technical problems lie in the adjustment of temperature, carbon and manganese, and oxygen to effect the desired distribution of bubbles giving the required pure outer shell and the level of the ingot top. These techniques naturally vary, but many are based on reactivation of the reaction in the ingot being teemed with oxidants and oxide fluxes and nucleants. The process will probably become more controllable when rapid oxygen determinations can be carried out.

Rejection of the finally formed product may result from defects originally in the ingot, or from processing defects. The grosser forms of these are described with their probable causes in I.S.I. Special Report No. 44 already referred to. Rejection may also occur because of major or blow-hole segregation detected in the finished product. Differential solidification gives rise as solidification proceeds to an increasing content of most of the constituent elements of steels, in the remaining mother liquor. Much higher contents of carbon, sulphur and phosphorus are then present in the metal which is the last to solidify and which surrounds the pipe. Whilst attempting to use the maximum pipe-free material, a segregated axial zone may be detected on sectioning the final product. The probable greater incidence of V- and A-segregates in big-end-down ingots should also be noted.

The extent of ingot segregation is usually mapped visually from a sulphur print of the prepared section surface where it is expected to be of importance. It is also evaluated chemically by the determination of the content of the elements most prone to segregation, usually carbon, sulphur and phosphorus across the section. These local results are compared with the cast analysis, or an analysis from a representative position in the ingot. Position "A", for example, was selected by the Ingot Heterogeneity Committee, because it more nearly represented the composition of the original steel.

But segregation can also be measured by Brinell hardness tests, so measuring the alteration in mechanical properties. The few examples of chemical analysis and Brinell hardness results, given together in the First Report of the I.H.C., show proportionality between these figures. If such tests are carried out the section should preferably have undergone a standardized heat treatment, e.g. tempering at 650°C. There is an almost straight-line relationship between carbon and phosphorus and the hardness and tensile strength. The increase in Brinell hardness is about two points for each 0.01% increase in carbon or phosphorus.

Inclusions in steel are a frequent cause of rejection. They are particles of non-metallic matter included in the metal. They may be large, up to 1 in. in length, and so easily visible to the naked eye, down to small specks just visible at a magnification of 1000 diameters. They are naturally the more readily observed if large and present at the finished surface of the metal. Magnetic-powder crack detection methods are employed to examine the surface and sub-surface more carefully. Ultrasonic methods are used to examine in depth. These techniques, and step machining with visual examination of the steps at each reduced diameter, are the usual inspection techniques.

Inclusion counting of many microsections, according to procedures established by ASTM, Fox, Jernkontoret, SKF, etc., are employed for the examination of special products such as ball-bearing steels, aircraft parts, etc., but are time consuming and present sampling difficulties, due to gross ingot heterogeneity.

Inclusions are deleterious when their presence would spoil the appearance of the metal, as for example in the case of most polished stainless steels, or prevent smooth enamelling in the case of sheet for domestic products, or spoil the covering ability of paints, etc., as in car panels. Mechanically they present a discontinuity, and so a source of weakness in the material from which a crack may more readily start. Surface stresses are usually higher than the nominal stress, so surface weaknesses are particularly undesirable. Normally inclusions run parallel with the principal direction of hot-working, and with micro-segregation give rise to longitudinal fibre and banding. The inclusions are usually malleable and so themselves extend in the direction of working, except in the case of fine dispersions of high melting-point oxides. This longitudinal fibre is reminiscent of wrought iron. Many of the excellent properties of the latter are due to its fibrous structure; at one time attempts were made to produce it on a large scale synthetically. But the properties across the fibre are impaired roughly in proportion to the amount of the inclusions, so this development of fibre is on the whole undesirable, especially in material to withstand transverse loads. The mechanical effect is not to alter the general shape of the load-extension curve, but to arrest it earlier at a thread of inclusions. So reduction of area first, then general elongation, and finally tensile strength are each reduced in turn. The effect can eventually be so bad that the inclusion acts as flake graphite, and even the yield point is not reached: the specimen breaks, parts. The higher limit of the temperature/toughness transition curve is lowered by inclusions, the transition occurs over a greater range of temperature and starts sooner. The fatigue strength is lowered.

Inclusions are desired in free-cutting steel, as the chips break easily at the inclusion. This breaking of the chip at an inclusion enables an inclusion to be distinguished from a "ghost" which is segregate only. Free-cutting steels do not contain hydrogen hairline cracks, as the hydrogen seems to diffuse into the inclusion or its interface.

Inclusions may occur as a dispersion throughout an ingot but are generally more prevalent in the bottom third, as a "sedimentary cone". They are also associated with the A- and V-ingot segregates, and at blowholes. They also occur on or near the ingot skin and care is taken in rimming steel to obtain a rim, sufficiently thick, of inclusion-free ductile iron.

Sulphur present in the steel (0·01–0·10%) normally, say 0·04%, occurs mainly as an inclusion, manganese sulphide, MnS. But as sulphur can be lowered to any desired level by many known techniques it is not considered here: neither are nitrides, borides, etc.

The inclusions are mainly oxides, as such, or silicates, or reaction products of oxides such as silicates, aluminates, etc. Probably a complex manganese iron silicate is the commonest, followed by alumina, and aluminium silicates. These arise primarily from the deoxidation products, i.e. the insoluble oxides arising from the metals added to remove the iron oxide, soluble in the molten steel, resulting from the almost inevitable oxidizing nature of the bulk steel-making processes. Owing to their size, surface properties, melting point, etc., they have not completely separated from the molten bath. They also arise from contamination with eroded materials, mainly oxides, from the entrained slag and all the refractories, with which the molten steel is in contact on its path from the furnace into the ingot mould. Reoxidation of the molten stream also occurs due to contact with, and entrainment of atmospheric air, and then reaction of the iron oxide with the dissolved deoxidants.

Generally, the inclusions resulting from deoxidation are small, and unless aggregated or segregated are only detectable under the microscope. They are called micro-inclusions. On the other hand, refractory contaminants are present as larger grains or glasses, and are called macro-inclusions.

The quantitative assessment of inclusion content is not easy, and is on the whole unsatisfactory. Certain steel products of homogeneous form are subjected to routine surface examination, and correlations and assessments can follow, e.g. sheet products, standardized shafts, tyres, axles and other forgings, and it is known whether the standard is improved on or worsened after a change in process.

Overall techniques for the elimination of inclusions from known areas are often employed as oxygen scarfing of slabs, and ingot turning for alloy steels. Probably the total evolution of carbon monoxide in the case of vacuum degassed steels would give the best idea of the cleanliness obtained. The oxygen in the carbon monoxide removed would mean that this quantity of oxygen at any rate could not be in the finished steel. A correlation could be gradually built up between the oxygen content of the ladle metal and the product.

Samples of the literature from the steel-making countries of the world reveal unanimity on the importance of the subject, and some agreement on the origin of the inclusions.

It is, however, concluded that the greater part of the inclusions come from sources other than entrained slag and eroded refractories. Reoxidation of the molten stream of steel and the meniscus rising in the mould are important secondary sources. Little is yet known of the paths of, and the part played by, ingot scums.

Improved deoxidation techniques, stirring, turbulence, injection in the ladle, prevention of reoxidation, and fluxing and gathering techniques in the mould are all methods of reducing inclusions.

Hydrogen in steel may also give rise to rejection. The defect is generally known as hairline cracking, and is revealed as colonies of fine radial cracks in the central zone of the product, if of medium or large section. Their formation may continue at normal temperature for a long time. Average contents for basic open hearth alloy steel are $4\frac{1}{2}$ p.p.m., which can result in cracking. This figure is partly dependent on the moisture content of the atmosphere, rising and falling with it. The figure is usually higher for basic electric furnace steel, and lower for acid open hearth steel. Special cooling treatments have been evolved which allow time for the hydrogen to diffuse outwards in critical temperature ranges. Some reduction may be obtained by gas washing and by surface jetting in the ladle. The present certain method of removal is by treatment of the molten metal in a vacuum. Various types of plant are in use, finality does not seem to have been attained and the hydrogen concentration may be reduced to below 2 p.p.m. in such plants.

The reaction between iron oxide and carbon is carried further *in vacuo*, hence some of the dissolved FeO can be so removed. The vacuum degassing process therefore permits deoxidation without the production of an insoluble reaction product (inclusion). Results show that the oxygen content of the vacuum-treated steel may be reduced by about one half its initial content. The yield of clean acceptable steel is thereby increased.

The losses of yield resulting from the adjustment in size of the finished worked product to the specified dimensions are capable of reduction by semi- or complete automation of rolling and cutting schedules. They may be also reduced to some extent by simple computation and semi-automatic techniques. Metallurgical factors do not here play a major part.

Continuous casting has not been considered here; it should possess many advantages over conventional methods from the point of view of yield. The increasing application of the process will doubtless soon give results of a quantitative nature and on which comparisons may be based.

The analysis of conversion losses attempted in this chapter cannot be complete or quantitative, but it indicates points where intervention leading to betterment of yield and quality may be made.

The writer has been influenced by Professors Turner, Wuest and Goerens and by the work of Professor Dürrer. He wishes to thank particularly the British Iron & Steel Federation for permission to reproduce Tables 2 and 3 and O.C.D.E., Paris, for permission to use the production statistics: the Metallurgical Society of India for permission to include a large part of the paper prepared for them in 1964.

REFERENCES

1. Organization for Economical Co-operation and Development, Statistical table.
2. *The Steel Industry, Stage* 1. Report of the Development Co-ordination Committee, British Iron & Steel Federation, July 1966.
3. Iron & Steel Institute Special Report No. 44, *Surface Defects in Ingots and their Products,* 1951.
4. BENEDICKS, C., and LOFQUIST, H., *Non-Metallic Inclusions in Iron and Steel.* Chapman and Hall, 1930.
5. FENTON, G., *Journal of the Iron & Steel Institute,* **186**, 1957, 396.
6. SCHOEBERL, A., *Stahl und Eisen,* **81**, 1961, 22.
7. STROGANOV, A. I., *et al., Stahl* (in English), 1 January 1962, 23.

CHAPTER 2

CASTING PIT PRACTICE

INTRODUCTION

J. G. Emmott

When fully killed liquid steel is cast into an ingot mould, it decreases in volume, the shrinkage being mainly attributable to three causes:

(a) Liquid shrinkage arising from fall in temperature before solidification.
(b) Solidification shrinkage occurring during the change from liquid to solid.
(c) Shrinkage in the solid state.

In the manufacture of ingots, it is essential that (a) and (b) are considered if a sound, homogeneous ingot is to be obtained.

In the solidification of a typical big-end-up (narrow-end-down) ingot the illustrations in Fig. 1 serve to show diagrammatically the balance of solid and liquid metal at various stages after teeming. This assumes that no heat is allowed to escape upwards to the atmosphere, the top surface remaining liquid throughout.

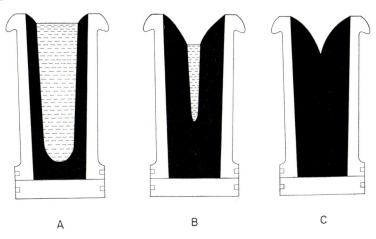

A   B   C

Fig. 1. Solidification of a big-end-up ingot.

After the formation of a thin layer of chill crystals at the ingot–mould/liquid–metal interface, columnar crystals are propagated, initially at right angles from the mould faces, towards the thermal axis. These columnar crystals continue to form until the thermal gradient within the remaining liquid metal is

19

insufficient to support further growth. The remainder of the liquid then solidifies in an equiaxed crystal formation.

The typical structure of such a sectioned big-end-up ingot is shown in Fig. 2. The inverted cone "sink" in the top of the ingot is termed "primary pipe".

In all steels there are constituents which have low solubility in liquid iron. The solubility is also a function of the liquid iron temperature and, as this falls, the constituents are largely expelled from the solidifying iron and pushed ahead of the freezing front, enriching the near zones of liquid metal. The enrichment becomes progressively greater as solidification continues, until the last metal to solidify is very rich in the expelled constituents.

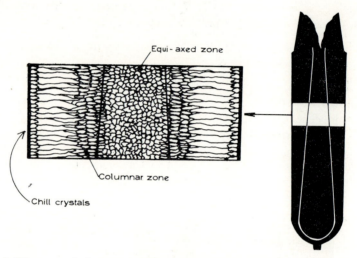

FIG. 2. Diagrammatical representation of a typical ingot structure (3-ton, bottom-poured, big-end-up ingot 0·25% C).

This phenomenon of zonal enrichment is termed "NORMAL SEGREGATION" but is more usually referred to simply as "SEGREGATION". Many elements segregate in solidifying steel, but, of these, carbon and sulphur are the most important, since concentrations result in changes of strength and ductility of the steel product. To a lesser degree the segregation of alloying elements and their compounds in high-alloy steels, especially high-speed qualities, can be troublesome.

Figure 3 illustrates the concentration of the segregated elements, particularly carbon and sulphur in the ingot section previously considered. It can be seen that the contour of the concentrated segregation is similar to that of the primary pipe, but penetrates deeper into the ingot body.

When such an ingot is subjected to hot working, the cross-sectional area is very greatly reduced and the length is proportionately increased. At an intermediate stage between the ingot and steel product, examination of the bloom or billet section will show the presence of segregation and pipe-discontinuities in much of the length. The cavities associated with the primary pipe are always deleterious to the quality of the final product and certain segregates, in sufficiently concentrated amounts, are also harmful.

It is thus of paramount importance to confine the cavitation and segregation defects to as short a length of the bloom or billet as possible. This must be attained by restriction of the pipe and segregation zone depths at the ingot stage, and is achieved by "feeding" the ingot from a "hot top", a process which is described in detail later.

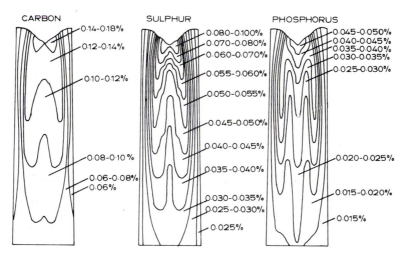

Composition: C=0.09%, Mn=0.49%, P=0.02%. (K.Möhl)

Fig. 3. Carbon, sulphur and phosphorus segregation in a killed steel ingot section (By courtesy of G. Guzzoni[1]).

Up to the present, only a big-end-up ingot has been considered. However, a large part of the world's steel output is confined to big-end-down (narrow-end-up) moulds where the area of the base section is greater than that of the apex. With such a mould the general considerations of solidification are the same as a big-end-up mould, but the final configuration of the solidification and segregation patterns are changed by the influence of the mould shape. Figure 4 illustrates three stages in the solidification of a big-end-down ingot, again assuming that no heat is lost vertically upwards to the atmosphere.

Chilled crystals form on contact between mould and liquid metal and columnar crystals are again propagated towards the ingot axis. Because of the taper or conicity of the mould, advancing freezing fronts meet at a point high on the axis whilst there is still metal in the liquid phase, at a point lower in the ingot. The point where freezing fronts meet is termed a "bridge". The liquid isolated below this continues to contract as it solidifies and becomes smaller than the zone occupied at the moment of bridging. The result is a cavity or cavities within the body of the ingot termed "secondary pipe". The isolated metal is also rich in segregated components and the cavity is associated with areas of segregate-rich metal. The surfaces of the cavity are not open to the atmosphere and do not become oxidized. During hot working the secondary pipe welds up to a large extent and the resultant billet is frequently axially sound but areas of segregation persist.

Above the bridge the liquid metal again freezes in a V-formation primary pipe, but the volume of metal thus solidifying is less than in the case of a big-end-up mould. The primary pipe is consequently slightly smaller in volume.

All the above examples have assumed that heat losses from the top metal surface by radiation to the atmosphere have been prevented.

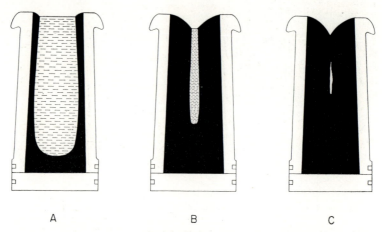

A   B   C

Fig. 4. Three stages in the solidification of a big-end-down ingot.

INGOT FEEDING

It has been described in the previous section how, when liquid steel is cast into a simple mould of conventional shape, primary pipe, secondary pipe and segregation may be found within the body of the ingot. After hot working these persist in the bloom or billet as axial imperfections, unacceptable in the majority of steel products. To improve the quality of steel ingots with respect to central unsoundness it is essential that as much of the shrinkage cavitation and harmful segregation as possible be positioned in the ingot in such a way that they can be removed by the discard of a small portion after hot-working.

During rolling or forging the deformation of surface regions of the ingot differs from that of the centre, resulting in areas corresponding to the top and bottom of the original ingot which are unusable. These end effects are termed "fish-tail" which are clearly seen in Fig. 4A. It is necessary to shear the billet and discard an appreciable length to eliminate fish-tail.

In recent years methods of casting, incorporating feeder heads, have ensured that primary pipe and associated segregation are raised into the region of the ingot where they can be discarded at the time of shearing. Secondary pipe and

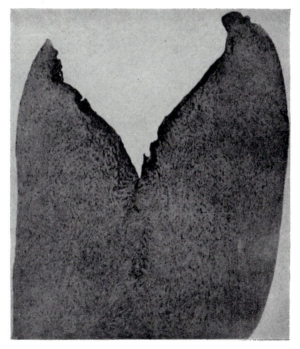

FIG. 4A. "Fish-tail" in rolled ingots.

B

associated segregation remain in the body of the ingot, but careful mould design can minimize defects resulting from these.

A feeder head can be described as a container, lined on the inner surfaces with some form of refractory, insulating or exothermic material and located at the head of the ingot mould. The metal content of the feeder head remains in the liquid phase longer than the metal in the body of the ingot beneath it and compensates for the total liquid and solidification shrinkages. The feeder head is also frequently referred to as a "hot-top" or more simply as a "head".

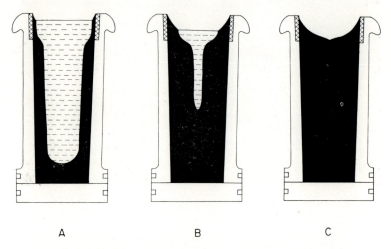

A   B   C

FIG. 5. Progressive solidification of a hot-topped big-end-up ingot.

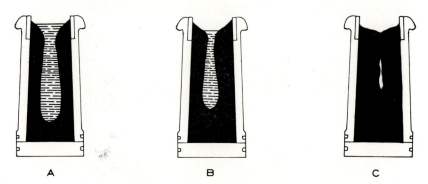

A   B   C

FIG. 6. Progressive solidification of a hot-topped big-end-down ingot.

To prevent heat losses by radiation to the atmosphere, the feeder head is covered by an insulating or exothermic powder layer, normally called "anti-piping compound", "topping compound" or "feeding compound".

When a reservoir of liquid metal is maintained at the top of the ingot by the use of a hot-top and anti-piping compound, solidification progresses much more slowly from the head lining. The final head contour of the ingot is in the form of a shallow dish, the progressive formation of which is shown in Fig. 5 for a big-end-up mould and in Fig. 6 for a big-end-down mould. Both illustrations assume that no heat has been lost to the atmosphere by radiation from the upper surface.

HOT-TOPPING METHODS AND THEIR DEVELOPMENT

J. G. EMMOTT

Hot-top Linings

In the early days of heading ingots, attempts were made to maintain an upper reservoir of liquid metal by fabricating a head box and lining it with refractory brick or clay, the contained metal being covered with such powders as vermiculite, kieselguhr, charcoal or coke dust. Improved yields were obtained when compared with unheaded moulds and the head boxes could be used many times with only minor repairs. Yields and feed patterns still left much to be desired, however, since refractory linings extract considerable heat from the head metal to satisfy their own thermal capacities. Also, when hot, both refractories and the inferior feeding compounds used are poor insulators and much useful heat is lost from the head by conduction. In addition to being thermally inefficient these hot-tops are cumbersome, needing cranes to apply them, and there are frequently problems of space requirements associated with preparation, repair and drying of heads before use.

As recently as 15 years ago expendable exothermic linings for ingot heads were being tested and during the ensuing years these have been improved in design and efficiency. They are widely used where yield increases of higher quality steels can support the extra cost. Figure 1 compares the axial sections of two ingots made in the same cast. The left-hand ingot used a partly inserted, preheated clay sleeve or "dozzle" whilst the right-hand ingot was made with a superimposed exothermic hot-top. Both heads employed the same feeding compound.†

The limitations of exothermic heads quickly became known. Although very effective in increasing the yield of small to medium ingots they are heavy and relatively difficult to handle on large moulds, are expensive when used on low-grade steel and do not give sufficiently good feed patterns on very large ingots.

The principal requirement for an exothermic head lining is that it burns raising its own temperature to that of molten steel or above, just before the head is filled. By this means, the head lining does not extract heat from the head metal and when exothermics with high calorific values are used, heat is given to the hot-top metal. However, exothermic residues after combustion have relatively poor insulation properties and the initial benefits are rapidly lost by conduction of heat to the head-box container or to the atmosphere. A much more efficient feeder head can be obtained by substituting a duplex form, where

† For details of the chemical reactions involved in the burning of these exothermic mixtures refer to page 445.

the head is made from a good refractory or insulating material faced with exothermic to prevent early chilling of head metal. Costs can also be greatly reduced by the use of a cheap backing material such as sand instead of part of the exothermic, but sand is not a good insulator and such heads are also unsatisfactory for use on large moulds.

A stage was reached in the evolution of hot-topping where ingots up to 250 mm square could be headed satisfactorily with simple exothermic shapes and larger ingots with exothermic-faced insulating material. However, even with cheap backing materials, problems remained with exothermic heads. Economics of application are frequently unsatisfactory on low-grade steels

Fig. 1. Sections of two ingots. *Left:* with preheated clay sleeve. *Right:* with super-imposed exothermic hot-top.

where the cost of the head may be greater than the value of the increased yield. Also exothermics fail to give satisfactory yield increases on very large ingots with long solidification times. Even refractory heads lined with exothermics were unable to maintain head metal in the liquid phase long enough to prevent primary piping and inferior segregation patterns in such ingots.

It was apparent that, when considering medium and large ingots, the insulation properties of the head were of the utmost importance in retaining head metal in the liquid state over a long period of time, frequently in excess of 3 hours.

If the high cost of exothermics was to be avoided, the thermal capacity of head-lining materials needed to be low and the linings themselves to be easily handleable and robust, to minimize crane usage and breakage in transit and handling. From these requirements was born the super-insulator.

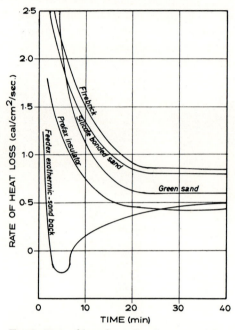

FIG. 2. Rate of heat loss of various materials.

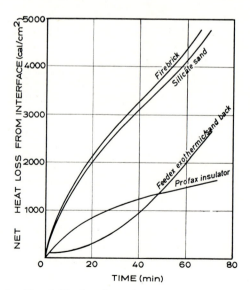

FIG. 3. Net heat loss of various materials.

If the thermal properties of various head-lining materials are compared, their ability to retain heat in a feeder head becomes apparent. Figures 2 and 3 show the "rate of heat loss" and the "net heat loss" with respect to time through 28 mm thick plates of firebrick, silicate bonded sand, green sand and a super insulator (Profax) when in intimate contact with a horizontal plate electrically maintained at a constant temperature of 1420°C. For comparison a 14 mm thick facing of good quality exothermic on a backing of 14 mm thick silicate bonded sand is also shown.

The following points should be noted from the graphs:

1. Firebrick, silicate bonded sand and green sand all extract heat from the plate very rapidly in the early stages, giving an initial chilling effect. Even when their heat capacities are satisfied they conduct heat to their environs rapidly.

2. A super insulator with low thermal capacity does extract some heat from the plate but this quantity is only small and is over a very short period of time. When the insulator's temperature has been raised to that of the adjacent plate highly insulating properties ensure that further heat loss is very slow.

3. The exothermic face of the sand-backed duplex assembly burns on contact with the hot plate and quickly reaches a higher temperature than the plate. For a short period of time after application, the lining contributes heat to the plate, raising its temperature. After approximately 7 minutes the lining and plate are at the same temperature and, from this time, heat flows from the plate by conduction through the relatively poor insulating layers of sand and burnt exothermic residues. After only 25 minutes heat is being lost more rapidly from the duplex-exothermic head than from the super-insulator and after 50 minutes the total heat lost from the exothermic head exceeds that from the super-insulator. Thus for ingots which take more than 50 minutes to solidify, the super-insulator can be expected to give better feed patterns and consequently better yields than exothermics. Although liquid steel is at a higher temperature than the 1420°C plate the relative performances of the materials may be anticipated to follow the same pattern.

Anti-piping Compounds

Earlier discussions of ingot solidification assumed that the top surfaces or feeder heads have not lost heat to the atmosphere by radiation. In fact Muellef and Bole[2] stated in 1949 and Fenton[3] confirmed in 1956 that 25 to 30% of total heat losses from a brick-lined hot-top without feeding compound cover are by surface radiation. In the case of an exothermic or super-insulator lined hot-top a larger proportion of the total heat will be lost in this manner, possibly exceeding 40%, since such heads tend to be wider and shallower than brick hot-tops and lateral losses are reduced by better insulation.

If no anti-piping compound had been used on the ingots illustrated in Figs. 1

and 4 of the previous section and the upper metal surface had been allowed to radiate heat freely to the atmosphere, the resultant feed configuration would have been different. Figures 4A and 4B show typical sections of big-end-up and big-end-down ingots which have used neither feeder heads nor topping compounds.

In each case the surface has frozen rapidly forming a skull bridge below which is a cavity. Other cavities have occurred, isolated from each other by successive bridges. After the formation of the first complete bridge, atmospheric pressure can no longer assist in the feeding of the ingot. When the cavities are divorced from the atmosphere their surfaces do not oxidize and, to a large extent, they weld up during hot working. Segregation associated with the lower cavities persists as an axial concentration.

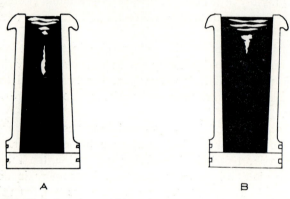

A   B

FIG. 4A and B. Typical sections of big-end-down and big-end-up ingots without use of feeding aids.

With certain low-quality steels where inferior structures and segregation patterns can be tolerated, or where the level of segregating components is low, ingots are sometimes cast without either feeder head or anti-piping compound. In order to ensure that the cavities do not become oxidized, it is a fairly common practice to apply copious quantities of water by hose or bucket to the mould after teeming is complete. This chills the upper surface and ensures a very thick top bridge. It should always be remembered that such practice can only be related to common, low-grade steel qualities and, even then the internal condition of the bloom or billet will be markedly inferior to that obtained from the efficiently hot-topped steels.

Before the use of exothermics and insulators for feeding ingots, such low-heat-capacity materials as vermiculite were used to cover the top metal surface in brick and clay hot-tops. With the advent of more efficient heads it was found that top covering compounds which were themselves exothermic enhanced the overall effect and gave improved feed patterns. Figure 5 compares the heads of three ingots made from the same cast of steel using refractory hot-tops. In (a) no anti-piping compound was used, in (b) the metal was covered by a 50 mm thick layer of exfoliated vermiculite and in (c) by a 25 mm thick layer of an exothermic powder.

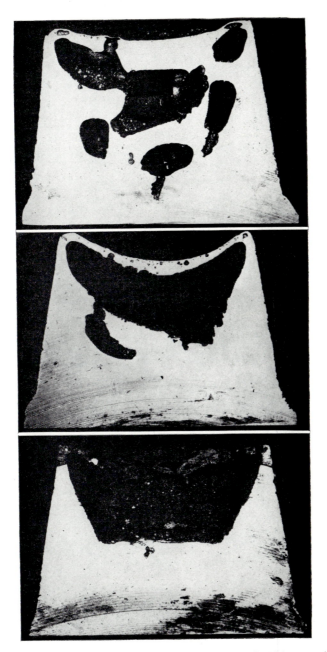

FIG. 5. Three ingot heads using refractory hot-tops. (Top) Without anti-piping compound, (Centre) With exfoliated vermiculite and (Bottom) With exothermic powder cover.

B*

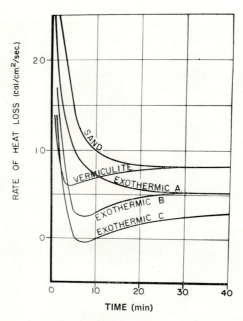

FIG. 6. Thermal characteristics of a variety of feeding compounds.

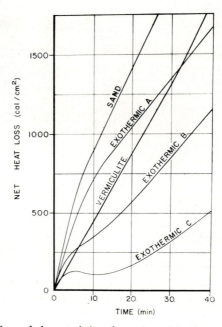

FIG. 7. Thermal characteristics of a variety of feeding compounds.

It can be seen that the tendency to form a top bridge is progressively reduced, and in the last instance bridging has been prevented by keeping the upper metal surface liquid until the ingot solidification is complete. If similar heads were made in either exothermic or super-insulator, the chilling effect at the sides would be much reduced and the upper contour of the residual head metal would be flatter.

The thermal characteristics of a variety of feeding compounds placed on a hot plate maintained at 1420°C are compared in Figs. 6 and 7. The rates of heat flow and nett heat flow with respect to time are illustrated respectively. The materials chosen are sand, vermiculite, and exothermic powders A, B and C. Powder A has low exothermicity and is very cheap, powder B is moderately priced and gives better heat output and insulation, and C is a compound with a high fuel content giving excellent heat output and heat retention but at a higher cost.

Again several conclusions can be drawn as follows:

1. Since none of the compounds is pre-ignited, each extracts some heat from the plate. Sand and exothermic powder A chill the plate heavily until they have reached the temperature of the liquid steel, then they permit fairly high heat losses by conduction through the layer.

2. Exfoliated vermiculite, because of its low density and conductivity, chills the plate only very slightly and prevents heat losses in the early stages. After a short period, however, the vermiculite starts to fuse in contact with the plate and its conductivity increases. From this point, heat losses become greater and after 30 minutes the cover is no better than sand and is inferior to the least exothermic of the powders.

3. Exothermic powders B and C, on burning, rapidly lower the chill effect and in the case of C, the feeding compound actually contributes heat to the plate. After burning, the powder residues are relatively insulating and heat is lost only slowly over a long period. At all times heat retention in the head is better with powder C, than with powder B. Again it can be concluded that a similar pattern of results would be obtained from application of the various powders to liquid steel in the feeder head.

FACTORS AFFECTING CHOICE AND DESIGN OF HOT-TOPS

Side-wall Linings

The exothermic materials available for heading steel ingots vary greatly in thermal output, sensitivity, rate of heat delivery and insulating properties of residues. Similarly insulators have widely different physical properties. It is thus essential that the correct exothermic or insulating head and anti-piping compound are selected for the size of ingot and quality of steel to be headed.

It is necessary to know the period of time during which the upper part of the ingot will remain liquid, so that the metal reservoir in the hot-top can be made efficient enough to feed the whole of the ingot. The time taken for an ingot to solidify is given approximately by one of the following formulae:

$$T = \left[\frac{D'}{2}\right]^2 \text{ or } \left[\frac{D''}{5}\right]^2$$

where T is the solidification time in minutes,

　　D' is the dimension of the ingot cross-section in inches,

　　D'' is the dimension of the ingot cross-section in centimetres.

(Note: If the ingot is rectangular in cross-section, D' or D'' is the smaller of the two dimensions.)

It follows that the greater the section of the ingot the longer is the time taken to solidify and the longer is the period during which the head metal must remain liquid. Thus for efficient feeding the design and composition of the hot-top to be used will be influenced by the size of the ingot being made.

Exothermic heads are extremely important in the small ingot, special steel field. It has been seen that exothermics are frequently unsatisfactory on large ingots due to the poor insulating properties of their residues. Further reference to Figs. 2 and 3 shows that the overall effect of a duplex exothermic/sand head, employing an exothermic thickness of 14 mm is similar to that of a 28 mm thick super-insulating head approximately 50 minutes after teeming an ingot. Up to that time the exothermic head is preferable, and for longer periods the insulator can be expected to give better solidification patterns. An ingot with a solidification time of 50 minutes has a section of approximately 335 mm (14 in.) square using the formula quoted earlier. If the exothermic thickness is increased or if the insulating properties of the backing material are improved, a duplex exothermic head can be used satisfactorily on much larger ingots. It is generally true to say that a duplex exothermic/insulating head is more expensive to apply than a simplex super-insulating one. In addition the heat retention of a duplex head is inferior over long periods of time unless the backing is a very good insulator.

It can safely be recommended that small ingots of up to 250 mm section can best be headed with exothermic lined reservoirs. These may be either simple exothermic shapes or units comprising exothermics backed by refractory or insulator. The former type can be used on any small moulds but the thickness necessary and the associated cost normally preclude their selection on moulds of more than 200 mm square. Duplex heads can be used on very much larger ingots providing that the insulation properties of the backing material are good enough to prevent high heat losses by conduction. In fact duplex heads have been used satisfactorily on ingots in excess of 20 tons, but the economics of such applications can be bettered by employing super-insulators.

When considering the application of hot-tops to ingots larger than 400 mm square, better feed contours can normally be obtained with the super-insulator than with an inferior insulator faced with exothermic since the total heat output of an economically acceptable quantity of exothermic is very small compared with the heat content of the feeder head.

Various applications have thus been defined. Simple exothermic or duplex heads can be recommended for all ingots up to 250 mm square and super-insulators for all ingots over 400 mm having considered thermal efficiency and cost. There remain, however, those ingots between 250 and 400 mm in section where the selection of head cannot be governed by a general recommendation. For all sizes of head within the group it may be anticipated that duplex, exother-mic-faced heads will give slightly higher yields than insulators, but at an appreciably higher cost. The choice of feeder head will depend upon whether the extra cost of exothermics is offset by the value of prime metal yield increase and some consideration will have to be given to speed and ease of application.

Covering Compounds

As in the case of side-wall linings the anti-piping compound should be selected to give the properties most desirable for a specific application.

For use on smaller hot-tops the powder should be sensitive and give a high and rapid heat output to prevent initial chilling of head metal and to contribute heat to the metal in contact with it. Since such ingots will solidify rapidly the insulation properties of burnt-out residues are less important than on larger ingots.

With large ingots it is only necessary that exothermic compounds, on burning, satisfy the heat capacity of the covering layer to prevent heat extraction from the head metal. They must, however, when burnt, have residues which are highly insulating, since they need to retain heat in the feeder head over long periods of time.

Application Rates

To a large extent the application rate of side-wall linings will be governed by their price. For a given heading cost, cheaper materials can be used in greater quantities. Thus it is possible to obtain a variety of side-wall linings with differing exothermic or insulating properties. Many of these can be expected to give similar results providing that the thickness of application is modified to give the same heat retention properties of the lining. However, excessive wall thickness of linings, whilst preventing heat losses through them to a satisfactory degree, may so alter the head volume or geometry that an inferior feed pattern results.

In the case of the best insulating side-wall linings currently available approximate wall thickness recommended by a typical supplier in the feeding of normal ingots is 18 to 20 mm on ingots with sections up to 550 mm square, 25 mm on sections of 550 to 700 mm square and 30 mm in almost all larger ingots. Figure 8 shows in graph form the total heat lost through varying thicknesses of a super-insulator widely used in the heading of steel ingots. It can be seen that an increase in wall thicknesses from 30 to 40 mm reduces only slightly the total heat lost, whilst increasing the application cost considerably. Thus it is not normally recommended that thicknesses in excess of 30 mm are used except in special circumstances. It must be stressed that such arguments can only be applied to materials with excellent insulating properties and inferior results will almost certainly be obtained if less efficient insulators are employed at the same thicknesses.

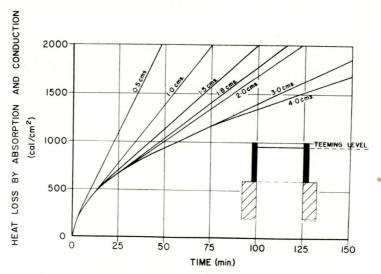

FIG. 8. Heat loss by absorption and conduction.

Anti-piping compound application rates also depend upon ingot size and the nature of the product to be used. The most important factor is to provide and maintain a continuous layer over the whole top surface of the head metal throughout solidification. This can be done by using either a compound which burns and forms a self-supporting crust or a material which, when burnt, gives powdery residues which feed down with the steel and form a coherent cover. Both methods are satisfactory, but those intermediate products which form a weak crust should be avoided. If the covering layer collapses, ruptures or contracts from the edges, allowing heat to radiate from the steel to the atmosphere, inferior results will be obtained.

It is not possible to generalize by relating application weights of anti-piping compounds to the size of ingots being treated since the variation in the length-to-breadth ratio of moulds differs widely and the density of covering powders varies greatly from supplier to supplier and from grade to grade. As a rough guide it will normally be found that a covering layer of 15 to 20 mm thick will be employed on ingots up to 300 mm square, 20 to 30 mm thick on ingots of 300 to 500 mm square and 30 to 60 mm thick on larger ingots, but this only applies where the correct grade of powder has been selected.

The Effect of Hot-topping on Segregation

It has been described and illustrated how the hot-top affects the feed characteristics of the ingot. Figure 9 shows the formation of segregation-rich areas in open-topped (unheaded), brick-topped and super-insulator- or exothermic-topped ingots. The freezing front associated with the unheaded or the inferior

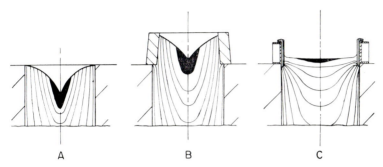

FIG. 9. Formation of segregation-rich areas in ingots. (A) Open-topped; (B) brick-topped; (C) exothermic hot-topped.

headed moulds results in the last metal to solidify being in the form of a deep inverted cone penetrating well into the ingot body. In the case of an efficient head the last, segregate-rich metal to solidify is in the form of a wide shallow dish and is wholly contained within the residual head. In all three cases the volume of liquid metal in which segregate concentration is at an undesirable level is probably similar. The distribution after complete solidification is, however, very different and can result in greatly varied prime metal yields.

In 1964 J. L. Duchateau[4] illustrated the segregation of carbon, sulphur and phosphorus in a 2·5 ton big-end-up ingot headed with a super-insulator. This is reproduced in Fig. 10. Twenty drilling positions were selected as shown and analyses were carried out. The results are given in the table shown in Fig. 11.

The steel quality specified a carbon level between 0·95 and 1·05% and it can be seen that, at positions 3 and 8 carbon segregation has been excessive. Sulphur and phosphorus also occur in greater concentrations in the upper, central part of the head.

When 140 mm square billets were rolled from identical ingots and analysed for carbon content of the surface and centre, at various distances from the head, carbon and sulphur variations existed. Carbon levels at corresponding positions in billets made from ingots with refractory, insulating and duplex-exothermic heads were compared graphically and are reproduced in Fig. 12.

Since the maximum carbon level acceptable in this instance is 1·05% it can be seen that the discard necessary to clear undesirable material is 7·8% with the exothermic head and 9·0% with the super-insulator. In the case of the refractory head, centre line carbon segregation is such that very large discards would be necessary and satisfactory heading of this grade of steel could not be achieved

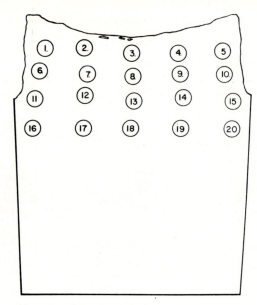

F<small>IG</small>. 10. 2·5 ton big-end-up ingots headed with super-insulator. Position of drilling holes.

Drill Analyses

Position		1	2	3	4	5	6	7	8	9	10
Carbon	%	0·98 0·92	0·98 0·98	1·15 1·15	0·95 0·95	0·95 0·94	0·95 0·95	0·98 0·98	1·02 1·04	0·98 1·00	0·98 1·00
Sulphur	%	0·008	0·008	0·016	0·008	0·007	0·005	0·007	0·008	0·007	0·008
Phosphorus	%	0·010	0·010	0·012	0·010	0·010	0·010	0·010	0·010	0·010	0·010

Position		11	12	13	14	15	16	17	18	19	20
Carbon	%	1·00 1·02	0·95 0·96	0·95 0·95	0·98 0·99	0·92 0·98	1·00 1·00	1·00 1·00	0·96 0·95	0·98 0·98	0·93 1·00
Sulphur	%	0·009	0·007	0·008	0·007	0·007	0·008	0·008	0·007	0·007	0·008
Phosphorus	%	0·010	0·010	0·010	0·010	0·010	0·010	0·010	0·010	0·010	0·010

F<small>IG</small>. 11. Analysis results of 20 samples of Fig. 10.
(By courtesy of R. Hallé of Société Ugine, France.)

economically. The choice between exothermic and super-insulator would depend on whether the 1·2% difference in yield is greater in value than the difference in the heading cost.

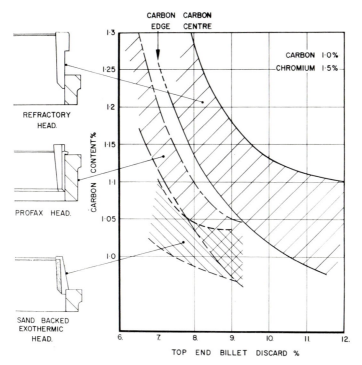

FIG. 12. Top end billet discard in %.

HOT-TOP VOLUMES AND FEEDER RATIOS

The hot-top volume, comprising that liquid metal contained within the exothermic, insulating or refractory-lined top reservoir, is normally quoted as a fraction of either the volume of the body below the shoulder of the ingot or of the whole ingot, and is termed a feeder ratio. The following formulae express net and gross feeder ratios.

$$\text{NET FEEDER RATIO} = \frac{\text{INGOT HEAD VOLUME} \times 100}{\text{INGOT BODY VOLUME}}\%$$

$$\text{GROSS FEEDER RATIO} = \frac{\text{INGOT HEAD VOLUME} \times 100}{\text{INGOT BODY VOLUME} + \text{INGOT HEAD VOLUME}}\%$$

The volumes contained by feeder heads vary greatly and are governed by mould sizes and proportions, steel qualities and the type and quality of hot-top to be used.

In the case of small moulds, especially those with a high height-to-breadth ratio, much of the liquid metal solidifies during the teeming period and, by the time the mould and hot-top are full, there remains perhaps only half the total contraction to be compensated by the hot-top. In such cases gross feeder ratios can be as low as 4%. As larger moulds are considered, the proportion of steel contraction to be compensated increases and the volume of available metal within the hot-top must increase.

Typical gross feeder ratios, when efficient exothermic hot-tops are used, vary between 4 and 10% depending upon ingot size, geometry and the type of steel being cast. In the case of refractory hot-tops gross feeder ratios are much greater, normally varying between 13 and 18% for mill ingots. It is much more difficult to quote feeder ratio for those ingots headed with super-insulating hot-tops since the insulation properties of heads from numerous suppliers vary greatly. It can, however, be said that a hot-top with very good insulating properties will usually have a gross ratio of 9 to 12%. An inferior insulating head with the same wall thickness would need a larger head volume of, say, 12 to 14%.

An inferior insulator with increased thickness can be anticipated to approach the thinner, better insulator in efficiency. Again the selection may be dictated by the application cost.

All previous comment has referred to either mill ingots or small forging ingots. There are, in addition, many large forging ingots where heading techniques are different. Frequently in such cases the residual head is used for holding and manipulating the ingot during hot working. Also with these ingots it is considered that a large safety margin is acceptable. As a result hot-tops in such applications do not aim at low residual volumes, but at achieving the correct head shape for ingot handling and a wholly sound ingot below the shoulder. For this reason the ingots are normally cast big-end-up and are fed by very large head volumes contained in refractory lined head boxes.

Yield Considerations

The benefits of hot-tops are seen principally as an increase of usable steel output, and better quality of that output. If a refractory hot-top using, say, 16% gross feeder ratio can be replaced by super-insulating hot-top of 11%, the saving is 5% of metal, providing both ingots are of equally acceptable quality below the shoulder. Since segregation patterns are better with super-insulators than with refractories, in addition to this 5% saving, the steel below the shoulder is normally of better quality.

On smaller ingots with exothermic replacement of refractory tops the saving can be even more advantageous. A 16% gross feed volume may well be replaced by an 8% exothermic head, thus saving 8% of liquid metal and giving a better quality product.

The saved metal can best be utilized in one of two ways. Firstly, from any given weight of ladle metal, for every ingot made with a refractory head, a longer chill length ingot can be made with an exothermic or insulating head. The increase in sound metal yield per ingot is equal to the increase in chill volume. Alternatively ingots can be made, replacing refractory heads as before, but maintaining the same chill length. After filling the same number of moulds there may then be 5 to 8% of the tapped metal weight remaining in the ladle. This can be made into additional ingots and represents an increase in yield.

In many instances a hot-top user makes a product or a multiplicity of products of specific length or size, from an ingot. Typical examples of this are rails from a square ingot and plates of prescribed dimensions from a slab ingot. In these instances it is unlikely that an increase in sound material in any one ingot will serve a useful purpose, so that the benefit of high yield must be gained from casting the same metal into additional moulds. However, setting aside for the moment discussions of yield increase, it is often desirable that a user be able to vary the sound yield from an ingot to suit certain end-products. For instance

Fig. 13. Eight-piece assembly of insulator tiles and corner pieces inserted below mould top.

the production of $4\frac{3}{4}$ rails from an ingot wastes approximately 15% of the sound metal available, since acceptable rails must be of a given length. If 5 rails per ingot cannot be made by increasing the yield with hot-tops, the chill length should be modified to give 4 sound rails so that the balance of the steel can be used for making additional ingots.

Although refractory hot-tops can be used for varying chill length, the range of insertion depth is relatively small and the nature of the refractory gives a lower yield than can be obtained from thermally more efficient heads. Insulating and exothermic hot-tops, properly designed, lend themselves admirably to fitting at any height in a mould where tiles can be wedged or nailed into position quickly and easily.

Figure 13 illustrates an 8-piece assembly of super-insulator tiles and corner pieces inserted below the mould top. To facilitate holding the tiles until the assembly is complete it is usual to use a jig, an example of which is shown in Fig. 14.

FIG. 14. Jig for tile assembly.

To summarize the comments on yield, it can be said that, from a specific weight of liquid steel cast, the replacement of refractory heads by smaller more efficient insulating or exothermic heads will result in an appreciable increase in good usable material. Similarly the reduction of chill length in certain instances will further increase usable material. In all cases where exothermic and insulating heads are used the quality of the bloom or billet product is improved with

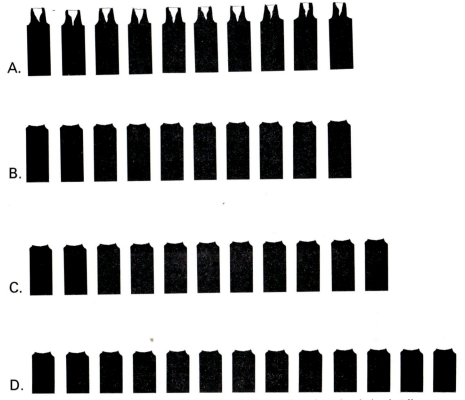

FIG. 15. Possibilities for using saved metal by exothermic or insulating heading.

A. 10 ingots made in moulds with brick hot-tops, chill length x cm.
B. 10 ingots made in moulds with efficient hot-tops, and chill length of 1·1 x cm.
C. 11 ingots made in moulds with efficient hot-tops and original chill length of x cm.
D. 13 ingots made in moulds with inserted, efficient hot-tops to reduce the chill length to 0·8 x cm.

respect to centre-line segregation. Some of the possibilities for using saved metal are outlined in Fig. 15.

Special Hot-top Applications

(i) When designing a hot-top for a slab mould, where the length-to-breadth ratio of the cross-section is large, a consideration of the mode of solidification can again be of use. Figure 16 shows in plan view three stages in the solidification of an open-top ingot. The freezing fronts tend to advance rapidly from both the sides and the ends of the section and the last metal to solidify lies on a central vertical plane, penetrating deep into the ingot. Similarly, if all four sides of the section were lined with hot-topping materials, the contours of the freezing fronts would not alter relative to the section, but solidification would be greatly retarded, the depth of the vertical plane would be much less than with an unheaded ingot and a better yield would result. With an efficient head all primary unsoundness would be contained in the residual head metal.

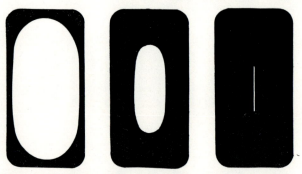

FIG. 16. Three stages in the solidification of open-top ingots.

FIG. 17. Solidification of ingot with only two sides lined.

If, however, only the longer sides of the section were to be lined with material, the mode of solidification would be different, as illustrated in Fig. 17. There would be a steep thermal gradient at the lined faces and a shallow one at the ends, caused by the mould chilling effect. With a correctly proportioned section this can result in an axial thermal centre, and good feed characteristics. It should be anticipated that such a two-slab assembly would give a yield 0·5 to 1·0% lower than a four-slab assembly in the same mould.

To be able to use only two insulating or exothermic plates on a large mould is obviously desirable to the user from an economic point of view. Whether or not this can be done depends upon the length-to-breadth ratio of the section and practical experience has shown that this should be in excess of 1·7 to 1·0.

(ii) In some instances, where the cost of 8-piece assemblies is too high for a specific application, it is possible to use four slabs without corner pieces. When thin slabs are used, the corner areas of the head will not be protected against heat loss by contact between mould and liquid metal. With larger ingots this effect is not great and satisfactory feed patterns can be obtained, although final yields are likely to be marginally smaller than where full insulating linings are used.

The Effect of Standing Time

It has long been realized that ingots must remain stationary during the whole of the solidification period to obtain best segregation patterns. This usually presents no problem except where moulds are cast on bogies. Since bogies are used in the casting of a very large proportion of all ingot steel, it is essential that, wherever possible, no external factors such as premature movement influence yields. The standing or "track" time can be defined as the period between teeming the last ingot of a bogie train and movement of the train to a railway siding or stripper bay.

The danger of moving ingots during the middle and late stages of solidification is proven by the recurrence of poor segregation patterns at the billet stage and consequent reduced yields. An investigation has been carried out by a steelworks on steel qualities of (A) below 0·40% carbon and (B) above 0·40% carbon, using both 12·4% gross and 10·6% gross super-insulator head volumes, and coverings of a good quality exothermic feeding compound. The ingots investigated were 5 tons, 570 mm square and big-end-down.

The graphs in Fig. 18 relate top-end billet discard to clear segregation with the standing time of the ingot train for the two carbon content groups. Where discard is recorded as more or less than a specific figure the graph location is accompanied by an arrow pointing in the appropriate direction. From previous considerations it is calculated that such an ingot would solidify in approximately 120 minutes. The graphs show clearly how the discard in each case falls suddenly when standing times of 115 to 120 minutes are used. Thus the benefit of allowing the ingots to stand until complete solidification has occurred is illustrated very positively.

In certain production circumstances it is impossible to leave the ingots at the teeming site until solid, in which case a compromise must be sought which will give the best results under prevailing circumstances. The best course of action is to move the ingot train as soon as possible after completion of teeming and place it in a sheltered railway siding where it will remain unmoved until solidification is complete. Great care must be taken to prevent splashing of the metal over the ingot top and loss or breakage of the feeding compound cover. Exposure to rain will also have some deleterious effect on the feed pattern, If such a train movement is made it is almost inevitable that the billet yield will be slightly inferior to that from a train which has solidified in the teeming bay. However, it will be appreciably better than the yield from a train moved in the later stages of solidification.

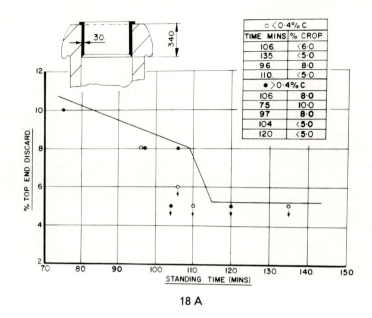

18 A

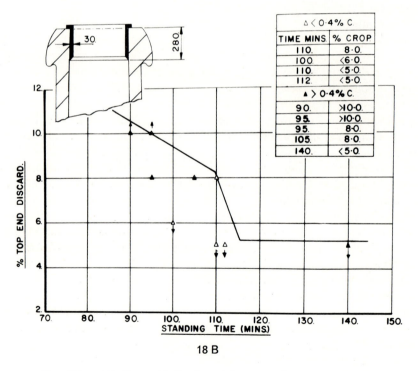

18 B

FIG. 18A and B. Top-end billet discard for two carbon content groups.

METHODS OF APPLYING HOT-TOPS

1. Refractory Hot-tops

The refractory head is well known and can take one of several forms. Figures 19A, B and C illustrate three forms in which it is commonly found. In most cases the refractory can be in the form of brick with or without a dried refractory facing layer, or can be of a monolithic rammed material. The life of such heads may be quite long if care is taken to repair the surface in contact with the steel when necessary. The high heat capacity of the refractory media used and their relatively poor insulating properties necessitate the use of high head volumes and feed and segregation patterns are usually inferior to those obtained from smaller super-insulator assemblies.

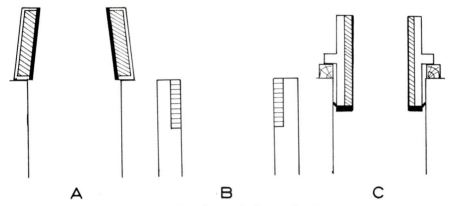

FIG. 19. Three forms of refractory heads.

Head A comprises a cast or fabricated metal container, lined with refractory bricks, and is placed on top of the ingot mould. It is essential that contact between head box and mould top is sufficiently good to prevent horizontal penetration of liquid metal. This would freeze in the form of a fin which would tend to support the ingot on contraction, resulting in metal tearing or hanger cracking.

Head B is a refractory material inserted in the form of either bricks or rammed monolithic into a preformed recess at the head of the ingot mould. There is again a danger of horizontal metal penetration at the base of the refractory and consequent hanger cracking.

Head C consists of a cast metal shell, lined with refractory material. The dimensions of the head are such that the outer surfaces are free to move vertically with respect to the mould and a degree of insertion can be obtained. The head, prior to use, is supported on wooden blocks which can be knocked out when the head is full. The head box then floats on the steel and can fall as the ingot body contracts. Penetration of liquid metal between head box and mould is prevented by the incorporation of a preshaped annular steel strip.

[*Note*: The comment on yield with refractory heads is valid only when the refractory is presented to the liquid steel. If these types of head are lined or faced with more insulating material, better yields and segregation patterns can be expected.]

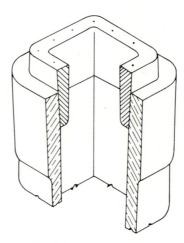

FIG. 20. One-piece sleeve.

FIG. 21. Four-slab assembly.

2. Exothermic Hot-tops, One-piece and Multiple-piece

Heads of this type usually comprise a simple exothermic shape for ingots of less than 150 mm square or exothermic backed by bonded sand or insulating material for ingots larger than 150 mm square. Whatever type is used, the head will displace a certain quantity of liquid steel and will consequently tend to float. The buoyancy forces must then be overcome by either a good friction fit or the use of a retaining/supporting ring.

The simplest fitting is a one-piece sleeve (Fig. 20) fully or partly inserted, and so shaped that it wedges into position in the top of a mould, support being given by the mould itself. On larger moulds this can be replaced by a four-slab assembly (Fig. 21) which is fitted with the aid of a simple collapsible jig or suspension straps and held in position by driving in wooden wedges. In both instances frictional forces prevent the head from floating.

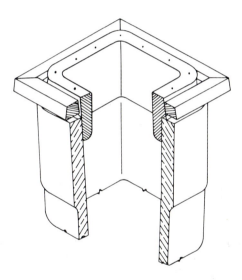

FIG. 22. Partly inserted exothermic head.

If an exothermic head is designed to be only partly inserted (Fig. 22) or superimposed (Fig. 23), support must be given to the exposed portion of the head and weight must be supplied to overcome buoyancy forces. This is conveniently done by forming the necessary counterweight into an annular ring which will fit closely to the outside of the head and give good support.

It should be noted that a small part of most superimposed sleeves is inserted into the mould. This portion is termed a skirt and prevents horizontal metal penetration between head and mould and resulting hanger cracking.

Fig. 23. Superimposed exothermic head.

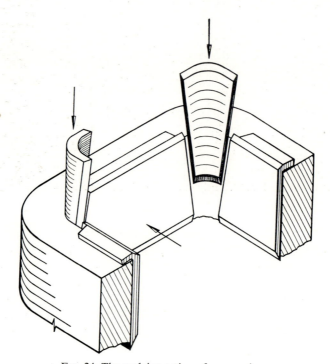

Fig. 24. The wedging action of corner pieces.

3. Super-insulating Hot-tops, Multiple-piece

Such assemblies comprise relatively thin tiles made with the correct curvature of the back faces to fit the camber of the mould or retaining head box. These are held in position by either wedging action, or nailing, or clipping to the supporting metal.

Wedging is best done by making an eight-piece assembly where correctly curved tiles are located centrally on the mould faces and shaped corner pieces are driven in to lock the whole head into a secure unit. The two-dimensional angle at which tile and corner piece meet is so designed that differences in mould size are compensated by a small vertical movement of the corner piece (Fig. 24). The wedging action also ensures that tiles are pressed back against the mould wall to prevent metal penetration between heading materials and mould. During the application of wedges it is necessary to support the tiles by hanging them on incorporated ledges, suspending them on straps or wires, or holding them in a jig. The assemblies may be inserted directly into the mould as

FIG. 25. Insulating tiles directly inserted into mould.

shown in Fig. 25 or be placed in a head box (Fig. 26). In the latter case the head box must conform in inner contour to the mould shape and the insulating tiles and corner pieces should cover the mould/head box joint.

In some assemblies the corner pieces are inversely tapered, the wider end at the bottom. These are suspended at the correct height in the corners of the mould and the slabs are pushed into position to obtain the correct wedging action.

It is also possible, especially with thicker insulating slabs used in moulds with

Fig. 26. Insulating tiles placed in a head box.

Fig. 27. Insulating tiles held by steel clips.

small corner radii, to obtain a satisfactory wedging action by shaping the ends of the tile to meet in a mitre or butt joint. Additional wooden wedges may also be used for extra security.

Steel clips may be used as a method of holding two- and four-piece insulating assemblies where no wedging action is employed. These can be in either of the two forms illustrated in Figs. 27 and 28. They are normally shaped from cold rolled steel bar and have sufficient resilience to retain a pressure on the tile after fixing.

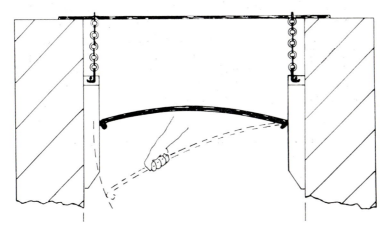

FIG. 28. Insulating tiles held by steel clips.

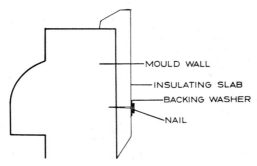

FIG. 29. Section through a securing nail.

A less expensive method of holding non-wedging assemblies in position employs the use of a metal nail. This is fired by an explosive charge from a specially designed gun. The nail penetrates the insulator and enters the surface of the ingot mould. A backing washer gives a greater area of support to the nail head. With such a method it is necessary to adjust the nail length and cartridge charge for differing thicknesses of insulating tile. If no other fixing methods are incorporated with the nailing technique, the nails should be placed

FIG. 30. Securing tile to mould with nail-firing gun.

FIG. 31. Typical nail pattern on large tile with hanging ledges.

in a pattern which will prevent metal penetration between tile and mould. Figure 29 illustrates a section through a securing nail, Fig. 30 shows a nail-firing gun being used to secure insulating tiles hanging on incorporated ledges and Fig. 31 a typical nail pattern on large tiles.

A number of proprietary guns is available most of which incorporate devices ensuring complete safety in use. It should be noted that nail residues must be removed before subsequent application of tiles.

4. Insulating Assemblies, One-piece

The multiple-piece assembly can be replaced by a single-piece feeder head, three types of which are drawn in Fig. 32. Typical examples of these used in production are photographed in Fig. 33.

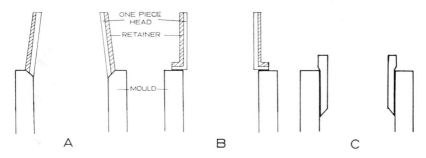

Fig. 32. One-piece insulating top.

33 A

Fig. 33A, B, and C. Typical one-piece insulating tops.

Type A is a simple thin-walled truncated pyramid which is superimposed on an ingot mould. A good union between mould and head is obtained by designing a heavily tapered section at the base of the hot-top. This fits into a machined mating recess at the top of the mould. The whole of the outer surface of the head must be supported by a head box.

Alternatively, Type B head may have a basal flange which fits between the head box and mould. This obviates the need for special mould shape and ensures a good shoulder seal.

C

33 B

33 C

Type C is a parallel-sided sleeve whose outer dimensions allow whole or partial insertion into the top of the mould. A thickening of the outer wall may be used to suspend the sleeve. With this type of head, unless used on a big-end-up mould where a wedging action can be obtained by external shaping, precautions must be taken to prevent the head from floating.

Insulating assemblies described in earlier sections are suitable for use on all

square and rectangular sectioned moulds. There are, however, many moulds
which are circular, hexagonal, octagonal, duodecagonal, fluted, etc., and the
use of wedging, clipped or nailed assemblies is difficult. In such cases a one-piece
assembly such as shown in Figs. 32 and 33 can be used, provided, in the case of
Type B, that the internal head dimensions at the base are the same as, or slightly
less than, the minimum corresponding mould dimensions. This is necessary to
prevent hanger cracking.

THE ECONOMICS OF HOT-TOPPING

There is little doubt that replacement of a refractory head by either exothermic
or insulator, or the employment of a hot-top on a previously unheaded mould,
will result in an increase in cost. It is anticipated, however, that the value of the
resulting increase in yield will exceed the extra application cost by a large margin
which will reflect the financial benefit to be gained. The saving (S) to be achieved
by sophisticated heading techniques is given by the formula

$$(S = V - C - W)$$

where $V =$ the value of the increase of sound metal yield at the shear,
$\quad\quad C =$ the *increase* in cost resulting from replacement of the existing
$\quad\quad\quad$ hot-top by exothermics or insulators,
$\quad\quad W =$ the scrap value of the increase in yield, since this quantity of metal
$\quad\quad\quad$ will not be returned for remelting.

The economics of replacing an existing head method by either exothermic
(Feedex) or insulator (Profax) have been illustrated by A. S. Godbehere[5]
in his study of typical product usage in the United Kingdom. His table, reproduced in Fig. 34, is based on the following assumptions.

(a) Number of ingot tons under consideration   10,000

(b) Billet yield from replaced hot-top   80%

(c) Cost of replaced hot-top per ingot ton   £0·15

(d) Cost of exothermic replacement head per ingot ton   £0·60
i.e. increase in cost due to exothermic head per
ingot ton £0·45, or per 10,000 ingot tons   £4,500

(e) Cost of insulator replacement head per ingot ton   £0·40
i.e. increase in cost due to insulator head per
ingot ton £0·25, or per 10,000 ingot tons   £2,500

(f) Cost of scrap needed to replace extra yield per ton £15

(g) "Break-even" refers to the point at which the extra cost entailed in
replacing an existing heading method by either exothermic or insulator
equals the value of the increased yield at the shear.

Col. 1 Possible billet yield from FOSECO heads (%)	Col. 2 Increase in yield from FOSECO heads (%)	Col. 3 Extra scrap to replace yield increase (tons)	Col. 4 Total cost of extra scrap needed (£)	Col. 5 Total extra cost with FEEDEX Column 4 + d (£)	Col. 6 Total extra cost with PROFAX Column 4 + e (£)	Col. 7 Weight of billets produced (tons)	Col. 8 Extra billet produced (tons)	Col. 9 Break-even price (£) Col. 5/Col. 8 FEEDEX	Col. 9 Break-even price (£) Col. 6/Col. 8 PROFAX
80	—	—	—	—	—	8000	—	—	—
81	1	100	1,500	6,000	4,000	8100	100	60	40
82	2	200	3,000	7,500	5,500	8200	200	37·5	27·5
83	3	300	4,500	9,000	7,000	8300	300	30	23·3
84	4	400	6,000	10,500	8,500	8400	400	26·2	21·2
85	5	500	7,500	12,000	10,000	8500	500	24	20
86	6	600	9,000	13,500	11,500	8600	600	22·5	19·2
87	7	700	10,500	15,000	13,000	8700	700	21·4	18·6
88	8	800	12,000	16,000	14,000	8800	800	20·6	18·1
89	9	900	13,500	18,500	16,000	8900	900	20	17·8
90	10	1000	15,000	19,000	17,500	9000	1000	19·5	17·5

FIG. 34.

(h) All tabulated values are quoted in pounds sterling and represent average figures which do not include feeding compound. (It is assumed that this commodity was also used in the replaced heads and is not, therefore, an additional charge.)

The figures contained in the table enable a graph to be drawn (Fig. 35) to illustrate more clearly the break-even point for a given increase in yield.

As an example, for an increase in yield at the billet stage of, say, 3 %, the steel-worker breaks even if the value of his steel, also at the billet stage, is £30 per ton. If he either increases his yield further with the same steel or maintains the yield with more expensive steel he has profited from the use of exothermic hot-tops.

With a less expensive insulating hot-top his break-even price would be lower, in this instance £23 for a 3 % yield increase. Thus he can profit from hot-topping cheaper grades of steel by changing from exothermics to insulators providing he can maintain the yield.

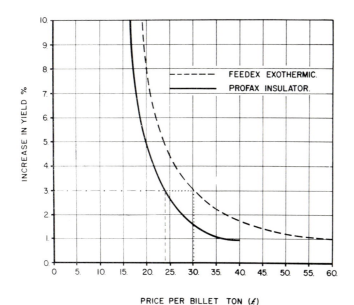

FIG. 35.

REFERENCES

1. GUZZONI, G., Estratto dal Bolletino Tecnico Finsider, numero 233, Luglio, 1966.
2. MUELLER, J. W., and BOLE, G. A., *Steel*, **126**, 80–82 (1950).
3. FENTON, G., *Journal of the Iron and Steel Institute*, **186**, 396–405 (1957).
4. DUCHATEAU, J. L., Le Massellottage PROFAX, Paper given to Commission des Ingenieurs des Acieries Electriques, March 1964.
5. GODBEHERE, A. S., Hot-topping steel ingots, Paper given to a meeting of the Ministry of Heavy Industry in Katowice, Poland, 1964.

HOT TOPPING OF ALLOY STEELS

J. G. Emmott

HISTORICAL

Mention has been made in the previous section of the use of exothermic hot-tops in alloy steel ingots. Preliminary work was carried out on small, high-quality steel ingots since these demonstrate poor feed patterns to a greater degree than the larger ingots, the risks in conducting pioneer work appeared to be less and the savings to be obtained greater than with the low alloy and carbon steel types.

In the late 1940's and early 1950's exothermic compounds had been used satisfactorily in the foundry industry in the feeding of sand castings. Up to 1952 many small ingots were being headed using preheated fireclay sleeves, frequently referred to in the industry as "hot dozzles", a practice which persists in a few steelworks today. A simple fireclay sleeve is heated in a conveniently situated gas furnace to temperatures in excess of 1000°C. It is then placed at the top of the ingot mould immediately before teeming commences. When filled with liquid steel it extracts less heat from the ingot head than warm firebrick and feeding metal is retained in the liquid phase for longer periods than hitherto. Consequently feed patterns obtained with hot refractory sleeves are better than with brick heads. However, the hot dozzles are still considerably lower in temperatures than the liquid steel when the feeder heads are filled and heat is extracted from the metal to raise the temperature of the refractory. In addition the fireclay is a poor insulator and permits rapid dissipation of heat to the backing mould, head box or atmosphere.

After the successful application of exothermics in foundries and the already demonstrated advantages of hot feeder heads on ingots it was reasonable to suppose that exothermic sleeves could also be used with benefit on steel ingots. Not only would they not extract heat from head metal but, if correctly selected they may even increase metal temperature. During 1952 and 1953 a great deal of test work was carried out with such applications in Europe leading to adoption in production. Figures 1 to 4 illustrate a very early production application on two small ingots in the United Kingdom. Figure 1 shows two clay dozzles previously used and alongside an exothermic and an exothermic/sand backed hot-top designed to replace them. Figures 2 and 3 illustrate the same exothermic sleeves in position in the ingot moulds and Fig. 4 shows ingots made with the various heads as follows:

(A) Ingot made with hot dozzle.
(B) Ingot made with simplex exothermic hot-top.

(C) Ingot made with hot dozzle.

(D) Ingot made with duplex exothermic/sand backed hot-top.

During the introductory period H. D. Shephard recognized the advantages of lining refractory dozzles and he reasoned that the burning exothermic would achieve higher temperatures (1500°C to 1700°C) than the preheated sleeves alone. Trials and early production work in the U.S.A. employed the method of

Fig. 1.

Fig. 2.

FIG. 3.

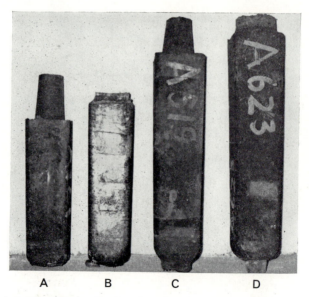

A   B   C   D

FIGS. 1–4. Very early application of exothermic sleeves on alloy steel ingots.

Fig. 5A. Refractory dozzles after casting.

Fig. 5B. Moulds with exothermic lined dozzles ready for casting.

c*

FIG. 6A. Ingot head from hot dozzle with side-wall skull to teeming level.

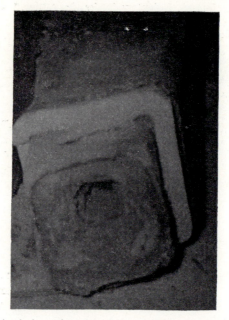

FIG. 6B. Central pipe in ingot from Fig. 6A, penetrating deeply below the shoulder.

lining fireclay dozzles and Figs. 5 and 6 illustrate the applications compared with normal preheated dozzles. Figure 5A shows dozzles after casting and Fig. 5B a pit of moulds with exothermic lined dozzles in position ready for casting. Figure 6 is an ingot head from a hot dozzle with side-wall skull to the teeming level and a central pipe penetrating deeply below the shoulder. For comparison in Fig. 7 is shown a stack of ingots made with exothermic linings to the dozzles. The residual heads are small and free from pipe cavitation. This work was published in 1956.[1]

FIG. 7. Stack of ingots made with exothermic lining to dozzles.

CURRENT PRACTICE

Although some feeder heads are still made by lining fireclay sleeves, it has become more usual to make either simple monolithic sleeves in exothermic materials or duplex heads which comprise a facing of exothermics surrounded by sand or refractory, moulded in the making of the sleeve. A large variety of such shapes is used today and a selection is shown in Fig. 8.

Most of the high-speed and special high alloy steels made are cast into small ingots which can be headed very satisfactorily with exothermic or exothermic/

FIG. 8. A variety of exothermic shapes.

TABLE 1.

Type	Steel quality	Price/ton in 5 in. billet £	Price/lb in 5 in. billet s.
1	Carbon Steel (0·12/0·32% C)	34·10	0·31
2	Tube Steel (up to 0·41% C)	41·80	0·37
3a	Alloy Steel 0·75% Ni	54·65	0·49
3b	3% Ni, 1% Cr	80·15	0·71
3c	3·5% Ni, 1% Cr, 0·3% Mo	96·20	0·86
4	Austenitic Stainless Steel	253·50	2·26
5	Martensitic Stainless Steel	138·00	1·23

refractory-backed feeding units. With austenitic stainless steel, however, much larger ingots are made and better feed patterns can be obtained using super insulators on the larger sizes.

It must be remembered that high alloy steels are of greater intrinsic value than low alloy and carbon qualities. The above table shows the cost of certain types of steel in 5 in. square section billet of various specifications as listed in publications in the United Kingdom in October 1968. These figures are given (Table 1) as a simple example and, whilst prices change from country to country and from time to time, a similar relationship of costs persists between the qualities of steel.

Considering a 1000 lb ingot, assuming that a brick or fireclay hot-top would result in a billet scrap level of 14% top-end discard, compared with 8% discard with one exothermic head, and 6% with another, the following observations can be made (Table 2).

TABLE 2.

Weight of ingot plus head	1000 lb
Discard at billet stage with fireclay head	140 lb
Discard at billet stage with 1st exothermic head	80 lb
Discard at billet stage with 2nd exothermic head	60 lb
Saved billet metal from 1st exothermic head	60 lb
Saved billet metal from 2nd exothermic head	80 lb

It now remains to compare the value of 60 and 80 lb weights of the various steel alloys from Table 2. If the value of the saved metal exceeds the *extra* heading cost attributable to the exothermic head and anti-piping compound an economical advantage will have been gained from the use of the more efficient head. The cost of applying exothermics to ingots varies greatly depending upon the size of mould being used. (It is cheaper to make, pack and transport one large sleeve for a 1500 lb ingot than to make 15 sleeves for 100 lb ingots.) In addition the quality of exothermics varies from supplier to supplier and consequently the cost for a given performance may also vary.

If 22 shillings is assumed as an approximate figure for exothermic heading costs per ton of steel a suitable duplex head for a 1000 lb ingot would cost 10 shillings. By comparison a clay dozzle for the same mould may be expected to cost in the region of 4 shillings per unit. Thus the *extra* cost of an exothermic head would be approximately 6 shillings per ingot, assuming that the same anti-piping compound is used in each case.

The value of the saved metal with each type of exothermic head is illustrated in Table 3.

TABLE 3.

Steel type	Value of saved metal		Difference in metal value between exothermic heads s.
	1st exothermic head (60 lb) s.	2nd exothermic head (80 lb) s.	
1	18·6	24·8	6·2
2	22·2	29·6	7·4
3a	29·4	39·2	9·8
3b	42·6	56·8	14·2
3c	51·6	68·8	17·2
4	135·6	180·8	45·2
5	73·8	98·4	24·6

Thus, it can readily be seen that either of the exothermic heads can be used with a significant saving to the steelmaker on any quality of steel providing that the ingot yield is required free from primary pipe and sulphur segregation. There is ample margin also for the use of an improved, higher-cost anti-piping compound.

However, when it comes to a choice between two exothermic heads circumstances could and do exist where the value of the yield increase is less than the increased application cost, especially with the lower quality steels.

Because of the saving to the steelmaker in hot-topping high alloy steels it is customary to ensure yield increases by using slightly higher head volumes than with plain carbon qualities.

THE EFFECT OF ALLOYING ELEMENTS ON FEEDING CHARACTERISTICS

Alloying elements affect the contraction characteristics of the steel during the significant periods of cooling in the liquid phase and during the change from liquid to solid.

During foundry experiments to measure volume change of steel during cooling Pilling and Kihlgren[2] expressed the contraction as a percentage of the mould volume and illustrated the significance of carbon content as follows:

Carbon content %	Total contraction %
0·10	10·4
0·45	11·3
0·70	12·1

The total contraction is the sum of liquid contraction, solidification contraction and solid contraction.

Other elements behave differently. Contraction of liquid steel is also increased by silicon, manganese and phosphorus and decreased by chromium and aluminium. The effect of these alloys on the liquid-to-solid shrinkage is uncertain.

When steels are to be hot-topped it should be remembered that high carbon, manganese and silicon steels will need a greater volume of compensating feed metal than low carbon, low alloy types. Feeder head volumes should be increased accordingly to supply this extra metal and increase the safety margin. Fortuitously the higher alloy steels can tolerate higher heading costs.

REFERENCES

1. SHEPHARD, H. D., New hot-tops save metal, make sounder ingots, *Iron Age*, **176**, November 24th, 1955.
2. Pilling and Kihlgren Second Report, Iron & Steel Institute, Special Report No. 15, 1933.

VACUUM TREATMENT OF MOLTEN STEEL

J. D. Sharp

It is well known that any liquid in contact with a gas absorbs some of the gas. This applies to liquid metals just as much as it does to other liquids, and owing to the nature of liquid metals the quantity of absorbed gas can be very large. Just as the quantity of gas absorbed in water depends on temperature, in the same way the solubility of gases in liquid steel are influenced by temperature. The hotter the liquid the more available positions there are in the structure of the liquid to absorb gas particles and therefore the greater the solubility of the gas in the liquid. In like manner, the pressure also influences very considerably how much gas will be absorbed. The greater the pressure the larger the amount of gas which can be forced into the liquid. Hence, by reducing the pressure above a liquid some of the dissolved gas will be extracted. Clearly if the steel is taken in a container the amount of removed gas will be proportional to the depth of the liquid owing to the excess pressure at the bottom of the vessel. It therefore follows that it is easier to remove gas from liquid in a shallow vessel than liquid in a deep vessel and this is of considerable importance when considering the correct design of ladle or actual degassing unit. Alternatively, if the liquid is poured into a vacuum the gas will have the optimum chance of escape as the pressure within the bubbles or droplets so formed will be at a minimum. Furthermore, the orifice through which the metal enters the vacuum can be so designed that the metal will break up into very small particles and every drop of the liquid concerned can be subjected to the vacuum. Clearly this has many advantages over the treatment of a metal in a container, as the degree of de-gassification depends to a large extent on the ability of the gas to redistribute itself rapidly in the bulk of the metal.

The undesirability of dissolved gas in steel has been known for many years and even Bessemer in about 1860 described the effect of a vacuum on steel held in a small crucible. Although Bessemer's experiments were very primitive, the very fact that he carried them out shows that even in his day vacuum was considered as a medium of improving the quality of the steel and removing the bubbles that were found on solidification. Engineers had already found these blowholes in castings and had concluded that these were caused by dissolved gases. In 1886 Mr. Russel Aitken, who was an engineer and not a steelmaker, described a process for the removal of absorbed gases. Although theoretically very far sighted it was many years before Aitken's system became reality as the necessary equipment was not available to create the vacuum.

69

THEORETICAL CONSIDERATIONS

The three gases that are of metallurgical importance and are readily absorbed by liquid steel are hydrogen, oxygen and nitrogen. In techniques designed to remove the gases efficiently, there are three main considerations. The most important of these is probably the required level of gas in the finished steel and in many cases a lower level of gas content is suggested than is actually required. Ease of application and cost of metal treatment are the other two important criteria and these will to a very large extent influence the choice of the process used. Each of the three gases mentioned above will now be discussed.

1. Hydrogen

The necessity of a low hydrogen content in forging steels has been known for many years. It was mainly for this reason that until relatively recently the acid open hearth was accepted as the only suitable furnace for making forging steels, because a selected metallic charge and very little slag making materials resulted in a final hydrogen content lower than alternative methods of steel-making. Even so, the steel produced still had to be treated in an annealing furnace for many hours to remove further hydrogen. It was the economics of this furnace cycle, which can run for many weeks, that gave the prime impetus to develop the vacuum degassing. Other benefits such as the possibility of improved cleanliness were not proved technically or economically feasible in the mid-1950's and it was the gain on cost and savings of heat treatment time for removal of hydrogen that prompted de Mare (1939) and all the pioneering work of the Bochumer Verein company lead by Adolf Sickburt. It is fair to say that it is the result of his early work that stimulated the growth of vacuum degassing.

Hydrogen is present in simple solution and it is probable that it obeys Sievert's Law, i.e.

$$\text{Weight percentage of gas in metal} = K \sqrt{\text{partial pressure of gas dissolved}}$$

This means that the variation of the pressure above the metal influences the equilibrium amount of gas that the metal can contain. Hence as the pressure is dropped the weight percentage of the gas dissolved is removed as a square function of this pressure. It has been shown that the origin of most of the hydrogen is from moisture incidentally charged to the furnace, e.g. as moisture in the lime, in the rust, on the metal scrap, or in the ferro-alloys. It is inevitable that as the furnace is tapped some hydrogen is picked up from the atmosphere and this can vary very considerably. Experience has shown that this is normally in the region of $0 \cdot 5 \text{ cm}^3/100 \text{ g}$ in a 100 ton heat. It is important to note that the amount of hydrogen removed during degassing is not dependent upon any other factor than the degassing pressure and the solubility coefficient; in particular it is not related to the initial hydrogen content. Thus in any particular set of conditions of temperature, etc., for a given final pressure, the hydrogen should be at a fixed level after degassing. This means theoretically that for a given pressure it is possible to forecast the hydrogen. In actual practice this is not always achieved and there is still a considerable amount of work to be done in discovering the reasons for the recorded discrepancies. Not the least of the factors concerned is the establishment of a fully proven system for determining hydrogen that is acceptable to all the authorities. This does not only refer to the actual analysis of the samples but also relates to the methods of taking them. There is already much literature describing the different techniques and it is not intended in this publication to discuss them.

For most purposes, residual hydrogen levels of under 2 cm^3/100 g are quite satisfactory, although for some special applications still lower values are necessary. Certain alloy steels used in very large forgings, e.g. rotor shafts, require the hydrogen to be as low as possible and if possible at least 1 cm^3/100 g. Most of the existing industrial processes can achieve 2 cm^3/100 g fairly readily and therefore in works where only hydrogen removal is required, factors other than this influence the choice and process. Following gas removal it is essential that unless a very low gas level has been achieved, care should be taken to minimize hydrogen pick-up during teeming.

2. Oxygen

Oxygen combines with the iron and is mainly present as some form of iron oxide. There will of course be other oxygen bearing materials present in the form of inclusions, i.e. silica, alumina, etc., and in particular, the refractory container which will normally be very largely silica. In practice, where a steel of more than about 0·20% carbon is degassed the amount of carbon monoxide evolved will be relatively small. Below this figure the equilibrium percentage rises very sharply and this means that if a low carbon steel is vacuum treated the amount of gas to be removed will be very large. This equilibrium is dependent on pressure and as the pressure falls the oxygen level becomes in excess of the equilibrium value, and the carbon–oxygen reaction commences. The carbon oxygen equilibrium diagram drawn for several pressures shows how the drop in pressure governs the volume of carbon monoxide to be evolved. If the pressure is dropped very rapidly the reaction can become very violent and cause very large increases in volume of metal in the containing ladle. Control of the rate of lowering of the pressure is required unless about 30–40% excess ladle capacity is available. If, however, the steel is poured into the vacuum the rate of pouring can be matched to the gas volume released; clearly this must be likewise balanced with the capacity of the vacuum pumps used to maintain the vacuum. The design specification for these pumps will be entirely different if they are required to cope with steel in a ladle being degassed as opposed to steel being degassed from a stream. In the former case the pumps must have relatively high duty at normal pressures to enable the ladle to be rapidly taken down to the degassing pressure. For stream degassing, as the vessel is already evacuated the pumps must be designed to match the flow rate required from the tapping ladle to the receiving ladle.

It has been shown that an active carbon–oxygen boil assists in removing hydrogen, especially from the lower parts of the containing vessel. In fact, techniques for artificially stirring the ladle, e.g. gas injection or electro-magnetic stirring are not necessary where a good natural boil can be created. It is, however, fair to state that in some cases it is not possible to create a very active boil especially where the steel contains a large amount of aluminium. Where a rimming steel or a semi-killed steel is degassed, very large gas volumes are concerned and in effect this can be used as a technique for also lowering the carbon content of the steel. This is used regularly in one of the installations and will be discussed under the Rhurstahl-Hereus process. It is quite easy to see how the oxygen level is lowered when an unkilled steel is treated under vacuum, but a similar reaction also takes place where the steel has been fully killed with silicon or manganese and, therefore, alternative oxygen reaction mechanisms must be in force. As the inclusion pattern of the degassed steel is certainly changed, the breakdown of the silica in the non-metallic inclusions is being considered as an obvious reaction, i.e.

$$SiO_2 + 2C \rightarrow 2CO + Si$$

There has been considerable discussion as to exactly how this reaction takes place, but what can be proved is (i), the size of the silica inclusions are considerably reduced and (ii) there does not appear to be a measurable increase in the silicon content of the finished steel. However, as has been reported by a number of workers in this field, although the size of the inclusions is considerably reduced, there is very little evidence to show that the total inclusion content is diminished. It has also been pointed out on many occasions that the walls of the containing ladles are highly siliceous and it seems surprising that reduction does not preferentially take place in these areas as opposed to the surfaces of the non-metallic inclusions. The degree of completion to which the silica reaction takes place is certainly a function of time. Unlike hydrogen degassing where the system chemically is trying to come to equilibrium, the chances of this happening during the removal of oxygen are very limited. It has been shown that the removal of oxygen is a function of pressure and the time that the liquid is treated. This makes comparison of results from one plant to another, extremely difficult as the exact pattern is rarely followed in two different installations. This conclusion is without regard to the wide differences between the different degassing techniques.

3. Nitrogen

Although nitrogen has for many years been known to have a significant effect on the surface characteristics and work-hardening characteristics of finished steel, particularly when considering material for deep-drawing operations, the degassing processes have not been considered important as a means of minimizing this effect. The results of degassing from plants all over the world have in the main shown extremely small nitrogen removal—only in the order of about 25 %. It can, however, be said that there is very little data available and it may well be that the pick-up of nitrogen is so fast when the steel is exposed to the atmosphere that some of the figures reported are not in actual fact correct. Nitrogen present in liquid steel forms chemical compounds, but these are not stable at steel-making temperatures. Nitrogen must therefore be considered to be present in the same form as hydrogen, but the variations between the calculated values and those found in practice for equilibrium clearly show that this state is not obtained because the speed of diffusion is much less than it is for hydrogen.

Steels containing aluminium, chromium or titanium are not likely to lose nitrogen because of their high affinity for this element. A further item which makes a precise examination of this problem more difficult, is the influence of the stirring action on the metal stream as a vigorous carbon monoxide boil, for instance, will not only lower the hydrogen but also present a greater proportion of the steel to the vacuum for the removal of nitrogen. At the time of writing, no steel is treated by vacuum for the removal of nitrogen, but it is possible that a modification of one of the existing techniques could result in this being done. The use of ferro-alloys with low nitrogen specifications shows quite categorically that there is a market for such a process.

DEGASSING PROCESSES

Broadly speaking, there are two types of degassing technique. The simplest and most widely used is to take a ladle of steel and simply submit it to a vacuum. This demands no particular skill or highly specialized plant in the steelworks, although it is clearly essential to have an efficient pumping system. By its very simplicity, there are many plants of this type. However, the degassing is dependent on gas moving from the bottom of the ladle to the top and the influence of this pressure gradient clearly shows the technique is not going to be as efficient as one in which a stream of metal is submitted to a vacuum.

As a modification, to this technique, in as much as a ladle of steel is taken and degassed portion by portion, the Dortmund-Hoerder process was developed. This process may be compared with ladle degassing as the degassing takes place from a relatively large surface and not from a metal stream. It could be likened to degassing from a very shallow ladle and allowing the degassed material to run back into the bulk of the metal and then withdrawing a second small sample for degassing. This will be discussed in greater detail in a later section.

The other type of equipment for degassing uses the idea of a stream of metal being shattered on entering a vacuum, thereby exposing an extremely large surface area per unit volume to the reduced pressure and degassing taking place drop by drop or rather bubble by bubble. In this instance the ladle of steel is placed on top of a chamber, in which there is an empty ladle held at reduced pressure. A most important feature of this process is the vacuum chamber being evacuated prior to the commencement of teeming. An aluminium disc is used to seal the chamber and when a vacuum seal has satisfactorily been made between the ladle containing the steel and the chamber, the aluminium disc is broken by the metal stream leaving the tapping ladle. Thus, every drop of steel is submitted to the vacuum. In addition, the steel collected in the receiving ladle is also exposed to the reduced pressure and further vigorous degassing takes place in this ladle. In a rather similar manner, the Rhurstahl-Hereus process or R-H process sucks a stream of metal through a nozzle into a vacuum chamber, where the upmoving stream is shattered into droplets in the same way as the falling stream is shattered into droplets in stream degassing. The degassed steel falls back into the original ladle. Two refractory lined legs are used to carry the metal from the ladle to the degassing chamber and from the degassing chamber back to the original tapping ladle. In this way, the process is continuous in as much as samples may be taken of the degassed steel and degassing continued until either the required analysis is achieved or the temperature falls to prohibit further degassing.

Each of the processes will now be discussed in detail, comparisons being drawn where it is felt necessary between them. Problems relating to the plants and the refractories will be discussed, but details of vacuum pumps are not included. These have been covered in considerable detail elsewhere and are readily available for those readers who wish to pursue this aspect of the topic.

LADLE DEGASSING

Without doubt this is the simplest of the degassing techniques and about half the total number of plants in existence are of this type. It is normally only possible to use this process with fully killed steels.

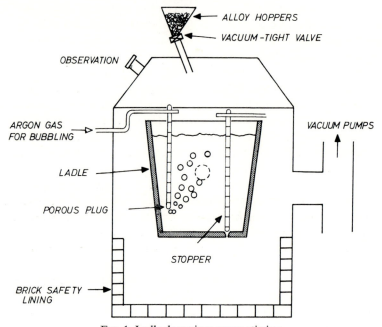

FIG. 1. Ladle degassing: argon stirring.

(a) The Plant

The plant consists of a large steel tank capable of containing the ladle of steel to be degassed, fitted with a vacuum-tight lid. This tank is connected to a vacuum system which is normally a set of steam ejectors. In most cases the tank is set into the floor of the melting shop, while the vacuum equipment is often outside the building close to the required supply of cooling water.

As has been mentioned earlier, with a deep ladle it is highly desirable to agitate the metal in such a manner that metal at the bottom of the ladle gets stirred to the surface and hence degassed. There are at least two techniques for artificial stirring in use:

(i) Agitation by an inert gas.
(ii) Electro-magnetic stirring.

Agitation by an inert gas usually consists of passing argon into the metal through a porous plug in the ladle bottom or down a dummy stopper rod. Relatively small quantities of argon are needed due to the very rapid expansion of the gas bubbles under the reduced pressure. The porosity of the refractory plug is such that the metal is unable to flow into the plug but the gas is capable of escaping quite readily.

Electro-magnetic stirring is only done in a few plants and is very much more expensive to install. This is made additionally expensive owing to the necessity of using austenitic stainless steel ladles. Also it tends to be vulnerable in normal industrial practice and can easily be damaged by molten metal. These stirrers were particularly fitted to plants capable of degassing at pressures of less than 0·050 torr.

(b) Process

The steel to be degassed is super-heated in the furnace some 80–100°C depending on the size of ladle and quality of steel. Small plants (say 20 tons) may have to have an even higher super-heat. The majority of the slag is removed and the furnace tapped as rapidly as possible into a ladle. This ladle is taken to the degassing pit, sampled and the temperature taken. The lid is placed firmly in position and the pressure in the tank reduced as rapidly as possible to its working level—say 0·5 to 2·0 torr.

The length of time the degassing continues is principally regulated by the initial temperature, as the rate of loss of heat will be well known in the particular plant concerned. Frequently ferro-additions will be made from automatic feed gear under vacuum to obtain the maximum yield, and the minimum of solid deoxidation particles.

When vacuum has been broken, the lid will be raised and the ladle taken and teemed in the normal manner.

It will be appreciated that good stopper and nozzle equipment are essential to withstand the erosive force of the metal in the ladle.

STREAM DEGASSING

This type of degassing was that pioneered by Bochumer Verein in the mid-1950's from a laboratory state to a fully developed process selling vacuum cast material. There is evidence to show that this is the most efficient system for the removal of gases from steel, but the heat losses are normally high. It is slightly more difficult to carry out than ladle degassing, requiring better maintenance of equipment, and a more controlled method of operation. Like ladle degassing it normally uses fully killed steels, but semi-killed and unkilled steels have been degassed by this process.

(a) Plant

The degassing chamber is the same type as that used for ladle degassing, but it is possible to use the plant in two ways. The metal may be run from the tapping ladle to a ladle placed in the vacuum chamber (ladle–ladle degassing) or from the tapping ladle direct into a mould (or moulds) placed inside the chamber (ladle–mould degassing). This latter type of degassing can also be correctly termed vacuum casting.

If a ladle–ladle practice is being worked, the vacuum tank need only be large enough to contain the receiving ladle. The lid, however, must be strong enough to bear not only the tapping ladle—but also the atmospheric pressure concerned. In some cases this atmospheric pressure can be greater than the weight of the full tapping ladle.

As in ladle degassing, the vacuum system is connected to the tank, but the equipment is such that it must be possible to create a vacuum in the tank prior to casting. This is normally done by fixing a thin aluminium disc over the hole in the canopy. There is a vacuum-sealing arrangement between the tapping ladle and the canopy (or vacuum lid) either on the whole of the circumference of the ladle or in a relatively small region round the nozzle. The metal stream bursts the aluminium disc on impact and the metal is sucked into the vacuum where the stream explodes into fine droplets. These droplets or bubbles must be contained and a stream limiter is necessary to contain the stream. The limiter is normally a steel shell lined with refractory.

(b) The Process

Steel is tapped from the furnace in the normal way—without a heavy slag-off. The ladle has the special fitting arrangement sealing it to the vacuum tank as mentioned above, and it is seated on the tank ready for teeming. Prior to tapping the mould or ladle will have been positioned in the tank and the pressure in the tank reduced to about 1 torr.

With the pumps running to maintain the reduced pressure, teeming commences through the aluminium disc. The rate of teeming should be controlled to maintain the minimum pressure. At the end of teeming ferro-alloys may be added from special hoppers isolated from the chamber by a vacuum seal.

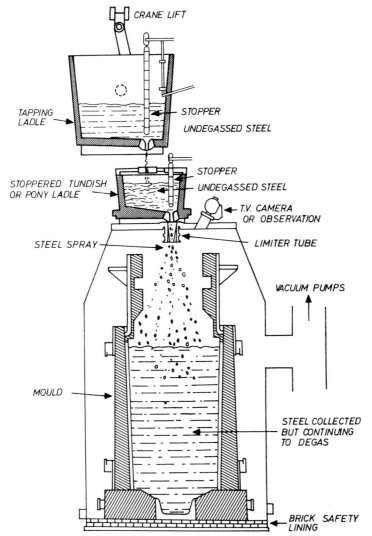

FIG. 2. Stream degassing: ladle-to-mould, two-ladle practice (Courtesy *Iron and Steel Engineer*).

The degassing period is clearly a function of the size of the nozzle, and in general, speed is an essential part of this operation if the teeming temperature is going to be correct.

Clearly, if ladle–mould degassing is being carried out a lower tapping temperature will be required than if ladle–ladle degassing is being used.

Normal ladle refractories are adequate, but if low carbon steels are being degassed at very high temperatures a special nozzle material will be required and a material other than fireclay used in the limiter. Unfired magnesite has been found satisfactory for these applications.

In stream degassing an enormous surface area is exposed to the vacuum, giving rise to excellent gas release. At the same time additional degassing takes place from the collected metal in the vacuum ladle—i.e. exactly in the same way as it does in ladle degassing. There is no slag present, however, and therefore less likelihood of slag becoming entrapped in the steel.

Following the completion of teeming, some slag is allowed to run through on top of the degassed steel *after* the vessel has been brought up to atmospheric pressure. The valve should be large enough to enable this to be done in about 20 sec.

In competition with the other vacuum degassing techniques available, stream degassing has one major drawback, and that is the temperature drop associated with the process.

TAP DEGASSING

As mentioned above under Stream Degassing the major disadvantage of that process is the high loss of heat due to the very nature of the process, i.e. the creation of a very large surface area all capable of radiating heat.

Although this heat energy cannot therefore be saved if the process is to be used, when a furnace taps normally in air into a ladle some 40–60°C are lost to the atmosphere, the ladle and in other heat losses. Bochumer Verein, therefore, decided to try and reduce the overall loss by using the ladle as the vacuum chamber—thereby saving an operation. This process has come to be known as tap degassing, and is now installed in at least three major steelworks.

(a) The Plant

The plant consists of a modified tapping ladle for the furnace concerned, having a refractory lined vacuum-tight cover. This cover also has the vacuum lead let into it, but as the volume required to be evacuated is relatively small, this is not as large as that required on either ladle degassing or stream degassing. Similarly the pumping system concerned is relatively small. One of the most difficult pieces of the equipment to design and maintain is this pipe which must be flexible to permit the required raising and lowering of the ladle as the furnace tilts during tapping.

A further item of importance is the maintenance of vacuum round the nozzle area of the ladle. A special disc must be fitted in position making a good seal with the ladle shell. Also the stopper gear must be such that a seal is still made by the cover all round the ladle rim.

On top of the ladle a small pony ladle is positioned and it is into this that the furnace is tapped. There must be a vacuum seal between the two ladles. The capacity of the ladle is only a few tons and as it is out of reach while under the furnace a remotely controlled stopper gear system is incorporated in the ladle design. An electric system is the one currently favoured—but an hydraulic system would be equally suitable.

(b) The Process

The degassing ladle, with the pony ladle in position is placed in the tapping pit and the vacuum line connected up. Near the time for tapping, the main ladle is evacuated, a temporary seal in the nozzle of the tapping ladle having been established.

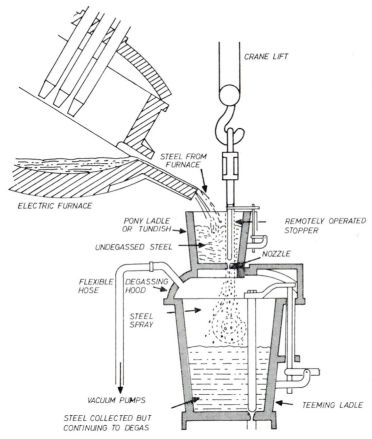

FIG. 3. Tap degassing layout as developed at Bochumer Verein (Courtesy *Leybold Eng.*).

The furnace is not slagged off, and the required superheat is about half of that required for stream degassing from the same furnace. Any necessary ferro-additions must be added to the furnace or put in the bottom of the tapping ladle before operations begin.

As soon as a small head of metal is established in the pony ladle, the pony ladle is opened and metal is sucked through into the main ladle—breaking up into the normal atomized stream as it does so. This process is continued until the furnace is empty, and then the tapping ladle is teemed in the normal way.

THE DORTMUND-HÖRDER PROCESS

The concept of this technique was outlined as early as 1886, but it was not until the early 1950's that suitable equipment was available to make the principles practical.

The process is described well, by its alternative title—degassing by parts. Steel is sucked from a normal ladle into a smaller vessel where it is degassed: after treatment the steel falls back and mixes with the undegassed metal. In principle, the benefits of degassing by parts are sound especially from the engineering point of view in that smaller pumps, ducting, etc., are necessary if part of the charge is degassed at a time. Also the fact that it is possible to sample the metal in the ladle, and take the temperature when required, has definite advantages.

(a) The Plant

Before the plant is ready to operate, the vessel must be preheated to nearly 1500°C either by gas or graphite resistor rods. The tapping ladle is normal and needs no modification. The actual degassing vessel consists of a brick-lined circular tank which has in its base a tube through which the metal can rise and fall into the ladle. The height of the tube must accommodate the height to which steel can be sucked up a pipe—i.e. about 57 in. As it is not desirable to make and break the vacuum in the vessel either the ladle is raised and lowered to empty and refill the steel in the degassing chamber, or the vessel is raised and lowered to have exactly the same effect. The tube is called a "snorkel" and the degassing vessel and the snorkel are well covered with high quality refractory, and in the past the necessity for high quality refractories on a routine basis has given rise to many problems. However, it must be said that many of these problems have now been overcome by the more general use of high alumina refractories.

A very valuable selling point for this plant is the ability to adjust analysis very accurately, and the use of equipment to weigh out and put alloys in the vacuum chamber through vacuum locks has given the D-H process great impetus. The particular possibility of fine adjustment to analysis after the furnace has been tapped has very great attractions for certain steelworks making special grades of steel.

The vacuum pumps are connected to the lid and in the cases where the vacuum tank is fixed and the ladle moves up and down, the connections are solid and non-flexible.

The snorkel, which has a refractory covering both inside and outside, has a cardboard or aluminium cone tied over the opening and this prevents even a fairly active slag getting sucked up into the vessel.

As an alternative to having to move the ladle or the vessel, a very powerful electro-magnetic coil round the snorkel has been proposed—but as far as is known, never yet used.

(b) The Process

The relatively low loss of temperature in this process makes it very attractive to open-hearth shops that desire to degass their product but have not the possibility of using the two early processes because of their heat loss.

After tapping the furnace the ladle is positioned under the degassing unit, and in a large plant is always placed on the carriage that is going to lift the ladle under the degassing unit. The snorkel is dipped into the metal and by switching on the vacuum pumps, steel is sucked up into the vessel. In the vessel the steel "boils" much as it does in the ladle during ladle degassing and as the reaction

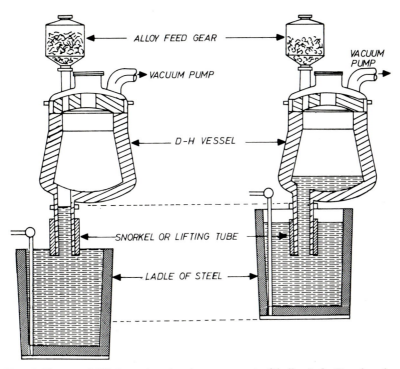

ALLOY FEED GEAR

VACUUM PUMP

VACUUM PUMP

D-H VESSEL

SNORKEL OR LIFTING TUBE

LADLE OF STEEL

FIG. 4. Dortmund-Hörder units, showing movement of ladle. *Left:* Vessel under vacuum: steel rises 1 atm equivalent. *Right:* Ladle raised so that steel rises into the vessel.

subsides the ladle is dropped so that steel runs back into the ladle. A point the critics of the process make, but which has never been substantiated, is that it is almost impossible to guarantee that ALL the steel is degassed, as the suction is made at the same point in the ladle every time. If dispersion of dissolved ferro-alloys is used as a measure of the uniformity of degassing, it is clear that excellent diffusion within the melt takes place.

The number of cycles usually employed on a particular plant varies according to the relative size of the degassing vessel and the ladle, but 25–30 cycles are typical.

Temperature dips and sampling proceeds during degassing to ensure that sufficient teeming temperature is maintained in the metal and that the analysis is correct. When degassing has been completed the ladle is lowered fully into its carriage and taken for teeming in the normal way.

THE RUHRSTAHL-HERAEUS PROCESS

Continuous processes have great appeal for those interested in the mass production of materials, and so a continuous degassing process obviously has considerable potential.

The British Iron and Steel Research Association have attempted the same exercise with some degree of success, but their system has never been put into operation on a large scale. The trials at Low Moor Alloy Steel Works run in conjunction with a continuous casting plant clearly set the pattern of the future, but so far the practical difficulties on a large scale have proved too great for a rapid expansion of this process.

The R-H process, as it is called, was developed about 1957 and has expanded very rapidly indeed. It has some of the advantages of the D-H unit—e.g. low heat loss, good alloy mixing potential, etc., but as it is a continuous process the load on the vacuum pumps is more uniform, there is less refractory wear and there are other advantages.

The principle of the process is that if a gas is introduced into a column of steel it can be sucked up more than the normal barometric height. A degassing vessel with two refractory lined legs is dipped into the ladle to be degassed. In one of the legs there is a flange that admits gas to the steel stream, and due to the vacuum that is created in the vessel the steel is sucked up the leg from the ladle and into the vessel itself. The height of the legs is such that they are over the barometric height for steel and so after degassing the metal falls back down the other leg into the ladle.

(a) Plant

The plant consists of a refractory lined degassing vessel in the form of a cylinder about three times as high as its diameter. The cylinder is closed at the top and contains two branches to the vacuum system. A hopper is fixed on the top of the cylinder containing the necessary ferro-alloys and fitments for TV cameras, etc., can also be made.

Steel only circulates in the bottom few feet of the vessel and therefore the area subject to heavy refractory wear is limited. Steel enters the vessel up one refractory lined steel leg and falls down a similar unit after degassing. The legs and whole vessel are preheated before use to prevent skulling, and this is normally done with gas burners. The lifter gas is normally argon, although a considerable number of gases have been tried—including carbon tetrachloride. The gas is introduced through fine orifices in a flange just above the level where the legs dip into the steel in the ladle.

It is possible also to dip the legs and operate the system in the furnace itself and this has been done in the United States.

The whole apparatus is arranged so that it can dip into the ladle or furnace and then readily be lifted and swung out of the way for easy access. This calls for excellent engineering and maintenance. Cyclone dust filters are used in each of the two vacuum lines from the unit, and these certainly make it possible for such an apparatus to be maintained in practical working condition.

It is certainly the most complicated piece of equipment used for degassing, although the volume to degas is relatively small. It may well be likened to stream degassing in that the actual treatment is on a spray of steel particles rather than on a complete molten metal surface.

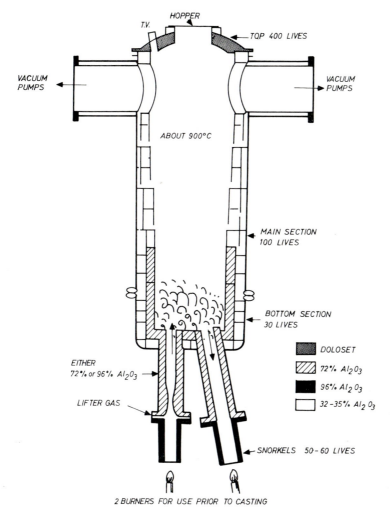

FIG. 5. Ruhrstahl-Heraeus process.

(b) The Process

Like the D-H process this degassing technique requires a low super-heat in the furnace, and again the furnace is not slagged off.

The ladle is moved to a cradle which exactly locates it under the degassing unit. The slag is broken up and the legs dipped into the metal. The temperature is checked and the vacuum pumps switched on—sucking the steel up towards the degassing chamber in both legs.

The lifter gas is then introduced into one leg and the steel gas mixture immediately surges upwards and is degassed. The steel falls back into the other leg of the system, and as the atmospheric pressure on the open ladle can only support about 57 in., as degassed material enters the top of this leg, steel runs out of the bottom back into the ladle.

The process can be kept running as long as there is enough temperature to permit it, and a constant watch can be kept on it, as the ladle is exposed. This is not possible, of course, when dealing with stream or ladle degassing where the ladle is inside the vacuum tank.

Similarly checks can be made on analysis at will, and with the aid of modern rapid methods of analysis, there is really no reason why any steel should be cast out of specification.

When degassing is complete, the ladle is taken from its fixed position and teemed normally in air.

THE ASEA/SKF PROCESS

This process was announced in 1965 and was designed as a technique to give a vacuum degassed, double slag, electric arc steel quality without completing all the steel-making in the furnace.

The key is the use of the ladle with two different lids. Firstly it is used to degass the steel and secondly with a different lid—it is used to carry out the reducing slag period.

Steel is taken either from an arc furnace or an L-D and the oxidizing part of the steelmaking is completed—i.e. the carbon, manganese, silicon and phosphorus reduced to the required levels. The ladle—without a stopper rod—is placed in an electro-magnetic stirring coil and a vacuum-tight lid placed in position. The steel is then degassed in the normal way, i.e. ladle degassing within the ladle—the relatively low temperature aiding the gas release. The stirrer is used to make sure that all the metal is treated. When a satisfactory low pressure has been maintained for long enough, the vacuum is broken and the vacuum lid swung to one side. The steel is now in a degassed condition with a relatively low oxygen content, without the use of solid deoxidants.

A reducing slag is now made up and added to the ladle to enable desulphurization to take place. A separate lid in the form of a miniature arc furnace roof is now positioned and the temperature of the steel raised to a temperature approaching the tapping temperature. Due to the depth of the ladle, the stirring coil is essential to stir the less dense metal down into the colder metal, away from the surface that will be absorbing the heat. The rate the system can absorb heat is entirely controlled by the power in the stirring coil. Before teeming, the stopper rod must be put in and this is achieved by using a special jig.

Although this system is relatively new, it is clear that for alloy steel-makers it is very attractive.

CHEMICAL DEGASSING OF STEEL

J. D. SHARP and K. STRAUSS

Whilst the degassing of steel in vacuum has now been established on a fairly wide scale, that of chemical degassing is only in its first stages. Non-ferrous metals can be degassed by the application of either inert gases or chemically reactive gases or substances which volatilize under the heat of the molten metal (see aluminium and copper alloys) and theoretically it should be possible also to remove hydrogen from steel by similar methods. Due to the high melting point of steel, and the attack on refractory materials, the practical difficulties of applying chemical degassing methods are, however, very much greater than with lower melting point metals.

According to hitherto unpublished reports[1] the Foseco organization have made tests using carbon dioxide gas CO_2 as the degassing agent. Blocks made from calcium carbonate (either limestone or marble) $CaCO_3$ were placed round the stopper rod of a steel ladle and the ladle was filled with molten steel. In contact with the metal the blocks started to decompose and to give off carbon dioxide which bubbled through the melt; after a few minutes the gas development slowed down or ceased.

A number of tests were carried out with amounts of steel from 500 kg to 5 tons; in many cases a satisfactory reduction of the hydrogen content could be obtained, ranging from 6·0 ml/100 g before treatment to 3·3 ml/100 g after treatment. The efficiency of the degassing depended on time of treatment and quantity of gas employed but results were not strictly consistent and reproducible and there were a number of failures. Although it is appreciated that 3·3 ml/100 g is not low enough for general commercial use, it clearly demonstrates the principle. For these reasons the tests were discontinued and it was concluded that for obtaining more consistent and better results, more chemically active gases must be employed.

Tests are also known to have been carried out using solid carbon dioxide blocks in an 8-ton arc furnace and liquid gases in a small laboratory furnace. These experiments were very dramatic and a very vigorous boil was obtained, but these materials at low temperatures attracted such large quantities of moisture that they had to be abandoned.

The problem of the injection of chemicals into molten steel has been studied during the last few years by H. Fiedler[2] and others, mainly for the purpose of inclusion removal from steel ingots. The materials used were manganese chloride ($MnCl_2$) and carbon tetrachloride (CCl_4) and there were definite indications of a reduction in the number of inclusions. With an improved technique of powder injection, and the development of highly resisting refractory material, it should become possible to inject chemically reactive materials into the steel which will volatilize and pass through the metal in the form of gas bubbles into which the hydrogen, which is dissolved in the steel, will diffuse and escape at the surface. More will no doubt be heard in future on this line of development.

REFERENCES

1. *Foseco Research and Development Report*, No. 123.
2. FIEDLER, H., *Neue Hütte*, Heft 1, 1963, p. 48; Heft 3, 1964, p. 131.

D

EFFECT OF MOULD ADDITIVES AND FLUXES ON INGOT SURFACE QUALITY

D. D. Beevers and J. J. Chappell

The introduction of mould fluxes for use on bottom-poured killed steel ingots represents a major step forward in ensuring superior ingot surfaces.

In the past the active and inactive mould dressings used were only effective under ideal teeming conditions and the many variables encountered at this stage of ingot production resulted in surface defects requiring considerable rectification treatment at the ingot or billet stage.

With the need to ensure defect-free billet surfaces, and the inability of any one mould dressing to compensate for these variables, it was found necessary at many companies to institute hot and/or cold deseaming operations.

In the worst cases rectification can involve inline deseaming, at an intermediate bloom stage, with a 3–5% loss in yield. This type of rectification is completely unselective, but is very often necessary on tonnage plants where selective deseaming is impractical and where subsequent surface defects cannot be kept within reasonable limits.

For these same reasons, but where in-line deseaming facilities are not available, cold deseaming plants have had to be commissioned at considerable expense in both equipment and labour.

The importance of controlling the factors which contribute towards these surface defects is therefore of considerable economic importance.

Before due consideration is given to the effects on ingot surfaces of mould dressings and fluxes, it is necessary to know the various types of surface defects experienced under normal casting conditions.[1]

Cracks

Cracks form one of the major defects in steel ingots and become obvious during hot deformation, although in most cases they are in fact present in the original ingot soon after solidification.

Cracks can broadly be divided into two distinct groups:

1. Cracks caused by obvious restriction of the ingot skin during cooling.
2. Cracks caused by factors not attributable to surface irregularities or by obvious restrictions.

Consider Group 1. Restriction cracks are directly attributable to some obstruction in the mould preventing the free contraction of the ingot. This group may be subdivided further:

(a) Hanger cracks—these either occur when the ingot is suspended in the mould due to badly worn or badly fitted head boxes or as a result of over-filling the mould. It is a notable advantage of an expendable type of head tile assembly that hanger cracks due to faulty head boxes have now been virtually eliminated.

(b) Fin cracks—these arise from cracks in moulds causing fins which will suspend the ingot during contraction. In much the same way badly worn moulds produce scabs which inhibit even contraction.

(c) Basal cracks—these result when fins or flashes of metal prevent free contraction of the ingot base.

Group 2 cracks result from completely different mechanisms.

As steel is poured the initial chilling effect arising from contact of liquid steel with the mould wall causes a solid skin to form. At this stage the mould expands at the same time as the ingot contracts, producing a gap between the mould and ingot skin. Subsequent rupture of this skin occurs if the ferrostatic pressure exerted by the liquid steel exceeds the strength of the skin layer. These inherent cracks then develop during mechanical working.

Not all the facts are completely understood in this class of non-restriction cracks but teeming speeds, teeming temperatures, mould temperatures, mould design and steel quality are known to be factors which can contribute towards cracking.

Clinking

Clinking is the term applied to internal cracks which result from too rapid heating or cooling of the solidified ingot. Such treatment produces severe internal stresses due to differential contraction and expansion limits which propagate along the coarse weak crystal planes, existing in the as-cast structure.

Alloy steels are particularly prone to this type of cracking.

Hairline

Hairline cracking is a term applied to those minute cracks which for so long eluded early detection and prevention. It has now been proved that hydrogen is the major contributory factor, being absorbed during melting and resulting in disruptive pressures being set up within the body of the ingot as the solidified ingot cools, and the solubility of the hydrogen decreases.

Double Skin

This defect can occur in both the bottom- and top-poured ingots.

In top-poured ingots the defect is caused by liquid metal splashing on to the mould walls at the commencement of teeming and also the surging action which swills the metal up against the walls. This metal immediately solidifies and contracts away from the mould wall leaving gaps which are subsequently refilled by the advancing metal level. Unfortunately surface oxidation occurs on the original solidified shell preventing eventual fusion.

In bottom-poured ingots the mould is protected from the initial force of the stream by teeming down the trumpet. However, uncontrolled teeming can still cause the metal to enter the mould via the runner systems at a rate sufficient to cause surging and splashes, with the consequent formation of double skins. Ferro-static pressure of the molten metal can cause a break-through from which bleeding of the ingot occurs.

Splash

Particularly prevalent in direct poured ingots.

Bottom-poured ingots enjoy relative freedom from this defect although slight effervescence can result in minor skin imperfections.

Rippled Surface

The rate of heat extraction by the mould on teeming and the surface tension of the metal influencing the meniscus are primary causes of this defect. The effect is characteristic of slow teeming speeds or where metal has been teemed at too low a temperature.

The rippled surface occurs as a series of peripheral undulations progressively forming as solidified surface metal forms and is engulfed by further molten metal as teeming progresses.

The defect can be regarded as small laps and in its slightest form no deleterious effects will result in the rolled product.

Lapping

A more severe form of rippling, and formed in a similar manner. Increased formation of oxide and solidified metal on the meniscus of the liquid steel, however, causes serious overlapping.

Double Teeming

The defect can be produced in bottom- or top-poured ingots when teeming is temporarily halted sufficiently long for the surface to freeze and oxidize.

Crazing

This ingot surface defect appears as a cellular network corresponding to the crazing effect found on the inside faces of worn moulds. In the early stages the resultant surface effect is not serious but subsequent defects will develop in the rolled products as mould wear increases.

Surface Porosity

These defects show up as massive pin-holing on the surface of the ingot. They are caused by gas evolution from the mould/metal interface resulting from moisture on the mould wall or even from the mould dressing used. These surface blowholes are oxidized during reheating resulting in longitudinal rokes or seams in the rolled product.

Slag Patches and Brick Inclusions

Any such foreign matter which finds its way to the surface and subsequently to the mould metal interface will become entrapped and form the basis of a surface inclusion. Even small areas of foreign refractory matter should be carefully investigated since deseaming or chipping can reveal more extensive trouble.

MOULD DRESSINGS

If one accepts at this stage that there are positive advantages to be gained from the use of mould dressings, applied by spray or brush, in improving the general surface appearance of the ingot skin, it is immediately apparent that ingot surface defects can occur as a consequence of factors completely uncontrollable by mere surface coatings.

Restriction cracks, double teeming effects, crazing, slag patches and brick inclusions under most conditions would not benefit from the application of mould dressings, and can be directly related to steelmaking practice and mould condition. In some cases volatile mould dressings could well aggravate the situation by becoming entrapped by capillary action in fissures present in the mould surface, causing subsequent blowholes as the advancing metal surface covers these regions. The same is true of thickly applied dressings which do not allow volatilization to take place ahead of the rising metal level.

Dehydrated tar for use as a mould dressing for killed steels has been in common use for years. More refined products have been designed to suit particular applications and many other dressings have been added to a constantly growing list of products.

In addition, both water and spirit based aluminium powder and graphite based products are used together with the reeking compounds.

Thornton at BISRA[2] clearly determined their mode of action and comparisons have been made on their relative properties and usefulness.

A perfect mould dressing should include properties to eliminate or reduce defects referred to earlier, so as to ensure a high standard of ingot surface and to allow a wider permissible range of teeming speeds and temperatures which cannot always be closely controlled in practice.

Even now opinions and preferences vary widely as to the choice of mould dressings thereby highlighting the varying conditions prevailing which lead to the relative success of different dressings.

Volatile mould dressings such as the dehydrated tar derivatives can be effectively used to minimize lapping, rippling, splashes and double skins.

As Thornton suggested, heat insulation properties alone in a dressing are not sufficient to prevent the sticking of splash or reduction of the chilling effect of the mould. He suggests that the degree of volatility of the dressing is a more important factor contributing to the following points:

1. Carburization of the steel splashes by the reducing gases evolved from the dressing, lowering the melting point of the steel.
2. Propulsion effect of gas evolution.
3. Prevention of direct contact between the steel and mould.
4. Exothermic effect of burning dressing.
5. Reduction in the incidence of lap formation by keeping the meniscus open and free from oxide.

The use of a volatile dressing is desirable to prevent lapping occurring when deliberately adopting lower teeming speeds and temperatures. This procedure is adopted to allow a sufficiently strong initial skin to form before the ferrostatic pressure has had time to increase to a point where a thinner skin would allow bleeding and a possible rupture.

Despite these considerations the non-active dressings such as soot-forming reeking compounds and various graphitic coatings still find favour. These provide a non-wet coating which discourages the adherence of splashes and laps and has none of the permissible harmful effects of a volatile dressing.

Aluminium coatings are very often used where possible pick-up from the coating in the ingot surface is undesirable.

Mould coatings appear only to deal with the mould/metal interface reactions which affect the final result and the major source of surface defects and indeed subcutaneous problems can never be dealt with either adequately or consistently by such ingot mould dressings particularly with respect to killed steel ingots.

Any pitside observer at some time will have watched with interest the appearance of teeming scum, consisting of refractory and metal oxides, on the surface of the metal during uphill teeming. Because of the central pouring system usually used, eddy currents are set up which tend to disperse these patches to the mould wall. With a volatile dressing the interaction at the metal/mould interface very often succeeds in rejecting these patches, forcing them away from the mould wall and preventing entrapment. This is particularly true where the metal surface remains sufficiently fluid. Non-active dressings, such as aluminium or reeking compounds are relatively ineffective in preventing this type of entrapment.

A common practice, which can still be seen, is the pitside personnel attempting to mop these patches up on the end of a metal bar before they have had time to reach the mould wall and become trapped. However, where several ingots are teemed together, only the moulds nearest the teeming platform benefit from this treatment.

The formation of this teeming scum and associated oxide is increased by low teeming speeds or low temperatures or by grades of steel containing readily oxidizable elements such as aluminium. The most serious aspect of this type of defect lies in the fact that entrapment invariably causes subcutaneous blowholes and associated inclusions, which can prove difficult to remove by even the most thorough deseaming operations. In addition these defects are not always obvious and can escape detection until later stages of manufacture. Despite the many precautions which are taken to prevent the entrapment of this teeming scum, a high proportion of ingots usually suffer from this defect.

As Rippon[3] stated at the sixty-second BISRA Steelmaking conference a successful product must possess the following properties:

1. Prevention of atmospheric oxidation of the steel and its alloys.
2. Should not contain volatile matter which can give rise to blowholes by becoming entrapped below the rising liquid level.
3. Prevent teeming scum from becoming entrapped in the ingot surface.

The use of inert gases to achieve the desired effects has been tried many times, but has usually failed due to practical difficulties nor did it satisfy all requirements.

In some instances chopped straw is still being used to provide an insulating blanket aimed at preventing surface oxidation, but it is relatively ineffective in preventing scum entrapment. Liquid slags have also been used but with limited success.

It has always been assumed that a material placed in intimate contact with the total surface of the rising metal would stand the best chance of preventing surface oxidation. In these last few years free flowing powder materials based on fly ash have been found to be effective. These products are added at a general rate of 3 lb per ton of steel and when placed in contact with the metal they partially fuse at the metal/powder interface providing a perfect means of mopping up and preventing scum formation. The insulating nature of such top covers prevents heat loss and provides protection from re-oxidation. The absence of volatiles completes the three requirements and thus provides a product giving considerable improvement over many of the best mould coatings used in any one application. To achieve the best results it is desirable to avoid turbulence of the slag cover during uphill teeming. This has resulted in teeming speeds being considerably reduced to about 8 in. per minute level. The almost complete elimination of heat loss from the metal surface permits this practice resulting in an ingot surface far superior to that obtained from a mould coating.

As Rippon[3] discovered, the physical damping of the metal surface turbulence by a slag layer allows surface tension to remain undisturbed, so that only minor blemishes, which are well rounded, appear on the ingot surface. He also found that a good ingot surface could still be produced from crazed mould surfaces, although badly crazed moulds would provide adhesion points for the powder which may result in the slag layer being drawn on to the mould wall to give a slag patch on the ingot surface.

While it is desirable to maintain a complete cover over the metal surface, on larger cross-section moulds a central area may result which is devoid of slag. This uncovered area (sometimes referred to as "eye") increases in diameter as the teeming rate is increased and where powder application rates are lowest. However, perfectly satisfactory results are obtained provided the mould/metal interface remains covered and undisturbed throughout the total teeming period.

The surface condition of the bottom six inches of the ingot is usually prone to a certain degree of slag entrapment. This is due to the initial turbulence experienced when first opening out. Many different methods of applying the powder have been tried to overcome this problem. Practices range from placing the bags of powder around the ingate prior to casting to suspending paper or polyethylene bags containing the powder some 18 in. from the bottom of the mould. Thus as the metal enters the mould the "bag" disintegrates and allows the product to fall on to the metal surface. Alternatively some users are content to throw the product into the mould on commencement of teeming. Despite these variations it is still true to say that this particular fault of dirty ingot bottoms has not been entirely eradicated to date.

Considerable work is now being carried out to perfect mould additives capable of being used on direct poured steels. Whilst suitable formulations have been devised the more intimate mixing of the slag and metal call for properties different from those required for bottom-poured ingots. Generally it is agreed that a higher degree of fluidity is required in a product for direct poured steel to avoid the obvious risks of product entrapment giving rise to non-metallic inclusions.

Future developments are aimed at producing synthetic formulations to perform the same function but to achieve consistency of product and performance that is not always found in compositions based upon "Fly Ash".

REFERENCES

1. Iron and Steel Institute Special Report No. 63, December 1958.
2. THORNTON, D. R., *J. Iron and Steel Institute*, **183**, 300–315 (1956).
3. RIPPON, J. R., The 62nd BISRA Steelmaking Conference, *Steel Times*, March 12, 1965, p. 369.

RIMMING STEEL

J. F. CARTWRIGHT, J. G. EMMOTT and E. T. R. JONES

AN INTRODUCTION TO THE PRODUCTION OF RIMMING STEEL

1. Steelmaking is essentially a process designed to minimize the content of certain undesirable elements in the iron from the blast furnace and to control the levels of other elements within desired specification limits. This is achieved principally by the oxidation of most of the elements to be removed and the absorption of others into carefully controlled steelmaking slags used in the operation. Blast furnace iron contains from 3·5 to 4·3% carbon, from 0·4 to 1·3% silicon and is very brittle. By a controlled oxidation process, these contents are reduced to varying levels depending upon such properties as strength and ductility required in the steel being produced.

2. The oxygen necessary for this control is introduced:

(a) As a blast of gas containing oxygen blown through or on to the surface of the liquid metal as in bottom- or side-blown converters. This may be air (containing 21% oxygen), oxygen-enriched air, steam and oxygen mixtures, etc.

(b) As pure oxygen gas blown on to the metal surface as in the Linz-Donawitz, Kaldo, Rotor and similar processes.

(c) As pure oxygen introduced through lances below the metal surface in open hearth and electric furnaces.

(d) As oxygen-containing solid additions to liquid metal, such as iron ore and/or mill scale.

Oxygen is dissolved in the iron and is in equilibrium with FeO in the slag. The oxygen normally reacts with the constituents in the steel according to the following equations:

$$FeO + C = Fe + CO_1$$
$$2FeO + Mn = 2Fe + MnO_2 \text{ (slag)}$$
$$2FeO + Si = 2Fe + SiO_2 \text{ (slag)}$$
$$5FeO + 2P = 5Fe + P_2O_5 \text{ (slag)}$$

In the first reaction carbon monoxide gas is evolved and gives rise to what is normally termed the carbon boil.

3. To obtain the reactions listed in (2) above, under normal production conditions there would invariably be present an excess of oxygen. The excess oxygen present is largely dependent on the carbon level of the steel. Figure 1

D*

illustrates typical carbon and oxygen contents encountered in different steel-making processes together with the equilibrium line at 1600°C and a pressure of 1 atm. This clearly shows the association of high oxygen contents with low carbon levels and vice versa.

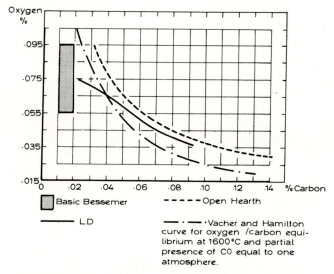

FIG. 1. Oxygen and carbon contents for different types of steel.

4. For a steel containing oxygen, if the temperature is reduced, the reaction $FeO + C = Fe + CO$ occurs and continues until equilibrium conditions are restored, or until one reacting component is exhausted. Similarly, a decrease in pressure would have the same effect.

Because over-oxidation of steel always occurs, some degree of deoxidation is necessary before pouring. Deoxidizing additions are made, wholly or in part, to the furnace, and/or ladle, and/or mould, to obtain the required level of oxygen. Depending on the final oxygen contents, three types of steel are produced.

5. *Killed steel* contains very low free oxygen levels, normally less than 0·003%. As a result, during solidification, there is no carbon/oxygen interaction and no evolution of carbon monoxide.

As the steel cools and solidifies, a contraction occurs which will result in internal cavitation (see Fig. 2) unless the ingot is "fed" by the use of hot-tops. In addition, certain elements of the steel which have low solubility in solid iron remain in the still liquid metal and progressively enrich it. As a result of this, the last metal to solidify contains a high percentage of these elements, which are normally termed segregates, and which are found in greater concentrations in close proximity to shrinkage cavities.

A hot-top is usually provided at the head of killed steel ingots in the form of a side-wall lining of insulating material or exothermic material and a cover of exothermic powder. These provide a reservoir of hot metal to ensure that the last solidification occurs in the head region. Thus, ideally, shrinkage cavitation and severe segregation are confined to the residual head where they can easily be removed during hot working. The same method can also be used for rim stabilized steel which is teemed in the rimming condition and killed with deoxidants in the mould after the formation of a pure iron skin.

FIG. 2. Killed steel ingot without hot-top showing internal cavity.

FIG. 3. Semi-killed steel. The evolved gas is entrained in the upper part of the ingot.

6. *Semi-killed or balanced steel* is one which, after partially deoxidizing with additions of ferro-manganese, ferro-silicon or aluminium, contains between 0·01 to 0·02% oxygen. This is sufficient to result in some reaction between carbon and oxygen to form carbon monoxide in the late stages of solidification. The evolved gas is entrained in the metal in the upper regions of the ingot

(Fig. 3) and segregation is located mainly around the area of cavitation. The production of semi-killed steel is confined to those compositions in which an excess of oxygen can exist. Thus, high carbon and most alloy steels cannot be cast with a semi-killed structure. In a correctly balanced steel, the volume of entrapped gas is equal to the liquid and solidification shrinkage and the ingot top is flat and very near to the teeming level. During hot working the cavities weld up and a high ingot-to-slab yield is obtained.

7. *Rimming steel.* The oxygen levels in rimming steel are in excess of 0.02% and are dependent to a large degree on the carbon and manganese contents. Some control is normally exercised in the ladle and/or mould by deoxidation with aluminium. Adjustment to the oxygen level in the steel prior to and during solidification can result in a variety of ingot structures. (Figure 4 shows one type, the subject being discussed more fully in para. 9.)

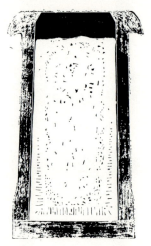

FIG. 4. Rimming steel with oxygen content adjusted.

Rimming steels, or effervescent steels as they are sometimes termed, generally contain from 0.05 to 0.20% carbon and 0.1 to 0.6% manganese to give a suitable range of hardness, ductility and strength, although higher levels of both carbon and manganese are sometimes encountered. The advantages of rimming steel compared with killed and balanced types are as follows:

  (i) High ingot-to-product yield.
 (ii) Lower prime cost due to absence of alloying additions and lower deoxidation costs.
(iii) Excellent surface quality in the finished product.
(iv) Improved cleanliness in surface and subsurface regions.

If properly controlled, the surface regions of a rimming steel ingot are relatively pure, and particularly with low carbon, low manganese qualities are a near

SCHEMATIC REPRESENTATION OF STAGES OF SOLIDIFICATION OF RIMMING STEEL

The solidification front (shaded) advances progressively inwards from each mould wall and the mould bottom, see right. The diagram below is a simplification of what happens at and behind this moving solid/liquid interface. It may be regarded as a vertical section, *not to scale*, mid-way up the ingot.

A layer of chill crystals forms immediately at the mould wall and selective freezing begins with the formation of columnar crystals. The Carbon/Oxygen equilibrium is disturbed in the liquid steel ahead of this freezing front.

The columnar crystals of high relative purity reject Carbon and other impurities (O, Mn, S, P) into the liquid steel already rich in FeO. At this early stage, the solidification rate is at its highest and the concentration of rejected elements at its maximum. The Carbon and Oxygen concentrations become sufficiently high to react and form CO gas bubbles at the solid/liquid interface.

With a high rate of gas evolution, bubbles rise rapidly up the solidification front and escape to the atmosphere. This vigorous sweeping movement displaces impure metal upwards which then circulates downwards in the centre of the ingot, as shown right. Under the ferro-static pressure of a full mould, CO gas evolution slows down as the Oxygen is consumed and the solidification rate decreases. At this stage, bubble formation is sufficiently slow to allow dendrites to entrap gas before it can escape upwards. The dendrite growth governs bubble propagation as shown left.

When the top surface of the ingot is sealed, either by freezing or capping, pressure builds up within the still liquid metal. The sweeping action of rising gas ceases and, as the temperature falls, the steel becomes pasty. Local concentrations of Carbon and Oxygen in this impure liquid result in CO formation and entrapment *in situ* of bubbles in globular form. The metal in the central zone is particularly rich in segregated Carbon, Sulphur and Phosphorus, more pronounced in the upper parts of the ingot.

Summary of the Transverse Structure of a Rimming Steel Ingot

1. Chill crystals at the mould wall and a rim of high relative purity enclosing an outer band of primary blowholes (elongated).

2. The transition rim/core zone and an inner band of secondary blowholes (globular).

3. The core heavily segregated. To prevent too heavy segregation, the rimming action can be stopped by chemical or mechanical capping.

FIG. 5. Schematic representation of the process of solidification of a rimming steel ingot.

approach to that of pure iron. These very ductile layers persist through to the final product making the steel suitable for thin section applications, such as deep drawing.

Producing a rim of adequate purity, together with a control of the internal segregation, is the art of making rimming steel ingots.

8. *Mechanism of the solidification of rimming steel*. The ingot structure is a function of the intensity of the rimming action and this is affected by many factors. Essentially the rimming action results from interaction between carbon and oxygen and the formation of carbon monoxide.

Figure 5 shows schematically the process of solidification of a rimming steel ingot. At the top of this figure, the triangle represents the formation of the solidification front which is moving vertically upwards and horizontally inwards as solidification progresses. Although this is related to time and temperature it is emphasized that the diagram is not to any scale. The main diagram below is an enlargement of the small triangle showing the progressive formation of the ingot rim. When the steel is poured into the mould, the latter has a cooling effect causing an almost instantaneous formation of a thin layer of chill crystals of the same composition as the liquid metal. Following this, columnar crystals begin to form perpendicular to the mould wall and solidification becomes selective. Due to the decreased solubility of certain elements in iron resulting from a fall in temperature, they are rejected from these columnar crystals into the remaining liquid metal. Thus, the unsolidified metal becomes progressively enriched by the rejected elements. The two most important constituents in rimming steel for the production of a satisfactory rimming action are carbon and oxygen. These are largely rejected by the solidified steel forming concentrations ahead of the freezing front. This and falling temperature create conditions for interaction of carbon and oxygen to form carbon monoxide. The gas bubbles are nucleated at the growing crystal front, the speed of their formation being proportional to the degree of supersaturation of oxygen in a liquid steel of a given carbon content.

As the bubbles form, they tend to rise and increase in volume due to bubbles joining together and reduction in ferro-static pressure. As the bubble size increases, the rate of rise of the bubble also increases. The result is a vigorous surging of gas bubbles upwards ahead of the solidification front which causes the formation of a current in the metal. These currents are upwards along the track of the bubbles and downwards again on the axis of the ingot. This manifests itself as a rolling action on the ingot surface which is a measure of the strength of the circulating current. This movement homogenizes the remaining liquid steel and also carries down with it, into the centre region of the ingot, oxygen from the atmosphere. As long as the rimming action proceeds, there is a progressive enrichment of rejected elements from the rim to the core. If the formation of carbon monoxide is prevented by chemical or mechanical means after the formation of the desired rim, the remainder of the liquid metal solidifies in a similar manner to a balanced steel.

During the rimming action, when the rate of solidification decreases after the formation of a skin, the rate of evolution of carbon monoxide is also reduced. A stage is reached when the reduced vigour of the gas evolution is such that bubbles can be entrapped by the advancing columnar crystals. In the lower regions of the ingot, the rate of gas formation is decreased sooner than in areas above, due to higher ferro-static pressures. Consequently, the entrapped bubbles are nearer to the mould surface at the bottom of the ingot than at the top. The gas cavities, which are termed primary blowholes, lie approximately perpendicular to the mould surfaces, their depth being determined by the rate of gas evolution. Solidification proceeds beyond this outer band of primary blowholes, the top surface of the ingot solidifies well before the interior, and there is an internal build-up of pressure which suppresses bubble formation. This state is usually accelerated by freezing the top surface using a metal cap and/or water. As the carbon monoxide evolution is suppressed, an inner band of secondary blowholes forms. As previously stated, the core of the ingot solidifies in a similar manner to semi-killed steels with randomly located blow-holes in the upper region of the ingot; the entrapped gas cavities, sometimes termed tertiary blowholes, compensate the volumetric shrinkage of the core. In larger ingots, solidification of the core progresses slowly and there is further segregation around the last metal to solidify.

9. *Oxygen content and ingot structure.* The rimming action is dependent on carbon and oxygen levels in the steel. Carbon and other elements are dissolved in the liquid iron. The oxygen present is in the form of dissolved gas in equilibrium with the iron oxide content of the furnace slag, and also in combination in stable compounds such as alumina and complex silicates. The level of dissolved oxygen in the steel is normally reduced by the additions of deoxidants as outlined in para. 7. For convenience, the dissolved oxygen which is free to combine with the carbon is commonly regarded as ferrous oxide (FeO).

Some carbon is rejected from the solidifying columnar crystals and forms a concentration ahead of the freezing front. It is, however, the total oxygen free to combine with this carbon enrichment which controls the rate of gas formation under equal conditions of temperature and pressure.

As carbon monoxide is formed, oxygen and carbon are removed from the metal in the approximate ratio by weight of 16:12. The result is that the concentration of oxygen in the remaining liquid decreases more quickly than does that of carbon. This also has an effect on the rate of gas evolution at a later stage. The rate of reaction, which is the determining factor in the type of rimming action produced, and the structure and distribution of blowholes and segregation, is dependent primarily on oxygen availability and to a lesser degree on the availability of carbon.

10. *Sinking rimming ingot.* This is also known as a box-hat or boot-leg ingot and has a structure (Fig. 6) associated with a high rate of carbon monoxide evolution. Due to the escape to the atmosphere of most of the gas evolved, the entrapment of primary blowholes is slight and these are deeply seated. They do not compensate for the liquid and liquid-to-solid shrinkages of the steel during the rim formation. As a result the liquid metal surface falls well below the teeming level. During subsequent rolling, a sinking ingot gives a low yield.

11. *Level rimming ingot.* The structure of such an ingot (Fig. 7) has resulted from a less vigorous gas evolution than with a sinking rimmer. Strong circulatory currents again homogenize the remaining liquid steel. Primary elongated gas

bubbles become entrapped perpendicular to the mould surfaces and are sufficiently deep-seated to prevent their exposure during subsequent scaling and hot working. During the later stages of solidification, secondary and tertiary blowholes again form. The total volume of the entrapped gas is equal to the combined liquid and liquid-to-solid shrinkage. The result is that the top surface of the ingot is flat and very near to the original teeming level.

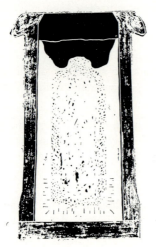 

FIG. 6. Sinking rimming ingot.   FIG. 7. Level rimming ingot.

This ingot shape gives a near maximum rolled product yield and deep-seated blowholes give optimum surface condition. For these reasons, rimming steelmakers normally aim to produce a level or near level ingot.

12. *Rising rimming ingot*. The structure of such an ingot is shown in Fig. 8. Due to insufficient concentration of interacting elements, the formation of carbon monoxide bubbles is considerably slower than in the other two types of ingots. Consequently, the gas bubbles are entrapped at an earlier stage during solidification and therefore are nearer to the ingot surfaces. Circulatory currents are relatively weak and there is less homogenization of the remaining liquid steel. Because of the slow formation of gas, much more is entrapped in the steel than in a level rimming ingot. The total volume of gas entrapped exceeds the combined liquid and liquid-to-solid shrinkage of the steel, and the upper surface of the ingot consequently rises to heights well above that of the teeming level. In addition, the tertiary gas entrapment is concentrated in the upper region of the ingot, resulting in a spongy top. During hot working of the ingot, the head contour gives an unfavourable yield compared with a level ingot. In addition the proximity of the blowholes to the ingot surface may result in their exposure by scaling during reheating and rolling. Bubble surfaces thus become oxidized, especially in the lower regions of the ingot, and the rolled product surfaces may be poor.

13. *Factors affecting the rimming action.*

(a) *Steel chemistry.* Certain elements in rimming steel affect the formation of carbon monoxide gas bubbles and so influence the structure of the final ingot. The effect of the initial oxygen content of the steel and oxygen pick-up from the atmosphere due to circulatory currents have already been discussed. It is essential that sufficient oxygen is available if a satisfactory rimming action is to be obtained. Certain other elements in the steel influence the availability of oxygen for the carbon monoxide gas formation. Figure 1 has shown how carbon levels of the steel control the oxygen content in equilibrium with it. As the carbon content increases, the oxygen levels in equilibrium decrease. However, very low carbon contents can adversely affect carbon monoxide bubble formation. Nilles has stated that the product of oxygen and carbon content expressed as a percentage must be in excess of 0·002 in order to obtain a rimming action. Thus it can be seen that with low carbon contents below 0·04%, and with higher carbon contents above 0·15%, some difficulties may be anticipated in obtaining satisfactory rimming action. These can be overcome by the use of additives to the steel in techniques which are to be described later.

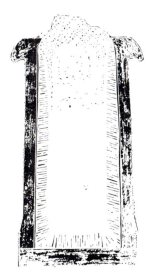

FIG. 8. Rising rimming ingot.

Manganese, which is always present in rimming steel, has deoxidizing properties and consequently affects the availability of free oxygen to the active carbon. At the time of teeming, manganese and oxygen in the steel are in approximate equilibrium. As the rim forms and both oxygen and manganese concentrate ahead of the freezing front, there is some interaction between these elements and some local deoxidization results. It should be remembered that, unlike the oxidation of carbon, the oxidation of manganese is not pressure-dependent and consequently the adverse effect which manganese has on the rimming action will be more pronounced in the lower regions of the ingot where ferro-static pressures are higher.

The oxidation of manganese occurs early during rim formation and results in a scum formation in the ingot mould. In certain circumstances this can be extremely heavy. Throughout teeming, the manganese in the liquid steel in the ladle is continuously being oxidized, especially if furnace slag is present. As a result of this, there is a progressive reduction in manganese in the ingots throughout the cast. As a general guide, the higher the manganese content of the steel, the less vigorous the rimming action. The combined effect of carbon and manganese can be seen in Fig. 9. For a given oxygen content of the liquid steel, an increase in the carbon content will result in an increase in the rimming intensity. Conversely, for a given oxygen content, an increase in manganese content will decrease the rimming intensity.

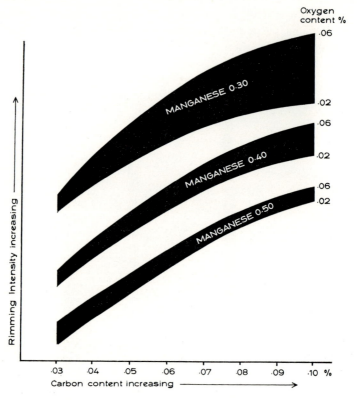

FIG. 9. Combined effect of carbon, manganese and oxygen on rimming intensity.

Silicon is a much stronger deoxidant than manganese and consequently it can only be tolerated in small quantities in rimming steel. The strong influence of silicon on rimming can be judged by the fact that it is sometimes used as a deoxidizing agent to terminate the rimming action by chemical capping.

(b) *The influence of physical considerations*. High solidification rates are conducive to the formation of a vigorous rimming action. Consequently low teeming temperatures are desirable from the point of view of initiating the rimming action, but it must be remembered that this practice could result in heavy ladle skulling and the need to teem quickly which, as seen below, can have an adverse effect. The selected teeming temperature and pouring rate are usually a compromise to obtain the optimum rimming action with the minimum extent of ladle skulling.

Since carbon monoxide bubble formation is pressure-dependent, increasing ferro-static pressure reduces rimming intensity. The greater the depth below the top surface of the ingot, the greater is the ferro-static pressure and hence the conditions for bubble formation and growth are less favourable. As a result of this, blowhole formation in the lower regions of the ingot is greater and nearer to the ingot surface, with a greater possibility of their exposure and oxidation during subsequent reheating. Since the rimming action starts early during the teeming of an ingot, a slow build-up of ferro-static pressure is preferable and consequently, for this reason, a low teeming rate is desirable.

14. *Adjusting the oxygen content for teeming*. There will generally be an excess of oxygen in the steel in the furnace and deoxidation is normally carried out in the furnace, or in the ladle, or both. Ferro-manganese is added in sufficient quantities to bring the manganese up to specification after allowing for oxidation losses. The final oxygen content in the ladle is frequently controlled by deoxidation with aluminium.

It is difficult to determine exact oxygen levels of the steel immediately prior to tapping, due to analytical problems. In addition, oxidation of the stream during tapping and teeming will change oxygen concentrations. This necessitates the final adjustment of the oxygen levels being made in the mould.

Oxygen levels and deoxidation practices vary from works to works and depend upon local conditions and plant. Most steelworks aim at a final oxygen content to meet their own requirements based on their own experience. This frequently means intentionally over- or under-deoxidizing the steel. Final control is then made during casting and the behaviour of the steel in the mould can be observed and corrected. The behaviour pattern, however, can vary within the same cast.

15. *Adjusting the oxygen content in the mould*. Corrective additions are made to the pouring stream in the trumpet or to the steel in the mould. Deoxidizing agents such as aluminium are introduced when a sinking steel (under-deoxidized) is encountered. With a rising steel (over-deoxidized), oxidizing additions are made.

16. *Deoxidizing additions to the mould*. Deoxidizing additions in the mould are normally in the form of pellets, sheets or rod. Aluminium is a very strong deoxidizing agent and great care should be taken to avoid over-treatment. Theoretically aluminium will combine with oxygen in the ratio $1:0.89$. Such efficiency is unlikely to be gained in practice due to such factors as impurities

in the aluminium and loss of metallic aluminium resulting from flotation and atmospheric oxidation.

Table 1 shows the variations in the oxygen levels between the types of ingots already discussed and the aluminium additions equivalent to their oxygen differences.

TABLE 1.

	Oxygen differences %	Aluminium differences %
Sinking rimmer to level	0·02	0·025
Level to rising	0·015	0·0175
Rising to fully rising steel	0·015	0·0175
Fully rising to semi-killed	0·007	0·0088
Semi-killed to killed	0·015	0·0175
Total oxygen differences	0·072	
Total equivalent aluminium additions		0·0863

The table shows that, when aluminium additions are made to sinking rimming steel, the first effect is to convert it to a level rimming structure. Larger additions would produce a rising rimming ingot with a typical cauliflower head pattern. Progressively greater aluminium additions would produce semi-killed and, finally, killed steels.

17. *Oxidizing additions to the mould.* The oxygen level of the steel can be raised during teeming by the introduction of gaseous oxygen on to the metal stream or by the introduction of simple solid compounds such as iron ore and millscale. However, other factors apart from the oxygen content also affect the rimming action. Reduction of surface tension at the liquid steel/bubble interface aids bubble formation and growth and promotes an early start to the rimming action. In addition there is an increasing demand for improved cleanliness of steel. To satisfy these two requirements and to provide, at the same time, readily convenient oxygen in a suitable form in a single addition, a range of rimming accelerators and cleansing agents has been developed.

18. *Control of rimming behaviour.* The main factors affecting the rimming action and the means available for modifying it have been outlined. The art of making rimming steel lies in the early assessment of rimming behaviour in the first ingot poured and knowing the remedial action to be taken in subsequent moulds. This depends upon trained observers with long experience in recording rimming behaviour and relating it to both the structure of the solidified ingot and the quality of the end product. Although general principles apply throughout the industry, pitside control is a matter of individual works' experience and conditions. Practices on identical steel qualities at similar types of steelworks may well differ in detail.

19. *Segregation.* It is the segregation of certain elements during solidification which initiates the rimming action. Strong circulatory currents in the steel resulting from the gas bubble movement homogenize the balance of liquid metal and a rim of high relative purity is formed. There is, however, a consequent increase in segregated elements in the core of the ingot. Of these elements, sulphur shows the greatest tendency for segregation followed, in order, by phosphorus, oxygen, carbon and finally manganese.

The distribution and extent of segregation are dependent on the intensity and duration of rimming. The shorter the rimming time, the less pure is the rim and consequently the distribution of segregation throughout the ingot is more general. The process of segregation does not end when the ingot top surface is closed. With cessation of the rimming action, segregation continues but to a considerably lesser degree until solidification is complete. Relatively pure equi-axed crystals sink to the foot of the ingot and heavy segregation is found in the upper central region of the ingot. The larger the ingot, the heavier is the resultant segregation and, for certain applications such as deep drawing, the top of the ingot is invariably of inferior quality and consequently down-graded.

20. *Suppression of the rimming action.* Gas evolution and the resultant rim formation are inherent in rimming steel and occur without external influences. To obtain better distribution of segregation in the core, the rimming action can be stopped by mechanical or chemical means after the desired rimming period. Bottle top moulds have been used for many years and, to function correctly, require the quality of the rimming steel to be slightly rising. The restrictive section at the mould top allows little pick-up of atmospheric oxygen and funnels the rising metal to accentuate the apparent rate of rise. Shortly after teeming is completed, a heavy metal cap is placed in the mould top. When the steel rises to touch this, mechanical capping automatically results. The duration of rimming in bottle top moulds can be varied by either adjusting the rate of rise of the steel, or modifying the teeming level relative to the base of the cap.

In other practices, moulds are used which are not restricted in the upper regions, these being termed "open top". When the steel has rimmed for the required time, a heavy metal plate is lowered on to the top surface of the metal, freezing it and subsequently inhibiting the rimming action. Thin metal plates with a following application of water are frequently used as substitutes for the heavy metal plate.

Another method of suppressing the gas formation and so obtaining a level rimming structure is the addition of strong deoxidizing agents such as aluminium or silicon following the required rimming period. These elements combine with the available oxygen in the remainder of the liquid steel, which then solidifies in a manner similar to balanced steel. This procedure is not to be confused with the method of rim stabilizing where the remaining liquid metal is fully killed by the introduction of aluminium after the formation of a thin rim. Such steel is subject to solidification shrinkage and ingots must be hot topped to obtain satisfactory yields.

21. *Cleanliness of rimming steel.* Inclusions in liquid steel are derived from external sources such as furnace, ladle and casting refractories. These are termed exogenous. Others result from the deoxidation process and reoxidation during teeming. These, together with sulphides, are termed indigenous inclusions. More than 80% of the total inclusions are indigenous. This is true of all qualities of steel, but the high oxygen levels present in rimming steel during solidification create even greater problems than with killed and semi-killed grades. There tends to be an enrichment of inclusions in the bottom regions of the ingot due to the sinking of equi-axed crystals which have been nucleated by non-metallic matter. Sulphide inclusions in particular tend to be present in greater quantities in the upper part of the ingot, but at a lower position than would be anticipated in a hot-topped killed ingot.

At the foot of the ingot, high ferro-static pressure tends to inhibit the CO bubble formation and manganese is preferentially oxidized. This leads to the earlier appearance of manganese-containing scum on the metal surface. The turbulence of the rimming action carries up the scum particles but further oxygen pick-up in the downward circulatory current may increase manganese oxidation. Ingot scum can be taken down into the ingot and entrapped during solidification.

Rimming actions which are either too weak or too vigorous can worsen the cleanliness of the ingot. The need for a strong but controlled rimming action is once again emphasized.

When the carbon/manganese ratio is high the carbon is preferentially oxidized. When the carbon/manganese ratio is low, the reverse is true and manganese is preferentially oxidized. This is a function of the oxygen level which is proportional to the carbon content. In general, cleanliness worsens with increasing manganese content. When the carbon/manganese ratio is low and in smaller ingots where circulatory currents are weak, this is particularly evident.

22. *Chemical aids to rimming steel production.* The previous paragraphs have outlined the many factors, the complicated simultaneous chemical reactions and physical phenomena affecting production of rimming steel ingots. Although the behaviour of a given steel composition can generally be predicted for a given set of conditions, the certainty of obtaining optimum results can be reduced by works and production variables. These prevent the adoption of identical practice by different works.

Two main factors decide the final yield of prime quality material, namely the rimming intensity and the cleanliness of the solid ingot. Both can be materially influenced by the addition of chemical rimming agents.

It has long been known that certain compounds, when added to rimming steel, have a pronounced effect in influencing the rate of carbon monoxide gas evolution ensuring a subsequent improvement in the ingot structure. These compounds function by either providing an element to react chemically with other elements already present in the steel to form carbon monoxide bubbles, or promoting early bubble formation by nucleation and reduction of surface tension at the bubble/liquid-steel interface. There is normally an added benefit in that the molten slag, formed from addition of these compounds, tends to agglomerate with, and absorb, inclusions from the steel, facilitating their rise to the surface and cleansing the ingot.

23. *Effect of rimming agents on ingot structure.* In the case of rimming steels which contain insufficient available oxygen to produce a satisfactory rimming action, additives contain oxygen-yielding components. Where the steel specification calls for very low carbon content, or where steel is cast in an underdeoxidized condition, rimming agents are available containing corrective carburizing or reducing media. When carbon and oxygen are at optimum levels, there is frequently a need to promote early bubble formation. This is achieved by incorporating materials which reduce surface tension.

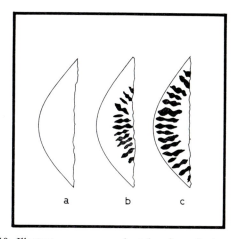

Fig. 10. Illustrates corner samples taken from the lower Third of:

(a) A sinking rimming ingot (Fig. 6) primary blowholes very deep-seated.
(b) A level rimming ingot (Fig. 7) optimum positioning of blowholes.
(c) A rising rimming ingot (Fig. 8) blowholes dangerously near the surface.

Treatment of sinking and rising ingots in "a" and "c" with suitable rimming agents will result in the formation of a level rimming ingot, "b", which is the type giving maximum yield of prime quality product.

TREATMENT OF INGOT MOULDS
AND BOTTOM PLATES

J. J. CHAPPELL

A great deal of money and research has been devoted over the last ten or twenty years to increasing the output and improving the quality of the steel as it leaves the furnace, and it is, therefore, essential that equivalent care should be taken on the pitside to ensure that the yield from the steel ingot is as high as possible. Because the steel discarded on the pitside can vary from 4 to 20% it will be appreciated that this represents a great deal of money and material.

It has been shown in some of the previous sections that by providing a correct feed to the ingot head the yield can be much improved, but the present section deals with the improvement of ingot yields by the correct preparation of the mould and bottom plate, and also gives some facts relating to yield and teeming control.

INGOT SURFACE

Poor ingot surface can lead to major defects in the final products; this is especially true of strip, sheet, bar or plate, and these defects may lie undetected until final inspection.

There are two major influences on the surface of the finished ingot. Firstly, the condition of the mould and secondly, the teeming control. It will readily be appreciated that the surface of the ingot mould is reflected in the surface of the finished ingot. Thus, a major defect on the mould may even result in the virtual scrapping of an ingot as it may be impossible to strip the ingot without breaking the mould. Alternatively, it may lead to serious rolling defects which again result in the discard of a considerable proportion of the ingot. There is a wide variation in the amount of work done on the mould from one works to another, but in most cases the ingot mould is cleaned and dressed before use. The decarburizing of the mould surface leads to crazing and cracking and it is the melting shop staff who decide when this is too bad to be further tolerated. Clearly, minor imperfections will be removed during the normal scaling of the ingot whilst in the soaking pits.

Secondly, the surface and general condition of the finished ingot will be reflected in the way in which it was teemed. Lack of teeming control results in many ingot defects and, to a large extent, this is in the hands of the teemer. The more common failures are lack of control of speed and lack of control of the stream itself. If the steel is teemed too fast, this leads to panel cracks, whilst too slow teeming leads to ripples, laps, etc. It is significant that during the last few years much work has been carried out in the U.K., U.S.A., Italy and other

110

countries to invent suitable devices to improve teeming control. Most of these, however, need sophisticated refractories which require careful and precise setting and only in melting shops where a high standard of technical supervision takes place, are these new devices going to be a practical proposition on a large scale.

Current Trends to Improve Ingot Surface

The effect of additives on the formation of the ingot skin has already been considered in the section on "Effect of Mould Additives and Fluxes on Ingot Surface Quality", but at least as much can be achieved by taking suitable steps to ensure that the mould condition is satisfactory and that the teeming control is as good as possible.

Unsatisfactory Mould Conditions

(a) Worn Moulds

There is really very little that can be done to produce a good ingot from a badly worn mould. However, there are suitable compounds already on the market to decrease the rate of oxidation of the graphite of the ingot mould surface; these coatings are applied when the ingot mould is red hot, this being the temperature at which the majority of the oxidation takes place. These compounds form an impervious layer over the whole of the inside of the mould, and are normally of a glassy nature.

For many years carbon has been sprayed or deposited on the inside of ingot moulds in an attempt to fill cracks and crazing marks with a deposit such that, as the rising metal comes into contact with it, it burns in such a way as to keep the steel meniscus out of the crack. It will be appreciated, however, that this does nothing to decrease the oxidation of the mould surface, as these materials cannot be applied over about 150°C as they burst into flames.

(b) Dirty Moulds

Unfortunately, after stripping, many moulds retain a ring of scum that was on the ingot surface at the time of solidification and if the next ingot is teemed past this ring, areas of entrapped slag oxide occur that may be rolled into the finished product. Clearly the only cure for this is to ensure that the moulds are clean, although there is considerable variation in the adhesion characteristics of this scum depending on the mould additives which have been used.

(c) Moisture in, or on, Moulds

Cast iron is semi-porous by nature and it is capable of retaining considerable quantities of moisture. Not until the mould has been used once will it be really dry and new moulds particularly give rise to pinholes and blowholes due to the water vapour expansion. Similarly rusty moulds will retain water vapour.

Whilst it is not possible to ensure that new moulds are free from moisture, the application of impervious layers to stop oxidation, as noted in (a) above, will likewise stop the absorption of water vapour. It should also be noted that many carbonaceous coatings used to improve the surface of ingot moulds contain water and it is essential that correct specifications should be used to control the supply of these materials.

(d) Poor Mould Design

If the corner radii is incorrect from the angle of the head tile (or head assembly of a forged ingot) and is too sharp, corner cracks or hanger cracks can result. Little can be done to improve these defects, except to improve the general level of work at the shop.

(e) Hot Moulds

If the moulds are above a certain critical temperature (determined by the ingot size and the quality of the steel being produced) the solidification of the ingot skin will not proceed normally and deep panel cracks will result as the ingot cools—particularly if the ingot is stripped after its normal standing time. This cracking is caused by the fact that the chilled "skin" is not strong enough to contain the body of the ingot as the ingot cools from the outside. Considerable improvements can be obtained under these conditions by the use of coatings on the ingot mould that, in fact, act as an insulating layer between the solidifying metal and the mould. This will enable moulds to be used more frequently and thus lower the stocks necessary for any given set of works conditions. The thickness of the layer and its physical characteristics will be critical for a given set of conditions of ingot size, quality of steel, etc.

TEEMING CONTROL

As already mentioned above, teeming control has a major influence on the quality of the finished ingot, even presupposing that the mould is in perfect condition. Many of the influences of poor teeming control are directly in the hands of the teemer and his equipment, so it is not possible to take remedial action for his errors.

Lack of control can be considered in two parts—lack of control by the operator or failure (or partial failure) of the equipment. The teemer may teem too fast or too slow or fail to teem in the centre of the mould. All these things lead to poor quality ingots and poor yields. If the equipment fails (say the refractories wear or crack) it is not possible to do anything about it after the occurrence, but by the use of coatings or special refractories, it is possible to minimize this wear. Some of the new teeming gadgets which have been developed over the last few years are now being tried in service. The most commonly used of these is "Autopour" which removes the teeming lever and replaces it by an hydrauelically operated arm which opens and closes the nozzle. It will be appreciated-that this does not overcome lack of teeming control associated with a badly set stopper and nozzle. More recently several devices have been tried to vary the rate of flow of metal from the ladle to the ingot. One of these has been developed by an Italian company. It is called the "Fischera" teeming control technique, and this consists of a refractory wedge which slides across the nozzle opening. However, practical difficulties are encountered in maintaining the sliding action of the wedge against the skulling tendency of the steel. A third design, at present marketed by the Vesuvius Crucible Company Ltd., which consists of a refractory needle valve is being used successfully in some works. However, it is fair to say that at present there is no piece of equipment which is being used on a large scale which takes the major part of teeming control away from the teemer.

It can thus be seen that although lack of teeming control is perhaps the most important single cause of poor ingot surface, very little can be done to minimize its effect.

BOTTOM PLATES

The wear on bottom plates as used in normal rolling mill ingots varies very considerably. The type of teeming used and the type of mould employed are two major influencing factors. Where the steel is teemed uphill or where closed bottom moulds are used, bottom plate wear is at a minimum. However, the teeming of large direct teemed slab ingots causes great damage to the bottom plate, and as this constitutes a large proportion of the total steel teemed, this is a major problem. It is not always appreciated that a badly worn bottom plate can influence the mill yield of the ingot very considerably and if the weight of metal lost in the crater so formed is considered the importance of this problem will be seen. On, say, a 15-ton ingot, these craters can be 2 ft × 10 to 12 in. deep and thus the steel filling this crater represents a very expensive yield loss.

There are two main approaches to this problem, which are either to allow the crater to form to predetermined depth and then completely refill the crater, or alternatively, to coat the bottom plate with a refractory dressing prior to every teeming. In certain circumstances both forms of treatment are combined.

Bottom plate patching materials are normally refractory based. Some are dependent on an hydraulic bond to keep them in position, i.e. using some form of "cement", while others use a chemical reaction to cause expansion and hardening of the whole mass. The most important criteria is that even under the condition of maximum stress large particles of this refractory should not become incorporated in the ingot. It has been shown that the effectiveness of this patch has only to be operational for the first few seconds of teeming, as after the formation of a pool of molten metal the wear on the bottom plate becomes negligible.

There are many proprietary brands of refractory coatings for bottom plates that are used as a wash. These fall broadly into two classes; either they are very cheap and large quantities are used, or alternatively, they are expensive with sophisticated fillers and binder systems and to ensure their economic usage, strict control must be exercised during their application. The first group usually comprises very cheap materials including fireclay grogs, sand, etc., while the second group uses special grades of silica, magnesite or chromite. There are many materials in both classes now on the market and clearly the local supply and demand will determine which is used at a particular plant. It has been demonstrated in several cases that it is not possible to use materials from the expensive group unless there is extremely good control of its use 24 hours a day.

In a few works, brick inserts on the bottom plate are used, and these are replaced from time to time. However, these still normally use some form of refractory wash and although the maintenance of these plates is relatively straightforward, the economics of the operation are such that this practice is gradually dying out.

ELECTROSLAG REFINING PROCESS

J. D. SHARP

INTRODUCTION

Although electroslag refining (E.S.R.) was first developed before 1939, it is only in recent years that the process has attracted significant commercial interest. The ever-increasing demands for high-quality clean steel have highlighted E.S.R. as being an attractive alternative secondary refining method to vacuum arc refining (V.A.R.). This is because equipment and running costs for E.S.R. are considerably lower than for V.A.R. and the process has a number of technical advantages.

R. Hopkins (M. W. Kellogg & Co., U.S.A.) first developed and patented the process in the late 1930's, but interest in it flagged during the war and it was only in the 1950's when E.S.R. was seriously pursued mainly in the U.S.S.R. where it was developed from an electroslag welding process at the E. O. Paton Institute of Electric Welding in Kiev whereby welding was carried out under a conductive pad of molten slag.

In 1959 Firth Sterling of U.S.A. acquired the Hopkins process from Kellogg and until 1964 used it exclusively for their own production which in 1966 was approximately 6000 tons by E.S.R. In 1964 Firth Sterling decided to licence other steel-makers and since then the process has made steady progress.

In the U.S.A., V.A.R. is very well entrenched (200,000 tons in 1966) and it is only within the last few years that E.S.R. has been seriously considered as a competitive method.

In contrast, the U.S.S.R. is by far the most experienced and extensive user of E.S.R. Many thousands of tons appear to have been made by the electroslag refining process and many types of steel have been examined. Most of the Russian production is in ingots up to 24 in. diameter weighing up to 1·5 tons, but there has now been considerable experience in producing ingots up to 36 in. diameter weighing up to 13 tons. A 56 in. plant capable of melting 30 ton ingots is under consideration.

Several French companies have licence agreements with the U.S.S.R. and Eastern European countries such as Czechoslovakia and Rumania are also using Russian know-how, but very few details have been published.

In the U.K. apart from the plant built under licence, some companies have built their own plants and a survey of the position indicates that by 1970 about 1% of the total U.K. alloy steel production will be E.S.R.

The main interest is now in ball and roller bearing steels, tool and die steels and the range is expanding to stainless, heat-resisting transformer and structural steels. Nickel alloys and titanium are also being refined in the U.K. by E.S.R.

and successful trials have been reported in the U.S.S.R. with copper and cobalt. Experiments are also being made with aluminium.

THE PROCESS

The metal which is to be refined is firstly formed into an electrode of suitable size which depends on the size of the required refined ingot. This electrode is either cast or wrought from primary material or scrap. A molten slag bath contained in a water-cooled mould is resistance heated by electrical current flowing between the consumable electrode and a water-cooled base plate. The slag temperature approaches 2000°C, and this causes droplets of the metal to melt off the tip of the electrode, fall through the slag and collect in a pool on the base plate. The consumable electrode is fed into the slag bath and a refined ingot, which now acts as the secondary electrode, is progressively built up.

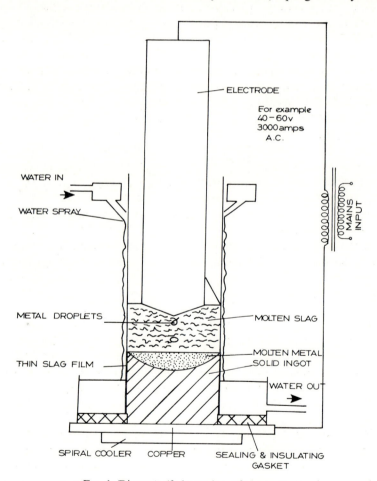

FIG. 1. Diagram of electroslag refining unit.

REFINING

Refining takes place by the reaction between the molten slag and the molten metal and E.S.R. proceeds from the molten metal to the solid ingot out of contact with air. The metal droplets fall through the layer of molten slag and collect in a pool at the tip of the already solidified ingot. It is important that stable current conditions are maintained throughout the melting period as fluctuations in current are reflected as ripples in the finished ingot. The use of alternating current assists the refining action of the slag by vibrating the slag-metal interface.

It is of the greatest importance that the slag has the correct electrical and chemical characteristics so that during the droplets falling through the slag certain reactions can be encouraged. As an example, the removal of sulphur to very low levels can be obtained by the addition of lime to the slag. Oxides and other non-metallic inclusions can be removed either by enveloping them into the slag or by chemical reaction with the slag.

A sound ingot is formed by near vertical solidification. Towards the end of the operation the current is reduced and complete absence of pipe and central porosity can be ensured.

Due to the method of operation, a thin film of slag solidifies between the mould and the ingot, and forms a smooth inner lining and an exceptionally smooth ingot surface is obtained. The layer of slag flakes off after stripping.

THE SLAG

From the previous remarks it can be seen that the slag itself is an essential part of the process. The basic requirements of the refining slag are that it should

(a) be molten at a temperature slightly below the melting point of the steel;
(b) be stable at temperatures up to 2000°C;
(c) be electrically conductive when molten but have sufficient resistance to ensure that the normal current is effective in maintaining a high temperature;
(d) have a low viscosity.

Calcium fluoride is a satisfactory basic material and the majority of slags used in practice for E.S.R. are calcium fluoride-calcium oxide-alumina systems. Calcium oxide is added in order to remove sulphur and phosphorus and alumina is added to increase the electrical resistance and give a hotter refining bath. The table on page 118 gives the compositions of a number of slags which have been used in practice and it also gives the melting range and application for particular compositions.

Some steel companies make their own fluxes and some buy them from outside suppliers. For obtaining consistent results it is necessary to prefuse the slag and then crush and grind the resultant material. Prefusion not only produces a more intimate mixture with a more definite melting range, but it also removes moisture

TABLE 1. *Table of Slags Used for Remelting*[1]

Details of slags used for remelting

Composition, %									Melting range °C	Application
CaF_2	Al_2O_3	CaO	TiO_2	MgO	BaO	ZrO_2	NaF	SiO_2		
70	30	—	—	—	—	—	—		1320–1340	Steels and alloys without Ti or B; secures desulphurization
80	—	20	—	—	—	—	—		1200–1220	Steels and alloys without Al, Ti and B; secures maximum desulphurization
60	20	20	—	—	—	—	—		1240–1260	Steels without Ti or B; secures maximum desulphurization
80	—	—	—	20	—	—	—		—	As above
80	—	—	—	—	—	20	—		1340–1360	As above, but secures a hotter slag bath
80	—	—	—	—	20	—	—		—	Dephosphorization
50	25	—	25	—	—	—	—		1220–1240	Steels and alloys with Al and Ti
95	—	5	—	—	—	—	—		1390–1410	Steels and alloys with Al, Ti and B
—	55	45	—	—	—	—	—		—	Steels and alloys without Ti or B; secures the hottest slag bath
80	—	—	—	—	—	—	20		1160–1180	Copper alloys
18	40	25	—	17	—	—	—		—	Steels and alloys without Ti or B; secures high output
—	60	35	—	5	—	—	—		1450	Steels and alloys without Ti or B; secures good desulphurization and high output
35	—	15	35	5	—	—	—	10	—	For use when slag conductive in the solid state is required

1. The SiO_2 and FeO contents of slags for ESR should not exceed 0·3%, sulphur content—0·05%.
2. The carbon content of slags for remelting low-carbon steels and alloys should not exceed 0·02%.

from the raw material which is always a source of hydrogen pick-up by the molten steel.

The main improvements required of the E.S.R. process are:

1. To avoid hydrogen pick-up.
2. To prevent loss of wanted elements.
3. To avoid contamination by unwanted elements.
4. To improve removal of sulphur, phosphorus and oxygen.
5. To promote good surface finish.

The degassing effect of E.S.R. is not very good except if the ingot to be refined has a high hydrogen content.

An equilibrium is reached between water vapour in the atmosphere, hydroxyl ions in the molten slag and hydrogen in the steel.

Quite small quantities of moisture in the flux can introduce considerable quantities of hydrogen in the first part of the ingot. This trouble is increased if considerable quantities of calcium oxide are added for good desulphurization because these slags absorb moisture readily and retain some of it even when molten in contact with the atmosphere. Prefusion of the slag, therefore, becomes necessary if severe gas pick-up is to be avoided. Another advantage is that the ratio of moisture pick-up during storage of prefused flux is very small compared with unfused mixtures.

A mixture containing 70% calcium fluoride, 20% aluminium oxide and 10% calcium oxide, picked up 1·85% of moisture after exposure to the atmosphere for eleven days. Exposure under similar conditions of a prefused flux did not show any increase in weight. With an increase of calcium oxide content the moisture pick-up of the unfused flux is considerably increased and for this reason for lime-bearing slags, prefusion is almost essential.

Moisture can be removed also by calcination at approximately 1200°C but this offers no appreciable advantages over fusion.

SULPHUR REMOVAL

TABLE 2

		Flux: CaF_2 with % CaO				
		0	10	20	30	40
Sulphur content %	Original steel 0·042	0·021	0·018	0·012	0·009	0·007

Calcium fluoride alone will remove sulphur from the steel droplets and additions of calcium oxide, barium oxide or magnesium oxide greatly improve desulphurization by the slag. Up to 40% of calcium oxide has been used but 10% or 20% is more usual. Sulphur contents down to 0·002% have been reported (Table 2).

E

PHOSPHORUS

Phosphorus can only be removed by either very basic or very oxidizing slags. Oxidizing slags are more efficient but cannot be used because of the removal at the same time of wanted elements. Fluorspar has no effect on the phosphorus content. The raw materials for the slag have to be carefully controlled for freedom from phosphates of any sort, because these are often reduced and then transferred to the steel.

Other contaminants in the raw materials for the slag are lead, zinc, tin, selenium, tellurium, arsenic, antimony, bismuth and silver and even if they are only present in traces either in metallic form or in compounds, they can cause serious deterioration in the properties of high quality steels and nickel alloys. As an example, the maximum tolerable amount of lead in nickel alloys and in certain steels is 0·0002%.

TABLE 3. *Analysis of fluorspar before and after Fusion.*

Element	Analysis		Element	Analysis	
	% before	% after		% before	% after
H_2O	0·15	Nil	Zr	0·005	0·005
CaO	0·015	0·067	Ge†	< 0·005	< 0·005
S	0·05	0·0025	Ta†	< 0·5	< 0·5
P	< 0·002	< 0·002	Mo†	< 0·005	< 0·005
Fe	0·80	0·07	Cr†	< 0·05	< 0·05
Si	0·75	0·30	Sn†	< 0·005	< 0·005
Al	0·10	0·01	Mg	0·01	0·01
Ti	1·00	0·50	Ga†	< 0·005	< 0·005
Mn	0·10	0·05	W†	< 0·10	< 0·10
K	0·14	0·028	Ni	0·005	0·005
Na	0·30	0·16	Be†	< 0·05	< 0·05
B	0·10	0·01	V	0·001	0·001
Pb	0·10	< 0·002	Cu	0·01	0·01
Sb	0·06	< 0·005	In†	< 0·005	< 0·005
As	0·005	< 0·0005	Cd†	< 0·05	< 0·05
Zn	0·005	0·0002	Ag†	< 0·001	< 0·001
Bi†	< 0·001	< 0·001	Co†	< 0·005	< 0·005

† The elements marked thus were examined spectrographically and could not be detected. The figures quoted indicate the lower limit of detection under the conditions used.

Great care must, therefore, be taken in the selection of raw materials for the slag.

Fortunately, most of the above elements or compounds are sufficiently volatile that they are wholly or partially lost during fusion of the slag and, therefore, prefusion represents a useful method of purification. The above list (Table 3) gives the analysis of a commercial grade of fluorspar before and after fusion.

Some of the alloying elements of steel will also be lost during the E.S.R. process because of the high temperature and the chemical reactivity of the slag.

The elements which are most affected are aluminium, titanium, vanadium, silicon, manganese and molybdenum.

Aluminium and titanium are most readily lost and this must be taken into account when transformer steels are refined because of the narrow and critical silicon range.

The loss of manganese may adversely affect the sulphur content because instead of the sulphur being present as manganese sulphide the formation of iron sulphide is favoured. This is particularly likely if the manganese content is less than ten times the residual sulphur content. Some of the losses of alloying elements can be prevented or diminished by the right selection of the refining slag. Other losses will have to be made good by alloy additions. The amount of slag used in the process varies greatly and depends on the size of ingot, melting equipment and other factors. Generally it fluctuates between 3·5% and 6·5% of steel remelted.

STARTING UP THE PROCESS

The process is started by striking an arc through powdered slag between the water-cooled copper base plate and the electrode tip. The copper base plate is often protected from the erosive effect of the arc by covering with a starting disc which is a slice of the metal to be refined $\frac{1}{2}$ to 1 in. thick.

The powdered slag around the electrode tip is melted and the electrode immersed into the fused slag as quickly as possible in order to quench the arc and to start resistance heating. Obviously the electrode begins to melt long before the slag is fully molten and during this time powdered or pasty slag may be trapped in the bottom part of the ingot. Also the ingot surface may be severely corrugated and the length of discard can be quite considerable.

Attempts have, therefore, been made to improve the starting-up operation and several ways have been tried:

1. A Russian method uses a simple exothermic mixture of one part metallic magnesium, one part metallic aluminium and two parts potassium nitrate and this is placed as a conical heap on the base plate and surrounded by powdered refining slag. The tip of the electrode is lowered on to the exothermic and passage of current ignites it and the heat of the burning exothermic assists in establishing a molten slag pool. In practice this method gives only a marginal improvement and there is danger of contamination of the ingot with alumina and magnesia.

2. Another Russian development is the use of a slag which is electrically conductive when it is solid. The compound which confers this property is titanium oxide, up to 40% of which is included in the starting-up powder. It is claimed that this gives a smooth start but there is a risk of pick-up of titanium by the metal. Therefore, the amount of titanium-bearing powder has to be kept to a minimum and the chilling effect of the normal slag powder leads back to irregular ingot surface at the base.

3. *Molten slag*—in this method the refining mixture is premelted in a separate furnace and the molten slag is transferred into the mould immediately before striking the arc. There are some difficulties in feeding the molten slag down the narrow space between the electrode and the water-cooled mould wall because the molten slag freezes immediately it comes in contact with the mould wall and a solid skin forms on the water-cooled base plate. Furthermore, the separate melting equipment for the slag adds to the cost of the plant but satisfactory results have been obtained on fairly short moulds with medium diameter.

4. *Exothermic starter tablets*—some satisfactory results have been obtained with exothermic tablets into which conducting wires are moulded and which pass from the top to the bottom of the tablet and project about $\frac{1}{8}$ in. each side. The tablets consist of exothermic mixtures which on ignition produce metal and slag and the composition can be so adjusted that the metal corresponds to the composition of the steel to be refined. The tablets are approximately $2\frac{1}{2}$ in. thick and the yield of molten slag produced after the thermic action is approximately 70%.

The exothermic reaction is started by lowering the electrode on to the surface of the tablet and switching on the current. The conducting wires become hot and ignite the mixture and the hot molten slag fuses the refining slag surrounding the tablet.

These starter tablets can also be used for nickel alloys if the exothermic mixture is suitably adjusted. So far only limited experience has been gained with this method.

THE ELECTRICAL OPERATION

The equipment consists of a mould, a transformer to provide the necessary electrical power and the means of supporting the feed electrodes with suitable mechanical gearing to raise or lower the electrode during melting.

(a) *The mould*—copper is the normal material employed for the mould although graphite and steel are also used. The copper mould is water-cooled by forming it into a jacket of the required shape, but this mould is expensive and heavy. Steel moulds must also be water-cooled and this is usually done by water spray. The choice of material for the mould depends very much on local conditions and the size of ingots, etc. The base plate to the mould is also water-cooled and it must withstand possible damage during the initial striking of the arc. The cooling must be so adjusted that the formation of a solid layer of slag between the base plate and the ingot is avoided.

(b) *Electrical equipment*—there are four modes of electrical operation:

  (i) Single phase, one electrode, one mould.
  (ii) Single phase, two electrodes, two moulds.

(iii) Three phase, three electrodes, one mould.

(iv) Three phase, three electrodes, three moulds.

There are certain metallurgical advantages in the use of three-phase current. The slag rotates on top of the solidifying ingot and this greater movement gives a better refining effect than with single phase. The power consumption can vary within fairly wide limits depending on the type of operation and is between 1000 and 1700 kWh per ton. The melting rate can vary between $2\frac{1}{2}$ kg and 14 kg per minute.

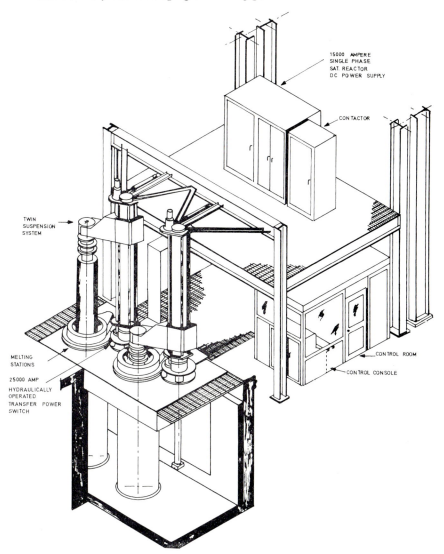

FIG. 2. Twin melting station furnace for electroslag refining. (By courtesy of Consarc Corporation, Rancocas, New Jersey, U.S.A.).

Figure 2 shows in diagrammatic form a twin melting station furnace producing 24 in. diameter ingots of approximately 7 tons. The power supply is single phase a.c. and the makers of the furnace are Messrs. Consarc Corporation, New Jersey, U.S.A.

(c) *The product*—the yield, of course, varies from plant to plant but it is from 92% upwards to nearly 100%. In 1967 only about 2500 tons of E.S.R. ingots were produced in the U.K. and the figure for U.S.A. was approximately 8000 tons but production figures are going up rapidly and by 1972 will be in the region of 100,000 tons in U.S.A. alone.

FIG. 3. One of the largest electroslag refining units in England, constructed by Firth Brown for melting of high tensile and stainless steels. (By courtesy of Firth Brown (Sheffield) Ltd.).

REFERENCES

1. *Metals*, 1967, **2**, March, page 46a.

CHAPTER 3

SPRAY STEEL MAKING

J. D. Sharp

1. INTRODUCTION

The steel-making processes are designed to remove unwanted elements from solutions of liquid irons, and to introduce alloying elements in carefully controlled amounts. The presence of these unwanted elements is due to the natural impurities in the iron ore and they can vary very widely—depending on the source of the ore. Carbon is the only exception, and this is present to its saturation limit; it arises from the use of carbon in the blast furnace—both as a refractory and as a reducing agent.

The rate of steel making is largely governed by the rate of elimination of these elements, and as most of them are removed by oxidation, clearly rapid mixing of oxygen and metal give high rates of steel-making. Bessemer in his original work (1856) did just this by blowing air into liquid pig iron, but clearly there are some disadvantages as very little steel is made by the original Bessemer process today. Spray steel making is even faster than Bessemer steel making, and so, as well as the potential advantages of speed, low capital cost, etc., the problems that remain still to be solved must also be considered.

2. BASIS OF THE PROCESS

Essentially the process consists in allowing molten iron to flow from a runner through an annular ring and drop into a receiving ladle. As it falls through the ring it is submitted to jets of oxygen and at the same time lime powder is fed into the system. The liquid metal stream is shattered to many fine droplets, thus exposing the maximum surface area of metal to the gas environment. In this way the chemical reactions are propagated in the most rapid way possible, but in this design of plant they must be completed during the falling of the metal from the nozzle to the receiving ladle. When the metal reaches the ladle it will certainly contain dissolved oxygen and also slag particles in fine dispersion—but the relative rate of reaction to that obtaining during the fall of the metal will be slow—and to a certain degree indeterminate.

It will thus be seen that in essence spray steel making is a very rapid method of removing oxidizable metalloids, but like all very rapid processes there are major difficulties of control. It is on the development of good control of element removal and temperature that the success of the technique depends. If there is good control of the analysis of the ingoing iron, it is quite feasible to make the necessary adjustments to obtain a range of, say, carbon analyses, but the process is so rapid that the ability to obtain carbon to, say, 0.02% C accuracy must be regarded as very difficult. The removal of each element will be considered below.

3. DEVELOPMENT OF THE PROCESS

Apart from the postulations of Bessemer, the first practical technique for a plant using this principle was put forward by the British Oxygen Company in 1955 (B.P. No. 790,790). The idea was to use a counter-flow technique, thus offering more chance of control as the gas/metal dwell-time would be longer than on a straight fall.

The B.I.S.R.A. patents are an advancement in that they provide for a physical breaking up of the metal stream by impinging oxygen jets, (B.P. Nos. 890,282, 949,610 dated 1957, 1959) and also do not have the great potential difficulties of preheating the tower and developing special refractories.

The original development was regarded as a method of desiliconization and this was carried out at the Hoyle St. Laboratories of B.I.S.R.A.; this was followed by trials at the Redbourn Works of R.T.B., Scunthorpe. The Hoyle St. experiments used a 10-cwt metal charge, but at Redbourn 16-ton batches of metal were treated. In this application the object was to prerefine blast-furnace metal before it was charged to the steel-making furnaces. In particular, the removal of silicon minimized the slag required on the open-hearth furnaces, thereby accelerating the consequent steel making.

After the experimental plant at Redbourn was closed down in favour of other prerefining techniques, the work concentrated on actually producing steel directly from iron. The concept of continuous steel making from a blast furnace runner with the minimum of capital expenditure had obvious appeal, but no further large-scale work was done until the Millom Hematite Ore and Iron Co.[1] decided to investigate it. It must be remembered that the conditions at this plant were exceptional, and in some ways ideal as a testing ground for spray steel making.

Firstly, the company had surplus iron and a method of converting it to steel had to be considered. The amount of iron was not large by normal steelworks standards—but large enough to demand that the surplus was not just put to solid pig iron. The very low capital cost of the equipment was an obvious attraction.

Secondly, the analysis of the iron was normal hematite analysis—i.e. it was relatively low in sulphur and phosphorus, thereby clearing what could only be described as a major area of uncertainty. Laboratory figures quoted for sulphur and phosphorus removal were sometimes good and sometimes not so consistent. The difficulties of coping with large amounts of foaming slag and guaranteeing consistent mixing of slag and metal are obvious.

So to prove the basic concept of spray steel making—without some of its major difficulties—Millom was an excellent site.

The first steel was put through the Millom plant on 18th May 1966, and since then the plant has regularly been making 10-ton batches of steel. Further development of this process is planned for Millom to give a production of some 3000 tons/week in conjunction with a continuous casting plant. Two other plants were commissioned in 1967 at Lancashire Steel (15/2/67) and Shelton works of John Summers. At the time of writing, both plants were having some difficulty with the high phosphorus iron they are handling, but if the same tenacity of purpose is shown with this problem as was shown earlier in the history of the process, it will be solved in due course.

4. DESCRIPTION OF THE PLANT

The plant described will be mainly that built at Millom as few details of the other plants have been released.

(a) Location

As part of the overall philosophy of making steel making continuously the Millom unit was built into the blast-furnace runner. This gives close proximity to the blast furnace without handling the metal—and at the same time provides a take-off for any metal that is not treated. Owing to the small size of the necessary equipment, it can readily be accommodated in the same building as the blast furnace and the minimum of steelwork is required as at no time is the hot metal being supported—except in the collecting ladle.

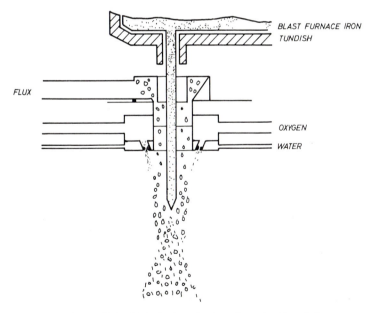

FIG. 1. The tundish which receives the iron from the blast furnace.

E*

(b) The Tundish

This is built as a section of the normal iron runner at Millom, and is so designed that it controls the head of metal and hence the flow of metal into the spray steel-making equipment. A short stopper rod is used to open and close the nozzle through which the metal flows into the reaction zone. The tundish is lined with normal refractories and the stopper is of very simple construction, as the temperatures concerned are only those of molten iron and not those encountered in steel making. The nozzle size varies from about $\frac{3}{4}$ in. to 1 in., depending on the flow required, but for the first Millom plant the published figures[1] are:

$\frac{7}{8}$ in. nozzle gives 24 tons/hour of iron
1 in. nozzle gives 30 tons/hour of iron

Clearly these figures will be subject to local plant conditions, but plans are in hand to design a 200 tons/hour plant. (This excludes the amount of scrap melted by the iron—i.e. the steel-making rates are considerably higher than the above.)

Comparable figures for the B.I.S.R.A. pilot plant[2], using a 9 in. head of metal are:

$\frac{1}{2}$ in. nozzle gives 6 tons/hour
$\frac{5}{8}$ in. nozzle gives 12 tons/hour

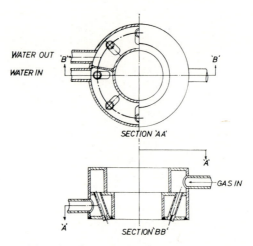

FIG. 2. Atomizing ring (from *Steel and Coal*, Sept. 14, 1962, p. 508).

(c) The Spray Ring

The spray ring is clearly the heart of the process and in some ways this addendum to the original patent is at least as important as the original idea. The ring has at least three functions to perform:

(i) To atomize the iron stream.

(ii) To provide the correct amount of oxygen to oxidize the unwanted metalloids.

(iii) To inject the required amount of flux to form a slag of the required composition which will retain the unwanted phosphorus and sulphur.

The ring is made in two parts, the flux injection and the oxygen jets being separate. The technique of using the oxygen to mix the lime curtain into the metal was initially devised at Redbourn,[3] and as the resultant metal/flux mixture is immediately oxidized, the exothermic reaction fluxes the lime so that it is in the correct state to absorb the phosphorus. Unfortunately, the very high metal temperature and the evolved gases can also give heavy foaming so that control at this stage is critical. The lime is fed either in a gas stream or mechanically. The quoted figures[1] are for the Millom iron 224 lb/ton of steel. This flux is a 70:30 lime–iron oxide flue dust mixture.

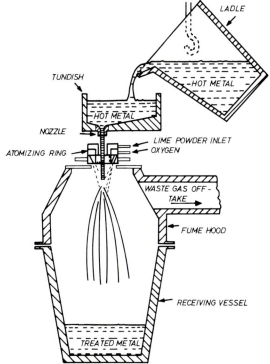

FIG. 3. Spray refining, pilot plant scheme (from *Steel and Coal*, Sept. 14, 1962, p. 509).

Below the lime ring is the atomizing ring and this consists of a water-cooled annular ring in which a number of oxygen jets are set. At Millom there are eight and it is claimed that the angle of convergence, internal profile and pitch circle diameter all have a bearing on the efficiency of atomization[1] (see Fig. 3).

The rate of oxygen flow and its pressure must clearly be controlled very precisely if the required carbon analysis is to be achieved. Again this must presuppose a known incoming iron analysis and temperature. Oxygen usage averages between 1750–2000 ft^3/ton of hot metal: as would be expected this is directly related to the carbon required in the finished product. The correct design of this nozzle undoubtedly stems from applying the established laws of fluid flow and jet design to molten steel as a normal liquid.

(d) The Reaction Zone

From the annular spray ring the metal falls freely into the receiving ladle. The time of reaction is governed by the time of fall and hence the design of this part of the plant is very important. As the reactions are taking place as the metal falls, the refractory problems in the process are relatively small. The top of the reaction zone is a flat water-cooled plate, with the metal entering the zone through a hole in the centre of the plate. A large off-take pipe sucks off the products of combustion to one side, this duct being refractory lined owing to the very high gas off-take temperature. The remainder of the gas off-take area is designed so that it is not struck by the falling metal—i.e. it is wide enough to avoid the metal spray.

The tundish, the rings, and this refractory lined combustion zone are all fixed and designed to enclose exactly the top of a receiving ladle—which is normally mounted on rails. The gap between the ladle and the fixed canopy is very important as the draught on the off-take fan and this gap regulates the amount of excess air. This control of the CO/CO_2 ratio is a very important part of the controlling process.

(e) The Receiving Ladle

This is a normal ladle, but it must have a great deal of free board for the required charge weight. It will also have a large lip so that the slag may readily be run off. In the Millom design there is no stopper assembly, and the ladle is emptied into a traditional steel teeming ladle when the cast is complete.

5. DESCRIPTION OF THE PROCESS

Iron runs from the blast furnace runner and into the tundish: when the tundish is full, metal overflows and continues down the normal line of the runner into the iron transfer ladle, that would normally take the iron to the steel plant or the pig-casting machine.

The receiving ladle beneath the spray steel-making unit has been filled with a precalculated weight of scrap. At Millom the scrap is preheated by an oil burner to about 300°C. Owing to the very high temperatures generated, this scrap can be a relatively large quantity. Figures of about 33% have been quoted in reports of the process, and this naturally has a very major effect on the economics. The lower the carbon of the finished steel, the greater the heat generated and hence the larger is the amount of scrap that can be used.

When the stopper in the tundish is raised the metal falls into the reaction zone. Here it is shattered by the oxygen and the added lime brought into intimate contact with the hot metal. Temperatures of about 2000°C are reported for this zone. The inclusion of iron oxide in the flux assists in the rapid fluxing of the slag. Temperature control of $\pm$ 15°C is claimed which is satisfactory for the steel qualities concerned.

Control of carbon is regulated by the draught and the rate of oxygen flow, and Dr. Davies[1] estimates that with an accurate method of measuring the iron flow to the unit the same level of control is possible as is obtained on a normal oxygen converter.

Silicon and manganese removal are expected to present no problem, but temperature will become difficult if the silicon analysis varies widely. Sulphur and phosphorus removal have been clearly demonstrated but some difficulties have been reported from plants where phosphorus removal is an important part of the process. Sulphur could be dealt with fairly effectively before it enters the tundish by adjusting the running conditions in the blast-furnace, and if this became a major problem this could certainly be done.

The collected metal is allowed to fill the ladle and when the slag has finished reacting it is run out and directly or indirectly teemed into ingots.

6. CONCLUSIONS

This process attempts to make steel from liquid pig iron by a very rapid mixing of liquid metal and oxygen and lime. It has the important advantages of low capital cost, high production rate, and very low refractory costs.

The success at Millom may be attributable to the very favourable conditions that obtain, and the process may find its best applications where conditions are similar. At those plants where a phosphorus bearing iron is used, the removal of the element still presents some problems.

NOTE ADDED

Since writing these notes it has been reported that the experimental plants at Millom Hematite Ore & Iron Co. had to close down due to lack of financial support and all work in connection with this process has ceased. This is rather unfortunate, but does not mean that technically spray steel making cannot be made to work satisfactorily or that economically it is not a sound idea. Two other experimental plants[4] working on similar principles but on a somewhat larger scale are still in operation, but these are shrouded in some secrecy and very little is known about them.

REFERENCES

1. DAVIES, D. R. G., RHYDDERCH, M. J., and SHAW, L. J., *J.I.S.I.*, August 1967, p. 810.
2. RHYDDERCH, M. J., *J.I.S.I.*, August 1967, p. 814.
3. *British Steelmaker*, Nov. 1966, p. 42.
4. *Steel Times*, Jan. 1968, p. 28.

CHAPTER 4

THE CONTINUOUS CASTING OF STEEL

R. H. HAMMERTON

INTRODUCTION

Continuous casting has become one of the most broadly applicable of the many cost-saving innovations of modern steel-making operations.[1] The history of the process dates from the time of Sir Henry Bessemer who filed patents on the idea in 1856. Bessemer planned to cast metal continuously by pouring the liquid metal between two water-cooled rolls and withdrawing a solidified sheet (see Fig. 1).[2]

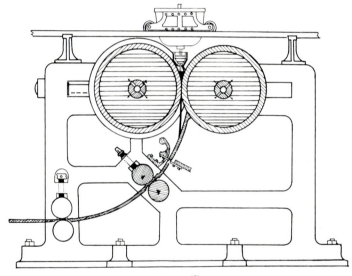

FIG. 1.[2]

Bessemer was able to produce a small piece of flat metal, but the effort was considered a failure. The process was difficult to control, the equipment expensive to maintain and the quality of the product poor. Unfortunately, Bessemer was ahead of his time and insufficient knowledge was available to solve these difficulties. Hence, the idea remained unexploited for many years.[3]

At the turn of the century the idea was re-examined and several new processes were patented, e.g. Evanov (Russia) 1905; Duter (Germany) 1909; Junghans (Germany) 1933. During the period 1930–40 the continuous casting of non-ferrous metals was adopted as a production method and the experience gained was applied to the casting of steel.[4] In 1943 Junghans demonstrated the new process by casting steel from a small capacity Bessemer converter into continuous strands on a vertical machine.[5] This valuable pioneering work was then examined by several other concerns (see below) and within 10 years of Junghans' first demonstration the process was attracting widespread interest.

TABLE 1.[6]

Country	Year work started	Company	Plant	Plant capacity (in 1951)
Germany	1943	Siegfried Junghans	Schorndorf	30 cwt
	1950	Mannesmann Hüttenwerke A.G.	Duisberg-Huckingen	5 tons
Austria	1947	Gebr. Böhler A.G.	Kapfenberg	6 cwt
U.S.A.	1946	The Babcock and Wilcox Company	Beaver Falls, Pa.	5 tons
	1949	Allegheny-Ludlum Steel Corp.	Watervliet, N.Y.	3 tons
U.K.	1946	Low Moor Iron & Steel	Bradford	2½ tons
	1952	United Steel	Barrow	5 tons

Today continuous casting is established as a production facility for casting a wide range of steel qualities in a variety of different shapes and sizes.

BASIC PRINCIPLES

The continuous casting of steel is based on the use of vertically supported, water-cooled copper moulds, with a means for pouring liquid steel into the top and extracting a continuous length of cast section from the bottom.

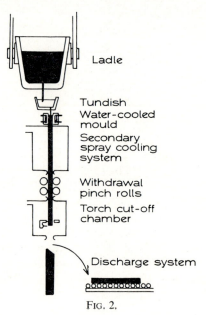

Ladle

Tundish
Water-cooled mould
Secondary spray cooling system

Withdrawal pinch rolls
Torch cut-off chamber

Discharge system

Fig. 2.

Figure 2 illustrates one of the early types of vertical casting machine which was developed mainly for the casting of sections of relatively large dimensions at relatively low speeds. Steel flows vertically downward from the tundish through an open bottom water-cooled copper mould where sufficient heat is removed to develop a solid skin on the outer surface of the steel being cast. The vertical movement is continued into a secondary cooling zone where solidification of the core is completed by a combination of radiative and water spray cooling. Below this zone the strand is gripped by sets of withdrawal rolls which control its descent and move it continuously to equipment of various types where it is cut to the required lengths. These lengths are received in a tilting basket which, by tilting upwards or downwards, can deliver the section to a roller path.

One of the main difficulties with the vertical machine is increasing its rate of output. In most cases, it is essential that the strand has almost solidified before reaching the withdrawal rolls. The time this takes to occur is restricted by well-known metallurgical and thermodynamic factors and therefore the only way to increase the production rate would be to increase the distance between the mould and withdrawal rolls. This would entail increasing the height of the machine which, in terms of the high capital costs involved and hazardous operation, would be undesirable.

The early machines, in particular large installations, were constructed at least partly below ground level. Such machines could be easily located under the crane of the existing melt shop thereby avoiding the necessity for a ladle transfer

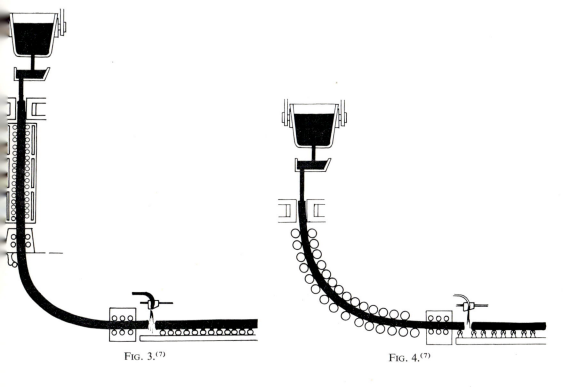

FIG. 3.[7]   FIG. 4.[7]

operation. Also, in the event of a breakdown of the plant or a running stopper the ladle could be conveniently removed and cast into ingot moulds at ground level. However, this type of plant is more expensive to construct than the tower type above ground level and is more difficult to maintain and operate. In particular it is difficult to keep the pits sufficiently dry to ensure satisfactory operation of electrical equipment.

The bending machine (Figs. 3 and 4) has the advantage of reduced height, and enabling the cut-off to be mounted horizontally. Also, the dummy bar can be disconnected at the inception of bending and lie in ambush until next required when it is immediately ready for insertion, so reducing the resetting time between casts.

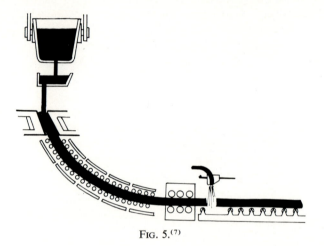

FIG. 5.[7]

The curved-mould curved-cooling machine (Fig. 5), first demonstrated at Von Moos, allows for a further height reduction.[7] The emergent casting retains the curvature imparted to it in the mould and follows a curved quandrantal path whilst solidifying. It is then straightened and cut to length as required. The curvature is designed to reduce the occurrence of sticking and the entire mould is situated at the end of a long arm that imparts a reciprocating action to it.

The most recently built plants are above ground and either follow bending or curved mould principles where simplicity of operation and speed of resetting offer very considerable advantages.

Figure 6 represents an early, somewhat different approach, developed in the U.S.S.R. (Goldoblin) and the U.S.A. (Hazlett) as an extension of Bessemer's original idea. The principle involves casting steel between two moving flexible metal belts which are subjected to intense water cooling.[8] Little has been published on the practical results achieved and there are few machines in operation. A report to the Russian All-Union Convention on continuous casting in 1963 concluded that whilst further development of this process would continue, more emphasis in the future would be placed on the modern vertical machines. The most recent development of the latter type of machine is by the Mannsemann Group and is the Elliptical machine. A curved mould is used, the strand being straightened by passing through a system of "withdrawal" rolls arranged in ever decreasing radius.

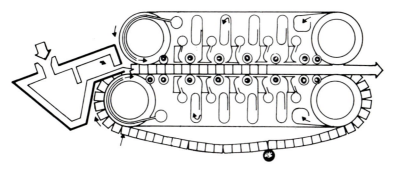

FIG. 6.[8]

PROCESS DETAILS

The continuous casting process begins at the melting furnace and requires more rigid controls than in orthodox practice.

The optimum temperature for continuous casting is less than that for conventional ingot casting of the same grade of steel. However, to avoid premature solidification and blockage of the tundish nozzle, the furnace is sometimes tapped at higher temperatures than normal determined by the normal heat loss occurring per cast. Oxygen steel-making processes can readily achieve the desired temperatures and ensure maximum utilization of the machine. Open-hearth furnaces are used by some companies but require a high standard of ladle preheating and hot metal handling practice.[9]

Deoxidation and alloying practice is a little more complex than in conventional practice. A high aluminium content adversely affects the fluidity of the melt and, with some steels, reactions at the nozzle result in premature cessation of the cast. For this reason deoxidation with silicon or other less deleterious elements is practised. If aluminium must be used, then continuous deoxidation in the mould by feeding aluminium wire appears the only alternative. The concentration of hydrogen and sulphur must also be closely controlled. Hydrogen is sometimes associated with the formation of subsurface pinholes in the cast product whilst sulphur is known to be a cause of longitudinal and transverse face cracks. A special desulphurizing compound is currently under development to overcome the latter problem.

Dimensions of Section

The choice of shape and size of the continuously cast product is determined by the subsequent hot working that is to be undertaken (rolling, forging, stamping and pressing), and by the necessity to avoid any appreciable defects during casting. If the work of the continuous casting machine is to fulfil the requirements of the subsequent hot reduction units (excluding cogging mills), the shape and size of the product cast on the machine must be within certain specific limits as defined by Boichenko.[10]

Ladle

There are three types of ladles in use today—bottom-pour ladles fitted with stoppered nozzles as in orthodox practice, syphon-pour ladles incorporating a tubular passage in the refractory wall to convey the metal to a pouring lip as the ladle is tilted, and lip-pour ladles (see Fig. 7).

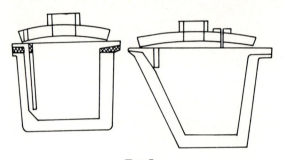

Fig. 7.

The choice of ladle depends on a variety of factors (see Halliday[6] and Morton[11]). Bottom-pour ladles can be handicapped by the durability of the stopper-rod assembly under the extended casting times of the process. However, independent research has shown that this problem can be partially overcome if a ceramic coating or a protective sleeve assembly is fitted around the stopper rod. Lip- and syphon-pour ladles provide a better control over the flow of metal between ladle and mould, but the capital cost of the installation (tilting equipment, etc.), maintenance requirements and inconvenience are generally greater than with a bottom-pour ladle. Part of this cost can be reduced using proprietary exothermic and insulation materials in place of the burner. A disadvantage of the lip-pour ladle is that unless a deep slag bridge is incorporated in the design, slag may be entrained with the stream of metal entering the tundish. This can be prevented by using a slag coagulating agent.

Heat losses in the ladle by conduction through the refractory lining or radiation from the metal surface can be reduced by any of a variety of methods. Ladles can be preheated prior to tapping using a removable lid fitted with a burner jet which can also be used during casting. Alternatively a deep layer of slag or proprietary insulating compound can be applied to the metal surface. The disadvantage of using slag is that it can erode the refractory lining of the ladle.

Tundish

This is an intermediate link between the ladle and mould and may be common to more than one mould. It has three functions:

1. Controls the flow of steel into the mould.
2. Accurately guides the molten steel into the mould.
3. In multi-strand plants, distributes the steel as required.

In early plants the tundish was basically a launder but modifications to meet its present-day functions have produced designs similar in principle to that shown in Fig. 8.

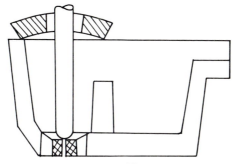

FIG. 8.[12]

Again it is necessary to minimize heat losses from radiation and conduction. Use of high quality refractory linings, a high degree of preheat, refractory lids containing burners and proprietary insulating compounds have been reported. Where an insulating compound is employed it is normally advisable to use a granular material which is less easily entrapped in the turbulent areas around metal entry and exit points.

Refractories

The extended casting times compared to orthodox practice require that refractories, lining designs, nozzles, stopper rods and heads are carefully chosen if successful casting is to be achieved.

The composition of ladle and tundish linings is made after considering the service conditions expected, i.e. temperature and quality of steel to be cast, slag practice, etc. Many articles have been written on this subject.

Both ladle and tundish refractories require some form of protection against attack by molten steel and slag, particularly if a deep slag layer is used for insulation purposes. Research on this problem has resulted in the development of special ceramic coatings which are not wetted by contact with steel and/or resist attack by molten slag. Their particular use is in facilitating skull removal from the tundish on completion of the cast so reducing the normal refractory wear.

The behaviour of refractory nozzles has an important bearing on the safety and control of the casting process and the quality of the strand produced. Willim[13] has published an excellent article describing the form, constitution and use of such nozzles in the continuous casting process. Although there is a wide divergence of practice in the selection of nozzle materials, in general, it seems that whilst zircon or zirconia is preferred for small throughput rates and for stainless steel grades and sequence casting, fireclay or fireclay with high alumina or magnesia inserts is used for larger rates and instances where economy leaves no alternative.

Many attempts by modifying both nozzle design and composition to overcome the tendency to nozzle blockage by alumina deposition have been tested. Non-swirl nozzles have been developed by Halliday[6] and special long nozzles (submerged nozzles) extending beneath the metal surface in the mould have been successfully demonstrated at Dillingen and S.A.F.E.[14]

Mould Design

Moulds for continuous casting are almost exclusively made of copper. There are three basic designs, namely tube, block and plate moulds. The tube mould is mostly used for smaller cross-sections up to about 100–200 mm square, while the block and plate moulds are used for larger square and rectangular sections (see Fig. 9). Prior to casting, the top of the mould is normally protected against adhesion of molten steel splash, with either a sheet of asbestos or a refractory coating.

Mould Cooling

Each mould is water-cooled (primary cooling) in order to accelerate the heat transfer and cause rapid solidification of the strand. The water maintains the mould temperature in a range of sufficient mechanical strength to withstand the ferrostatic pressure and cooling water pressure.

The mould cooling water should be in chemical balance or slightly corrosive;[18] if the latter, a corrosion inhibitor must be used to protect the piping and mould lining. Deposits of calcium carbonate restrict the heat flow on the coolant side of the mould which, in addition to reducing mould life, can cause billet deformation and an increase in the possibility of breakouts. Appropriate measures are also taken to minimize organic growth in the emergency water tank and cooling tower. Jaeggi[18] has evaluated the specific requirements for the mould cooling water and has represented the results in graphical form (see Fig. 10).

Mould Reciprocation

The solidification rate is determined by the geometry of the mould, the cooling characteristics of the steel being cast and the rate of withdrawal from

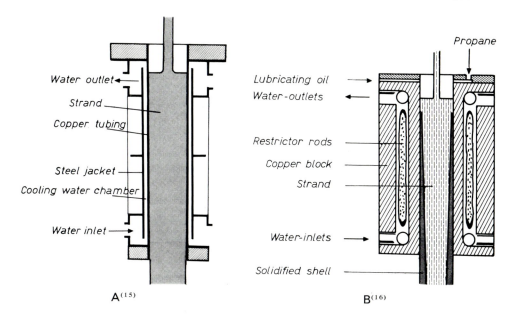

A(15)   B(16)

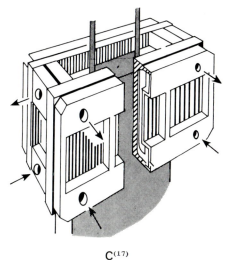

C(17)

FIG. 9.

the mould. Any given section size has a critical rate of withdrawal which, if exceeded, can result in a "breakout". In the event of such an occurrence, molten steel drains from above the fracture and solidifies on to the superstructure in the secondary cooling zone. This steel is more easily removed if the superstructure has been prior coated with a release agent that prevents welding.

The rate of withdrawal is also limited by the tendency of the solidifying strand

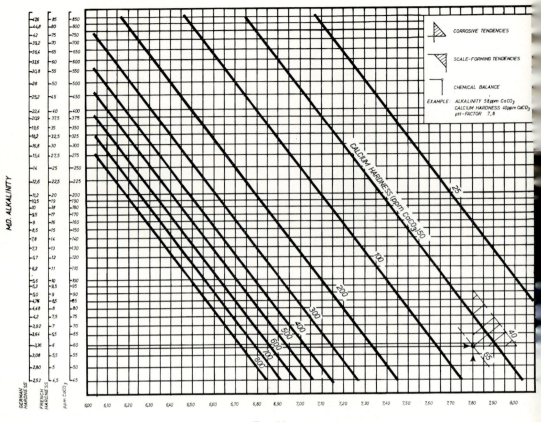

FIG. 10.

to stick to the mould wall, an occurrence which can be minimized by reciprocating the mould. This idea was first patented by Junghans in 1933. The mould
moves a short distance downwards at the same speed as the billet and then
upwards at a higher speed. The downstroke prevents constant friction of the
newly formed shell with the mould wall after which the upstroke strips the skin
from the mould surface in a more effective manner than the sliding and dragging
that occurred previously. Several modifications of this principle are now in
operation.

1. B.I.S.R.A have developed a spring-mounted mould. When sticking occurs,
the mould moves downwards until the force of the springs effects the upward
release. Modern versions of this technique ensure that release occurs in a
downward direction tending to compress the billet skin rather than tear it as
before.

2. United Steel Ltd.[6] discovered that moving the mould downwards faster
than the hot metal subjected the newly formed skin to a slight compressive
force, which would help an incipient crack in the solidifying skin to heal itself
before it left the mould. This method also reduced sticking and subsequent

cracking since at no time did the mould travel at the same speed as the billet. This "negative strip" technique has permitted speeds of up to 37 ft/min for a 50 mm square billet.[4]

Mould Lubrication

The tendency for the solidifying steel to stick to the mould can be further reduced by applying a lubricant to the mould walls. The lubricant may be drip-fed on to the meniscus near the mould wall although more generally the oil is supplied above the meniscus level by one of two principal methods (see Fig. 11).

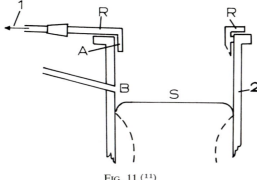

FIG. 11.[11]

A number of evenly placed small-diameter holes (B) are drilled through the mould wall at a position a few inches above the meniscus level. These holes are connected by means of small-bore pipes which pass through the cooling jacket to the oil supply. Alternatively, the lubricant is applied through an oil ring main (A) clamped on top of the mould. This main feed ring is drilled with regularly spaced small-diameter holes (R) through which the oil is forced under pressure, so that it runs as a continuous thin film along the mould wall.[11]

The lubricants most commonly used are rape seed oil and its derivatives and, in Russia, paraffin wax or transformer oil. The rate of supply to the mould varies for different machines but a usage of 1 ml of lubricant per minute per inch of mould perimeter is a general average.

The role of the lubricant as a possible cause of subsurface defects and carbon pick-up is the subject of frequent comment and has led to development of special mould fluxes. The practical requirements of such a flux are that it should lubricate the forming strand, prevent surface oxidation in the mould and absorb non-metallic inclusions. In general, particularly with smaller sections, mould fluxes are handicapped by being too easily entrapped within the body of the strand. This problem may be overcome by use of the submerged pouring tube (sometimes referred to as "nozzle") practice referred to in an earlier section. The above factors have been examined by several investigators.[19,20,21]

Dummy Bar

Prior to the beginning of a cast a dummy bar of approximately the size and shape of the casting is inserted into the mould from the bottom. The space between the dummy bar head and the mould is packed with asbestos rope or a proprietary sealing compound to form a temporary mould bottom. The head of the dummy bar is provided with a suitably shaped casting known as a "key" around which the first molten steel solidifies. Thus, as the dummy bar is withdrawn on commencement of the cast so is the newly formed strand.

The dummy bar key is normally a cast iron "fir tree" shape or one or more stout bolt heads for casting small sections. For large slabs, a small bolt is insufficient and the use of several bolts inconvenient, hence a short section of rail or a special casting is used (see Fig. 12).

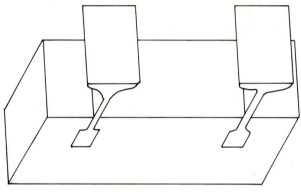

FIG. 12.[22]

The larger the slab, the more valuable the "key" becomes, and some operators retrieve the key from the solidified strand. Such retrieval, using oxyacetylene torches is easier when the key has been prior coated with a ceramic coating which prevents both erosion and welding of the key to the strand.

Secondary Cooling

As the casting emerges from the mould, solidification of its core is completed by water cooling. This "secondary cooling" is required for the following reasons.[11]

1. To increase the rate of solidification below the mould so reducing the distance of the withdrawal rolls, this distance being one of the major factors in influencing overall plant height.
2. To obtain an improved as-cast structure.
3. To prevent excessive bulging of the casting below the mould.
4. To dissipate the heat otherwise generated in the plant components and structure.

The secondary cooling must be regulated so that the influence of those factors causing the formation of bridges and axial porosity is reduced. Similarly, the control must be such that the deleterious effects of internal stresses in the section, crack formation and changes in shape do not occur. Speith and Bungeroth[23] have reported that water usage exceeding 0·9–1·0 litres/kg of steel causes internal cracks and other defects to occur in the cast product.

The method of distribution of secondary cooling water varies widely. Three main types of secondary cooling nozzle are in use, producing either laminar or hollow-cone jets, or air-atomized sprays. To ensure that these jets are not blocked during operation, some operators treat the recirculating water with proprietary flocculating agents. This ensures complete precipitation of any finely suspended matter.

Withdrawal Rolls

The withdrawal rolls fulfil a three-fold function. In preparation for casting, they support and guide the dummy bar upwards into its position in the mould. When casting begins, the rolls withdraw the dummy bar together with the cast product at the desired casting rate. As casting proceeds, the weight of the cast product progressively increases until the function of the withdrawal rolls can to support this weight and to control the speed of withdrawal.[11]

Cut-off

When the cast ingot has been drawn through the withdrawal rolls and the dummy bar sections have been disconnected, the strand is cut into the required lengths.

An oxyacetylene torch is mainly used for cutting carbon steel billets, whilst iron powder injection is often an essential extra for the cutting of stainless and other high alloy steels. Manual operation is customary for small sections and low casting speeds, but faster rates of withdrawal necessitate automatic cutting. Removal of the waste products after cutting is facilitated by prior coating all metal work with a refractory based release agent.

After being cut, the strand lengths are delivered to a roller path and thence to stock or further operations.

Machine Ratings

These are dependent on the section size cast and are governed by the casting speed per mould or strand. The larger the section size or the less the reset time between casts, the higher is the rating.

The graph shown in Fig. 13 relates casting speeds and mould tonnage rates for square sections from 50 mm to 300 mm in size, assuming equivalent casting conditions.[24]

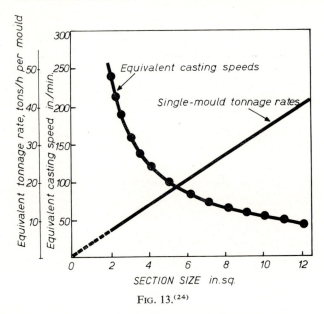

Fig. 13.[24]

The speed and tonnage parameters above are based on the operating performance of machines recently built by the Concast group. This group has published the nomogram[25] shown in Fig. 14 relating the size, casting speed and mould throughput capacity when continuously casting slab sizes.

METALLURGICAL ASPECTS

The main metallurgical advantages of the continuous casting process enumerated in the patent literature and in the results of experimental work are:[10]

1. A better quality of cast product because of greater homogeneity, less segregation, a finer-grained primary structure and, in the case of hard steels, a more homogeneous distribution of carbides.
2. A reduction in the amount of hot-working.
3. An increase in the yield of sound metal because of the considerable reduction in top and bottom discards.

The metallurgical character of continuously cast steel is that of a cast product which still requires hot-working. Its structure consists of a thin chilled layer, columnar grains which have grown from the perimeter inward, and a central zone of equi-axed crystals. For low carbon steels the columnar grains extend to within half to three-quarters of the distance to the centre depending on section thickness. Higher carbon steels give columnar growth to a progressively shorter distance, with a correspondingly larger central zone of equi-axed crystals. As the carbon content increases, the structure of both columnar and equi-axed crystals becomes finer.

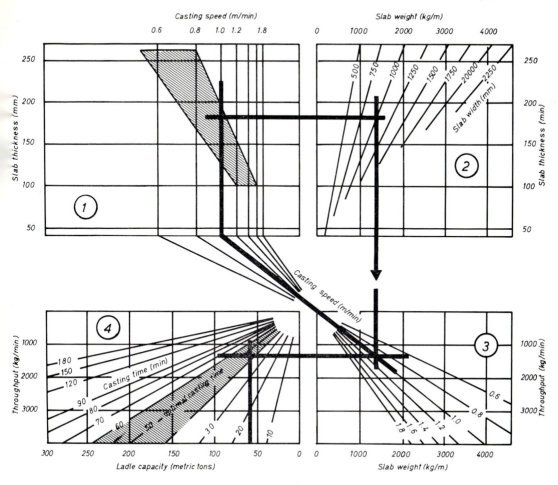

FIG. 14.[25]

In general, steels which solidify characteristically in orthodox ingots with a coarse grain structure are found to solidify with a relatively coarse structure when continuously cast, and those which are usually fine-grained when conventionally cast also exhibit a fine grain structure when continuously cast.[26] However, continuously cast rimming steel differs fundamentally from the conventional product. The structure is more uniform—the solid rim has a carbon content as high as that in the central zone, i.e. there is no low carbon ferrite rim, and transverse analysis shows negligible differences in carbon, manganese, phosphorus and sulphur contents.

Generally speaking the quality of continuously cast steel is good. The physical and mechanical properties of the "as-rolled" material are satisfactory and indistinguishable from those obtained from a good quality product derived from conventional ingots.

ECONOMY

The economics of continuous casting are based on two main factors:
1. The actual costs (conversion costs) of the machine.
2. Costs involved with production.

Conversion Costs

Halliday[26] has represented conversion costs in graphical form as shown in Fig. 15.

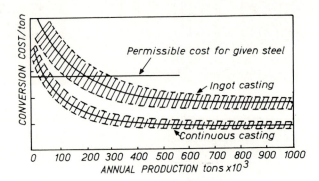

FIG. 15.[26]

As would be expected annual tonnage output has a major bearing on metal conversion costs and therefore on the comparison between the two processes. Mr. Halliday's conclusions were:[26]

1. The cost of producing semi-finished sections of any given size by continuous casting is generally appreciably lower than that of producing the same annual tonnage of such sections by orthodox means.
2. For all tonnage outputs, the costs for conversion by continuous casting tend to bear roughly a similar ratio to those for production by the orthodox methods at similar annual tonnage levels. The initial capital cost for the machine is usually offset by the elimination of the heavy consumption of ingot moulds. The consumption costs of continuous casting moulds are much less than ordinary moulds and there is also a saving in space.
3. The production by ingot casting of a given steel with a given permissible conversion cost is likely to be uneconomic below a certain annual tonnage level, while its production by continuous casting would still be economic down to a much lower annual tonnage level.
4. For high tonnage outputs continuous casting generally provides a distinct saving on the normal overall conversion cost from liquid metal to semi-finished sections.

Production Costs

Savings in production costs provided by continuous casting are derived from the inherent technical advantages and reliability of the process and suitability to production operations in a steel plant. These advantages are represented in graphic form as shown below in Fig. 16 and are listed in Table 2 (from Halliday[26]).

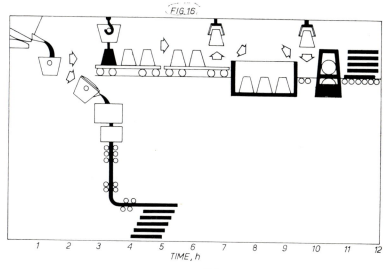

FIG. 16.[26]

TABLE 2.[26]

Process

More standardized, less an art
High average yield (6–12% gain)
Reduces fuel requirements and recycled scrap
Eliminates heavy primary rolling mills
Greatly reduces overall throughput time
Allows reduced manning—greater productivity
Often reduces works area required
Allows casting of special shapes

Cast Product Quality

Uniform and good/excellent quality
Dressing reduced, frequently eliminated
Freedom from macrosegregation
Grain size control easy
Equal, if not greater, cleanness
Withstands severe primary hot reduction
Spreads more uniformly in rolling
"Difficult" steels produced more easily
Facilitates production of new types of steel

TABLE 2—*contd.*

Rolled Product Quality

Reductions in area in rolling reduced
Mechanical properties equally good
Improved transverse properties
Cold rolling possible to thinner gauges
Steel with enhanced properties expected

Costs

Capital cost of machines relatively low
Overall capital cost for new works greatly reduced
Reduced investment on materials in process
Operating costs notably lower
Small-tonnage plants made feasible
Major savings on large-tonnage plants
Progressive plant installation to match expanding markets

Integration Aspects

Scheduling of casts in detail more reliable
Operational flexibility quite feasible
Diversification of production possible
Process can be automated
Highly suited to integrated automation
Closely linkable with steel-making (sequence casting)
Combinable with vacuum degassing
Closely linkable with rolling (direct rolling)

TABLE 3.

	Number of Machines†	Capacity†
End of 1948	1	—
End of 1953	8	0.28×10^6
End of 1958	21	1.3×10^6
June 1963	61	5.5×10^6
June 1964	105	10.5×10^6
June 1966	173	26.5×10^6
End of 1968	243	35.8×10^6

† Including machines on order or under construction.

WORLD-WIDE SITUATION

Growth of Continuous Casting

Over recent years the continuous casting of steel has undergone very rapid development as outlined in Table 3.[27]

The distribution and density throughout the world of these machines in 1967 is shown in Fig. 17.

Malcor[27] has shown that the growth rate of continuous casting runs parallel to that of LD steel-making (see Fig. 18).

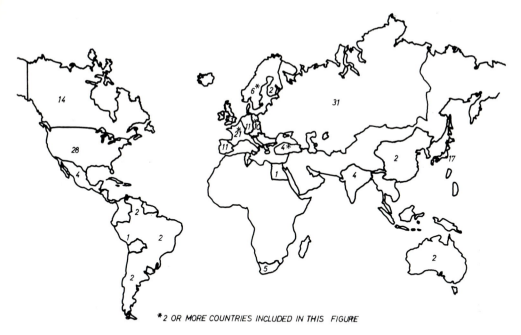

*2 OR MORE COUNTRIES INCLUDED IN THIS FIGURE

FIG. 17.[34]

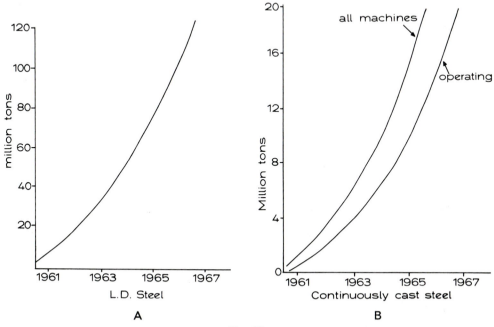

FIG. 18.

F

Dr. Tanner[28] of Concast A.G. has predicted that providing developments in the iron and steel industry follow their present course, continuous casting capacity will exceed 50 million tons per annum by 1970.

Current and Future Developments

A few years ago the bulk of continuously cast steel was billet section, but today the emphasis is on slab casting machines. The immediate priority in all section sizes is towards quality improvement, developments being led by partial or complete automation, i.e. the use of computers for continuous control and regulation.[29]

Already partial automation is in evidence on several commercial machines. Examples include automatic control of metal level in the mould, automatic guidance of the steel stream, electromagnetic stirring, self-start and stopping and automatic control of ladle and tundish stopper rods.

A further approach to improved quality is the use of degassing techniques, the principle being to degas and cast in a continuous operation. Many solutions have been proposed as indicated in Fig. 19[30] and already some operators are using ladle-to-ladle degassing (e.g. Paderwerk Gebr. Benteler and Dillingen and Tube Investments), or tap degassing (e.g. Brierley Hill Laboratory—Round Oak).

The continual need to reduce costs is requiring higher casting speeds which in turn demand improved cooling and the development of new techniques:

1. Concast have developed a pseudo-mould extender[32] which consists of a set of cooling plates extending beneath the mould. These support the strand and hence reduce the chance of breakouts arising from the thin walls rupturing at the higher casting speeds. Research on an alternative solution has led to the development of a special additive which when added to the mould produces a wider dendritic zone in the solidified shell so providing "reinforcement".

2. A multiple section mould[31] has been developed by B.I.S.R.A. which enables high casting speeds of small billets. The mould is designed to cast three square billets joined along their diagonals and is in production usage at Western Canada Steel (see Fig. 20).

Another promising development is the B.I.S.R.A. "H" section (or "dogbone") shaped mould.[33] The principle which was researched jointly by B.I.S.R.A. and Algoma Steel is to cast a specific shape such as the "H" section and then directly roll the product in the withdrawal rolls or an adjacent mill.

The development of special products for the continuous casting industry to improve quality and/or facilitate casting is being carried out by several companies.[34] Already a special inclusion remover composition has been developed that is selective to alumina, silica and alumino-silicate inclusions. Special

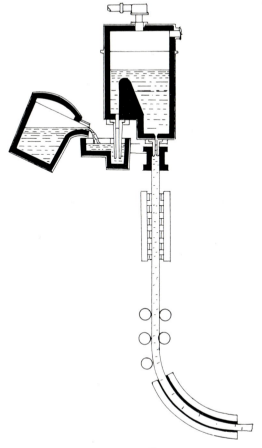

Fig. 19.[(30)]

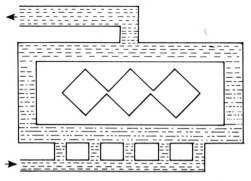

Fig. 20.

rimming agents are also being developed. Other research and development topics that have been reported to date include additives to water to create more efficient cooling,[36] special low density fibrous sleeves to protect the stream of metal between tundish and mould against surface oxidation[37] and chemical coatings to prevent atmospheric oxidation of the cast strand.

Recently considerable progress has been made in adopting the practice described as "sequence casting", i.e. continuous-continuous casting—the uninterrupted casting of several ladles in succession on a given casting machine. The importance of sequence casting lies in the increased output that can be achieved by eliminating reset times which normally account for 25% of the process time.

An alternative, less publicized means of increasing output is the "casting-on" technique referred to by Summers.[35] The principle is to use the previous cast as the dummy bar for the next cast. This idea becomes quite attractive with large bloom and slab sizes where the dummy bar key is an expensive item. Dummy bars have short lives and easy disconnection is never reliable. A special refractory shape is inserted into the mould just before the end of casting and the remaining metal is poured into same. This develops a "key"-shaped casting on the end of the strand which is suitable for use in ensuring satisfactory union with the next cast.

CONCLUSIONS

Continuous casting is not just a substitute for conventional ingot casting and primary rolling to semi-finished products. It is a process in its own right, with its own limitations and criteria, many of which require adjustments of procedures and even rethinking of steel-making practice. Neither is continuous casting a panacea for all ills. It is a very useful steelworks tool with a wide application but there are some situations in which it offers no advantages.[26]

REFERENCES

1. READ, Dr. R., and BRIEN, H. L., *Blast Furnace Steel Plant*, 689 (Aug. 1965).
2. B.I.S.R.A. Special Report 89, 1, A (1965).
3. *B.H.P. Rev.* **43**, 6–10 (1966).
4. MIDDLETON, R. H., *Steel Times*, 808–812 (June 1965).
5. WESTON STARRATT, F., *J. Metals*, 861 (Aug. 1965).
6. HALLIDAY, I. M. D., *J. Iron Steel Inst.*, **191**, 121–163 (1959).
7. GALLAGHER, L. V., and OLD, B. S., *Scientific American*, 75–88 (Dec. 1963).
8. HAZLETT, R. W., *Iron Steel Engineer*, 106 (June 1966).
9. JACOBS, W. W., A.I.M.E. Met. Soc. Conf., 26, 477 (1965).
10. BOICHENKO, M. C., *Continuous Casting of Steel*, Butterworths, London, 1961.
11. MORTON, J. S., *Met. Reviews*, 9 (87), 121–177 (1964).
12. BAUER, GERLING, BOCK, A.I.M.E., Symposia, 155–176 (1961).
13. WILLIM, F., *Concast-News*, **4** (3), 4 (1965).
14. GOSSELIN, C. M., Experience at S.A.F.E. with submerged nozzles. Mannesmann Symp., 1967.
15. PRECHT, H., *Concast-News*, **2**, 1 (1962).

16. PRECHT, H., *Concast-News*, **3/4**, 4 (1962).
17. BUCHER, W., *Concast-News*, **1**, 1 (1963).
18. JAEGGI, W., *Concast-News*, **5** (3), 5 (1966).
19. HALL, D., and McHUGH, T., B.I.S.R.A. Special Report 89, 1, 103 (1965).
20. HALLIDAY, I. M. D., B.I.S.R.A. Special Report 89, 1, 14 (1965).
21. ELDRIDGE, P. E., *Concast-News*, **5** (1), 3 (1966).
22. Design used A.G. der Dillinger Hüttenwerke.
23. SPEITH, K. G., and BUNGEROTH, A., *Stahl und Eisen*, **52** (72), 869–881 (1952).
24. HALLIDAY, I. M. D., *Iron and Steel*, 475 (Sept. 1965).
25. *Concast-News*, **4**, 314 (1962).
26. HALLIDAY, I. M. D., B.I.S.R.A., Special Report 89, 1, 9 (1965).
27. MALCOR, H., *B.I.S.R.A. Trans.*, 3978 (Nov. 1964).
28. TANNER, Dr. H., *Iron and Steel Inst.*, 199–208 (Nov./Dec. 1966).
29. KOENIG, Dr. P. J., *Concast-News*, **4** (1), 5 (1965).
30. Canadian Patent No. 682859.
31. SAVAGE, J., and PEARSON, J., *Iron and Coal Trades Review*, 725–728 (April 1961).
32. Austrian Patent 259,780.
33. MARR, H. S., *J.I.S.I.*, 1186–1193 (Dec. 1966).
34. Technical Marketing Bulletins, *Foseco*, 1–13 (1965).
35. SUMMERS, W. H., *Steel Times*, 639 (Nov. 1966).
36. British Patent App. No. 3985/67 and 16123/67.
37. British Patent No. 1,096,706.

B. TREATMENT OF CAST IRON

CHAPTER 5

CAST IRON

J. L. FRANCIS

THE CUPOLA FURNACE

The essential parts of a cupola furnace are indicated in Fig. 1. In its simplest form it comprises a vertical steel cylinder or shell, lined with refractory. This shell is supported on four columns and is provided with a number of openings. At the top is the charging door. Nearer the base are openings for the blast tuyeres, a slag outlet and back and front breast openings. The latter contains the tap-hole and the former facilitates making up the sand bottom and lighting the furnace. Two hinged, semicircular steel doors close the opening in the base plate when the furnace is operating. A steel prop or locking device holds the doors in position when closed, and allow of their easy release when it is required to empty the furnace at the end of a "blow".

Much study has been devoted to the melting of cast iron in the cupola during recent years. This has resulted in improvement in design and operation. The introduction of hot blast and water-cooling have enabled a higher degree of control over both the chemistry and the quality of the molten iron. Water-cooled, hot-blast cupolas are able to melt continuously for a number of days because there need be no refractory lining in the melting zone and therefore no burn-back problems.

Absence of refractories in the melting zone helps in maintaining a constant melting rate and to produce molten iron of more accurate and constant chemical composition at the furnace spout. Also, the liningless melting zone allows of either acid or basic fluxing as desired. With basic fluxing a high level of desulphurization and recarburizing can be achieved. This in turn allows the use of cheaper raw materials for the furnace charges.

Cupola Charging—Acid-lined Cupola

Normal charge materials include coke, pig iron, cast-iron scrap, steel scrap and ferro-alloys. There must also be a fluxing agent and an air blast to provide the oxygen for combustion of the coke. Apart from the air blast, all the materials charged into the cupola are solids and, while it is in operation, nothing other than liquid material and effluent gases can come from it. So the non-metallic substances such as coke ash, sand and rust on the scrap and any other non-volatile, non-metallics must be rendered liquid so that they can escape from the furnace via the slag hole. It is the fluxing agents which must perform this function.

F*   159

Cupola furnaces will melt iron over a wide range of adverse conditions of operation. However, it is only under standardized methods of correct operation that they give their best performance.

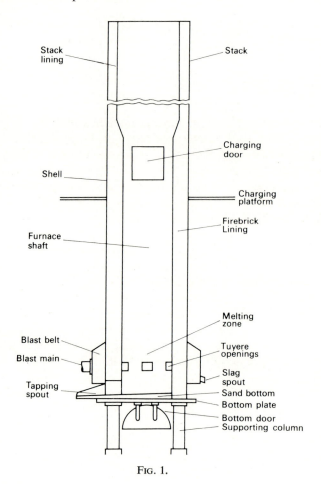

FIG. 1.

Coke Bed Charge

This is a most important item. Coke of varied size or small coke should not be used for the bed charge. It is the function of the coke bed to support the metal charges at the correct height in the melting zone of the furnace. Therefore, it must be charged to this height before any metal is charged.

Correct coke-bed height is really a function of the blast pressure and increases with it. Under average conditions the bed should extend 3 feet or 1 metre above the level of the top tuyeres, where more than one row of tuyeres is installed. After the coke bed has been ignited and thoroughly burned through, it should be gauged to ensure correct height before charging commences.

With too low a bed the free oxygen in the blast is not all burnt to CO_2 or CO before it contacts the iron melting in the melting zone. As the iron melts in drops and trickles it offers a large surface area to the furnace gases and at a temperature which is well above its melting point. Thus any free oxygen present at this stage rapidly combines and oxidizes the silicon and manganese as well as the iron, thereby causing high melting loss and molten metal of inferior quality, high in sulphur and likely to cause hard spots in the castings.

Metal Charge

There is an optimum charge weight depending on the size of the cupola. It is related to the melting rate in tons or kilos per hour. Thus, if a cupola has a melting rate of 6 tons or 6000 kilos per hour the unit charge weight should be between one-eighth and one-tenth of this quantity. That is from 12 to 15 cwt or from 600 to 750 kilos. It is always better to use charge units of the smaller than than the larger limit as this induces a more uniform working of the furnace and a better mixing of the different components of the metal charge.

Coke Split

This is the quantity of charge coke which separates the metal charges. It can be calculated in two ways. A coke-to-metal ratio can be used to determine it. The textbook ratio is usually given as 1 lb coke to 10 lb metal or a weight ratio of 1:10. In this case the weight of coke split is one-tenth that of the metal charge.

Another approach is based on the rule that the melting of one charge of correct weight will consume coke from the bed equivalent to a layer 6 in. (15 cm) in depth. Thus if a circle of cupola firebricks is built equal to the diameter of the furnace and 18 in. (45 cm) high, it can be filled level with coke. Remove the bricks, weigh the coke and divide by 3 to obtain the weight of the coke split.

Fluxing Materials—Acid-lined Cupola

Limestone is the universally employed cupola flux. Marble chips, marine shells or coral would serve equally well. Dolomite or magnesian limestone is another alternative. With standard cupola practice and coke of 10% ash content it is used at a rate of 3% of the charge weight or 60–70 lb/ton, provided that it contains about 96% of calcium carbonate.

The calcium carbonate in the limestone is calcined, inside the cupola, to produce burnt lime (CaO) and this forms the fluxing agent.

$$CaCO_3 + heat = CaO + CO_2$$

Lime is a basic material and will combine chemically with the acid oxides of iron and silica to form a slag which can be run off from the furnace.

In some cases it becomes desirable to intensify the fluxing action of ordinary limestone. This can be done by adding along with it a quantity of fluorspar or of soda ash. The fluorine from the fluorspar or the sodium oxide from the soda ash energize the ordinary limestone flux.

Thus the fluxing action can be rendered more efficient in removing the silicate films of ash from the incandescent coke. By so doing the oxygen of the air blast can react more quickly with the carbon and hotter and faster melting takes place. With faster melting, less sulphur is picked up from the coke and molten iron is produced with improved quality, a higher total carbon and greater fluidity.

In the case of the fluorspar addition, its action can be explained simply, as follows:

$$SiO_2 + 2CaF_2 = SiF_4 + 2CaO$$

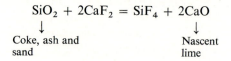

Coke, ash and   Nascent
sand   lime

It will be recalled that lime (CaO) is the active fluxing agent derived from limestone. The equation above also shows CaO as one of the end products. However, there is a difference. Lime from calcium carbonate is formed by driving off the carbon dioxide (CO_2) by calcination. In the second case, lime is formed as a result of a chemical reaction at high temperature and the lime formed is nascent lime which is much more reactive. With soda ash, the energizer is sodium oxide (Na_2O) which is a more powerful base than lime.

With normal limestone fluxing in acid cupolas the quantity of slag produced is about 80 to 100 lb per ton. It is of no value as a by-product and the fuel consumed in its formation is not available for melting iron. Moreover, the carting away of the slag and its dumping costs money. Thus any means of reducing the volume of slag will be advantageous.

If 60 lb (28 kg) of limestone is used per ton of iron charged, the quantity of flux materials used can be reduced by 25% when fluorspar is used. The 45 lb (21 kg) of flux is then made up from 30 lb (14 kg) of limestone and 15 lb (7 kg) of fluorspar per ton (1000 kg) of iron charged. Slags from acid cupola melting have chemical compositions between the following limits when melting white and grey iron:

Silica—SiO_2	44–55%
Lime—CaO	15–30%
Alumina—Al_2O_3	5–12%
Ferrous oxide—FeO	1–10%
Magnesia—MgO	Trace—3%
Manganese oxide—MnO	1–5%
Phosphorus as P_2O_5	Trace—2%
Sulphur	Trace—0·5%

Fluxing Basic Cupolas

The quantity of fluxing material used in basic cupola operation is greater than in acid practice. It depends on the degree of desulphurization required and is of the order of 175 to 200 lb/ton (80 to 90 kg per 1000 kg) made up as follows:

> Limestone 30–40% of weight of coke
> Dolomite 10% of weight of coke
> Fluorspar 25% of weight of limestone

Instead of 25% fluorspar a 50–50 mixture of fluorspar and soda ash can be used.

A typical slag analysis from a basic cupola melting charge for nodular graphite cast iron, quoted by W. R. Perry[1] is:

	%
Silica—SiO_2	20
Lime—CaO	56
Alumina—Al_2O_3	6·5
Ferrous oxide—FeO	0·64
Manganese oxide—MnO	0·20
Calcium fluoride—CaF_2	15·00
Sulphur	0·85

For efficient desulphurization the ratio of ferrous and manganese oxide should not be greater than 1·5% of each, for above this level the slag will undergo a progressive colour change from brown to black. Examination of basic slags offers a much more positive guide to melting conditions than a similar examination of acid slags as in the latter case changes in composition are not so readily indicated. A slag of the composition given is white to light grey in colour and is known as a "falling" slag because it disintegrates to a fine powder when cooled in the air.

Slag Basicity

The chief components of basic cupola slags are lime—CaO, magnesia—MgO (if dolomite is used) and silica—SiO_2. Also there may be some alumina (Al_2O_3). The lime and magnesia are basic oxides and silica is acid. Alumina is often regarded as a neutral agent. In fact it can act as both acid and base depending upon the basicity of the slag. That is to say, in acid slags it functions as a base and as an acid agent in basic slags.

Slag basicity can be assessed very simply for cupola slags by relating the basic and acid constituents thus:

$$\text{Basicity ratio} = \frac{CaO\% + MgO\%}{SiO_2\%}$$

The basicity ratio, calculated in this way should be in the range 1·3 to 1·6 if satisfactory desulphurization together with sufficiently high carbon is to be achieved from low carbon charges.

Slag basicity	S content of metal
0·8 or lower	0·08% or higher
0·8 to 1·3	0·08% to 0·04%
1·3 or higher	0·04% or lower

The basicity ratio is not the sole determining factor. Carbon also plays a part in producing the desired reaction:

$$FeS + CaO + C = Fe + CaS + CO$$

If ferrous oxide or other acidic oxide is present in excess of the ratio mentioned above, sulphur will pass from the slag and return to the metal as shown by the equation:

$$CaS + 2FeO = CaO + FeO + FeS$$

Fluxing the Bed Coke

The desulphurizing slag should be formed right at the commencement of the melt by charging flux materials in the coke bed. Charge the bed coke to about two-thirds of its full weight and then add limestone, fluorspar and soda ash in a 5:1:1 ratio. The total quantity of flux materials to add to the bed coke is about 40 lb (18 kg) per ton (1000 kg) of hourly melting rate of the cupola. Thus a cupola rated at 8 tons/hr requires $40 \times 8 = 320$ lb (145 kg) of flux materials included in the bed charge. The remainder of the coke bed is placed on top of the flux.

Because of the additional flux materials and the need for hot melting, the full bed height will be beyond that normal for an acid-lined furnace. It should extend to not less than 4 ft (120 cm) above the tuyeres. The weight of flux materials between the charges will vary between 7·5 to 10% of the metal charge weight.

Calcium carbide is favoured as a fluxing material by some operators, especially when charged with the bed coke. When used it can replace the sodium carbonate in both the bed coke and between the metal charges. Its cost and the special storage conditions required are a disadvantage.

Use of Steel Scrap

Steel scrap can form part of the metal charge for either acid or basic cupolas and ratios as high as 60% are used, especially in basic melting because of the much higher recarburization which is possible. When steel is charged higher coke to metal ratios are needed. This applies particularly to basic operation where because of the larger slag volume even higher coke-to metal ratios are needed than for acid practice and 1:6 or 16% is not unusual.-

The melting loss of silicon is at least double that for acid melting and may rise above 25% of that charged. High-temperature melting favours lower silicon loss. On the other hand, the loss of manganese on melting is less than in acid cupolas being 15% as against 25%.

Weighing Charge Materials

All the metallic constituents of a cupola charge must be weighed with reasonable accuracy. Coke can be weighed also provided that it is kept dry, but charging by standard volume can be permitted. Also it is preferable to weigh the fluxes. Where fluidizers such as soda ash and fluorspar are added, the weights required are relatively small compared with those of the metals and coke, and may require the provision of a smaller and more sensitive weighing device. To avoid this, these additions and ferro alloy additions can be added in the form of standardized briquettes. For many years the Foseco Company has engaged in research and development in fluxing and processing molten metals and alloys covering the entire field of casting production. The result is that a whole range of products is available which facilitate the standardized application of these essential treatments.

Air Requirements

It is not always realized that on a weight basis, air is the major item entering the cupola. Accurate measurement of air supply is a most desirable feature. Air is supplied to provide oxygen necessary for combustion of the coke and the quantity will vary according to the coke to metal ratio at which the cupola is operated. In some circumstances over a ton of air is required to burn the carbon of the coke charged to melt 1 ton of iron.

Approximately 140 ft^3 (4 m^3) of air are required to effect complete combustion of 1 lb of good-quality coke. However, in practice combustion is not complete and so 115 to 135 ft^3 per lb of coke is normal (7·0 to 8·5 m^3 per kilo), or on a weight basis 10 lb air per 1 lb coke. This ratio takes no account of the presence of water vapour in the air. Where a constant quantity of oxygen is required, the air delivery must be weighed and its volume regulated according to the variation of relative humidity. Some cupolas are fitted with an automatic air weight blast-supply system. The quantities are calculated from the chemical equation, as follows:

$$C + O_2 = CO_2$$

The atomic weights of carbon and oxygen are 12 and 16 respectively. Therefore 12 lb carbon requires $16 \times 2 = 32$ lb oxygen for combustion to carbon dioxide.

32 lb oxygen is the 1 lb molecule.

Air contains 21% oxygen, by volume.

Therefore 32 lb air at n.t.p. will supply $\frac{21}{100} \times 32 = 6·72$ lb oxygen.

From the above equation, 1 lb carbon requires $\frac{32}{12} = 2 \cdot 66$ lb oxygen for combustion to carbon dioxide.

$$2 \cdot 66 \text{ lb oxygen is contained in } \frac{32 \times 2 \cdot 66}{6 \cdot 72} = 12 \cdot 7 \text{ lb air}$$

Thus 1 lb carbon requires $12 \cdot 7$ lb or 156 ft^3 air at n.t.p. to burn it to carbon dioxide.

If good-quality coke contains 90% carbon then
1 lb coke requires $11 \cdot 4$ lb or 140 ft^3 air at n.t.p.

Combustion is seldom, if ever, complete and the heat generated will vary accordingly:

1 lb C burning to form $\frac{44}{12} = 3\frac{2}{3}$ lb CO_2 evolves 14,550 B.t.u.

1 lb C burning to form $\frac{28}{12} = 2\frac{1}{3}$ lb CO evolves 4400 B.t.u.

1 lb CO burning to form CO_2 evolves 4350 B.t.u.

$\therefore 2\frac{1}{3}$ lb CO burning to form CO_2 evolves $4350 \times 2\frac{1}{3} = 10,150$ B.t.u.

Air is supplied to cupolas by either a centrifugal fan or a positive blower. The volume of air delivered by a fan is affected by the resistance to the passage of the air by the piled material of the charges in the furnace. If this resistance increases, then the volume of air delivered by the fan will decrease. It is essential, therefore, that the fan be designed to deliver sufficient air against any resistance the charges are likely to offer. Where a positive blower is used, a definite volume of air is swept forward per revolution. Thus, the supply of air is controlled by the speed of the blower and any obstruction is signalled by the opening of the safety valve provided.

DESULPHURIZATION OF MOLTEN CAST IRON

Removal of sulphur from molten cast iron during its passage through the cupola is possible only when it is operated under basic slag conditions. The difference in fluxing technique as between acid and basic melting has been discussed. With cupolas lined with acid or silica-based refractory materials an acid slag only can be used and no desulphurizing action takes place. Sulphur removal from the melt occurs with basic fluxing and for this a basic or dolomite lining is necessary unless a liningless water-cooled melting zone is fitted which allows of either acid or basic fluxing as desired.

The tonnage of cast iron melted in basic lined cupolas is small compared with that melted in acid cupolas. The latter increases the sulphur content of the iron every time it passes through the furnace. Sulphur is picked up from the coke and cast-iron scrap used in the charges. Its presence in grey cast iron is disadvantageous as it makes machining more difficult by stabilizing iron carbide and it decreases tensile strength.

Basic lined and liningless melting-zone cupolas are more expensive to operate and maintain than the acid-lined furnace. So, as specifications for the mechanical properties of grey cast irons became increasingly demanding and exacting, attention was given to the possibility of removing sulphur from acid cupola melts after the molten iron had entered the ladle. Pioneers in this field were Colbeck and Evans.[2] They demonstrated that sodium carbonate, used at a ratio of about 1% of the weight of metal to be treated, could function as an efficient ladle desulphurizing agent.

The technique is simple and inexpensive. Anhydrous soda ash is placed in the bottom of the ladle and the molten iron tapped on to it. The following desulphurizing reaction ensues:

$$Na_2CO_3 + FeS \rightleftarrows Na_2S + FeO + CO_2$$

$$CO_2 + Fe \rightleftarrows FeO + CO$$

$$Na_2S + MnO + FeO + Na_2CO_3 = 2Na_2O + Mn + Fe + SO_2 + CO$$

From this it is seen that some of the CO_2 evolved from the soda ash is reduced by the action of the molten iron to CO and burns as carbon monoxide gas "candles" on the surface of the melt.

It is also evident that the desulphurizing action is reversible so that sulphur can pass back again into the melt. This reversion in the presence of the acidic oxides SiO_2 or FeO has already been mentioned. Therefore foundry sand must not be used to thicken up the slag prior to skimming. Limestone chippings are employed for this purpose which is very necessary as the soda slag is extremely fluid and must be removed from the melt to keep it from entering the moulds.

Efficiency of Sulphur Removal

The amount of sulphur which can be removed from molten iron contained in a static, acid-lined ladle by a 1% addition of soda ash is limited. It depends upon the intimacy of contact between the metal and the soda ash slag, and also upon the level of sulphur in the iron. Under normal conditions between 50% and 60% of the sulphur present is removed. When the initial sulphur content is already low it is more difficult to reduce it further. A higher degree of sulphur removal can be attained by using a basic-lined ladle in which to conduct the treatment; also by pouring the desulphurized iron from the treatment ladle into a second ladle, thereby increasing the contact of the molten iron with the soda slag. Unfortunately, this double-pour treatment results in a loss of both time and temperature.

Where the molten iron is being desulphurized in the transfer ladle prior to conversion to steel in a converter, this additional treatment can be carried out. If the iron is for pouring into moulds the loss of time and temperature is a much more serious matter. Even so, the time that must be allowed for the desulphurizing action to complete itself is always a drawback in casting production. The time varies with metal temperature and volume of molten iron. A casting poured before the action is complete will show a profusion of blow-holes and slag inclusions.

With ladle desulphurized metal there is always the danger of slag inclusions when the iron has to be poured into moulds. This risk is overcome to some extent by the use of teapot-spout ladles. However, the soda slag tends to form build-up on ladle linings and pouring lips. Slag inclusions in castings are more detrimental than sulphur.

On the other hand, desulphurized iron has improved properties. With castings for chemical plant, failure by pitting corrosion is far less likely because not only is the sulphur level reduced but inclusions of manganese sulphide are removed and these, when present, form locations from which corrosion pits commence. Furthermore, grey irons of low sulphur content are more easily machined, particularly where casting sections are thin. In the case of ingot moulds their life is longer when cast from iron of low sulphur content.

Desulphurization by Injection of Calcium Carbide into the Cupola

Injections of calcium carbide have been made into molten cast iron in the cupola and in the ladle. H. J. Leyshon and M. J. Selby[3] of the British Cast Iron Research Association have carried out tests using powdered calcium carbide at a rate of 2% by weight of metal treated. This was injected below tuyere level but above the slag line, into the carbon-lined well of an acid cupola. The sulphur content was reduced by 0·04%. With a well lining of graphite-bearing ganister the sulphur reduction was 0·03% for the same injection ratio. The average, original sulphur values in these two cases were 0·128% and 0·112% respectively.

Injection of the same quantity of carbide through the tuyeres of the furnace was less efficient in reducing the sulphur level. This method of reducing sulphur in the well of the cupola has the advantage that it results in no loss of temperature of the molten iron. Neither does it cause delay, as it needs no extra time. The fall in sulphur is caused by an increase in basicity of the cupola slag and a reduction of its iron oxide content.

Ladle Injection

Up-grading of cast iron by ladle injection treatment of the molten metal has been practised for more than a decade. Powdered materials such as calcium carbide, graphite, ferro-silicon and other ferro alloys have been blown through graphite lances extending to near the bottom of the ladle, by nitrogen gas under pressure. The turbulence generated by the gas stream allows the necessary contact of liquid metal and powdered agent. With calcium carbide injection a dry slag is thrown up on to the metal surface from whence it can be removed with comparative ease.

M. N. Volianik[4] has developed a technique for avoiding the tendency of the carbide to clog the graphite gas lance. A predetermined amount of calcium carbide is placed on the surface of the molten metal in the ladle. Acceleration of the reaction is achieved by turbulent mixing created by nitrogen blown at high pressure through the metal bath. A cover fitted to the ladle prevents splashing and acts as a support for the lance(s). Gas is blown for 3 to 4 minutes.

To attain sulphur removal efficiencies of 85% the iron must have a temperature of 1400°C or more. The temperature drop is from 25° to 70°C, depending on the volume of the iron in the ladle. The process is inexpensive, uses rugged equipment and has particular application to the production of nodular graphite cast iron.

The Porous-plug Ladle Technique

In the production of nodular graphite cast iron (S. G. iron) a low sulphur content of the order of 0·02% is necessary to economize in the use of the nodularizing alloy and also to minimize the incidence of dross defects. From the early work of Norbury at the British Cast Iron Research Association we know that manganese can be used to balance the ill effects of sulphur in grey cast iron. The ratio given is that the minimum manganese content should be equal to 1·7 × sulphur + 0·3%. However, there is a limit to how far manganese can be used to balance sulphur if subsurface blow-holes associated with manganese sulphide segregation are to be avoided.[5]

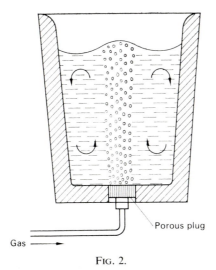

Gas

Porous plug

FIG. 2.

For these reasons much work has been done with the object of developing a simple, practical and inexpensive means of removing sulphur from molten iron obtained from acid-lined cupolas. The lance injection method has been mentioned and the porous-plug ladle using compressed air or nitrogen may be regarded as a simpler application.

R. B. Coates and H. J. Leyshon[6] have given a detailed description of it. The diagrams and assembly layout (Figs. 2 to 4) are taken from their published work.

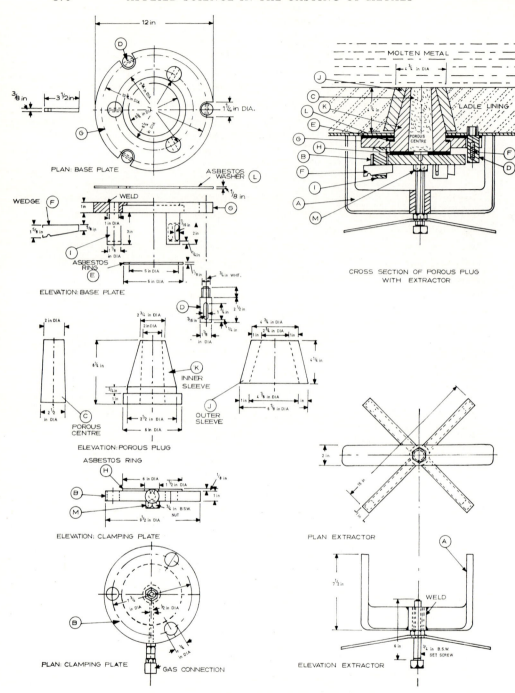

FIG. 3. Porous plug clamp assembly and extractor.

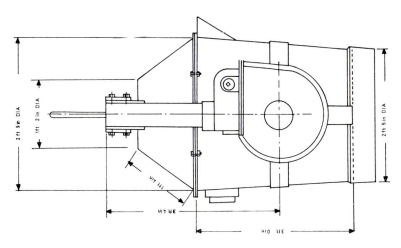

2 ft 9 in DIA

1 ft 2 in DIA

2 ft 5 in DIA

1 ft 4 in

3 ft 4 in

3 ft 0 in

KEYING BOLT
1½ in LONG AT RANDOM
TO SUIT

SALAMANDER
BASIN

GRAPHITIC
GANISTER

STEEL FLANGE

CUPOLA CIRCULAR
BRICK

LADLE TILES

SAND

CIRCULAR BRICK

GRAPHITIC GANISTER

POROUS PLUG

STEEL SKIRT
welded onto base
of ladle

FIG. 4. 30 cwt porous plug treatment ladle.

Calcium carbide was the preferred desulphurizing agent used, because of its dry slag formation. Using this at a rate of about 1·5% of the weight of metal treated, very low sulphur levels can be achieved with metal temperatures running no higher than 1350°C and air flowing for 2 min at a flow rate of 9 ft³/min. The method is equally applicable to carburizing with graphite additions or simultaneous desulphurizing and carburizing.

As with all desulphurizing processes carried out externally of the furnace, there is a temperature loss which depends mainly on the bulk of molten metal treated. Less loss is claimed for the porous plug method and the actual figures given were 55°C for a 4 cwt melt reducing to 25°C for 1 ton.

The Shaking Ladle

This device (Fig. 5) makes use of a rotary-reciprocating motion applied to the molten iron in the ladle so as to induce it to swirl with a parabolic meniscus. The ladle does not revolve about its own axis. Kalling developed the method in Sweden in 1958 and it produces a very efficient mixing which is essential to these methods of ladle desulphurization. The principle is discussed by Sven Eketorp.[7] With this apparatus pig iron has been desulphurized to 95% using lime at a rate of 1·5%. It is also employed for simultaneous desulphurization and carburization.

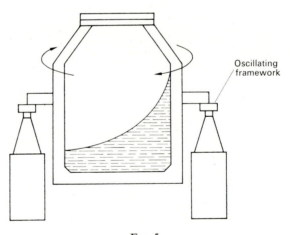

Oscillating framework

FIG. 5.

The equipment requires a capital outlay of up to £25,000 and is suitable for treating large melts of several tons at a time for which the temperature loss is about one-quarter of that for small charges. In fact it is more likely to find application at blast furnace plants and steelworks than for the average iron foundry.

Continuous Desulphurization

From the foregoing it is apparent that an inexpensive, simple method of continuous desulphurization of molten iron from an acid-lined cupola would be of great advantage. Here again, a process has been developed by the British Cast Iron Research Association, which is not difficult to operate, requires no additional labour and a limited amount of supervision. It employs soda ash as the desulphurizing agent and is applied to a continuously tapped, acid cupola capable of delivering molten iron at above average temperature so as to compensate for the inevitable temperature loss incurred by the process.

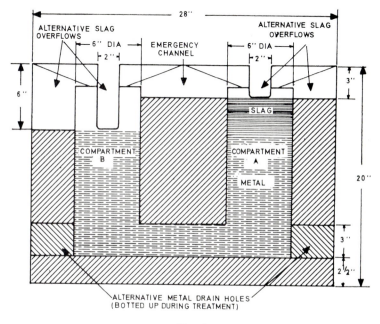

FIG. 6.

The acid cupola slag is taken off at the side of the cupola spout and soda ash is fed, from a vibratory feeder, on to the stream of molten iron as it emerges from under the slag dam. This is carried out in a manner such that mixing of the metal and soda ash slag occurs prior to the stream entering a slag separating box constructed of a carbonaceous refractory ramming material as detailed by R. B. Coates and H. J. Leyshon.[8]

Figures 6 and 7 are taken from the reference quoted and show the design and general appearance of a prefired block suitable for treating up to 10 tons of iron per hour. It has a weight of about 4 cwt and data from a light castings foundry, melting 30 tons of iron per day, and desulphurizing with soda ash addition of 17 lb per ton, indicate a life for the carbon block equivalent to a throughput of about 2500 tons. The capital cost of the blocks, together with routine maintenance charges, are said to amount to a cost of about 2 shillings per ton of iron treated.

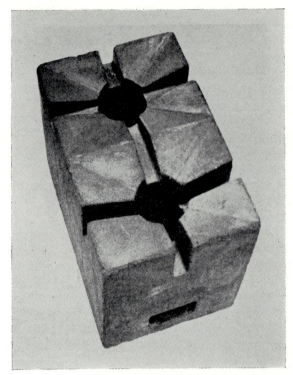

FIG. 7.

Ladle Slags

Slag may be present on the surface of the molten iron in a ladle from several sources, apart from ladle desulphurization treatments. It is always a potential cause of defective castings. If it enters the mould with the metal stream it forms obvious inclusions which may also be accompanied by gas holes. The slag contains ferrous oxide which can react with the carbon in the iron to produce carbon monoxide gas according to the simplified equation:

$$FeO + C = CO + Fe$$

There is no good reason why cupola slag should ever be allowed into a pouring ladle. Slag formed from ladle refractories is more insidious. These refractories normally contain high contents of silica and clay. The presence of clay makes complete drying out of linings difficult. In the presence of moisture steam will be formed which will oxidize the iron to produce a fluid surface slag, rich in oxide and very reactive.

Ladle linings of naturally bonded sand milled with a proportion of ganister or fireclay are a source of ladle slag when the iron enters at a high temperature. In these circumstances slag forms continuously on the skimmed surface of the iron while it stands in the ladle. Another source of ladle slag is manganese sulphide which separates from irons of high sulphur and manganese content to float on the surface of the melt. Such irons will also form slag inside the mould while they remain liquid. Cast irons in which the silicon and manganese contents approach each other in value likewise tend to encourage the formation of low melting point manganese silicate slags.

Subsurface blow-holes are frequently associated with these slag inclusions especially when pouring temperatures are low enough to permit precipitation and segregation of manganese sulphide in the ladle and in the mould according to the reversible reaction:

$$Mn + FeS \rightleftharpoons MnS + Fe$$

When the surface oxide dross becomes enriched in manganese sulphide its melting point is lowered and may fall to 1050°C which is below the eutectic temperature for cast iron (1125°C). It is, therefore, liquid during the formation of the eutectic graphite. The iron oxide present in the slag can react with this graphite to produce carbon monoxide according to the reaction already shown.

Chemical analyses of ladle slags vary between the following limits:

Silica (SiO$_2$) 32·5–41%
Manganese 10–27%
Sulphur 0·085–2·2%

Much of the knowledge of this type of defect is obtained from the work of Morgan[9] and Tonks.[10] Figure 8 is a photomicrograph illustrating a typical defect due to segregation of manganese sulphide slag accompanied by blow-hole type porosity from the evolution of carbon monoxide gas. Experience shows that balancing the sulphur by adding manganese, as discussed earlier, is not a solution to this problem. It is necessary to reduce the sulphur level to that compatible with a manganese content of 0·7%. Another preventive step is to ensure that pouring ladles are maintained free from slag build-up on lips and linings. This entails efficient skimming and cleaning of slag from the surface of the molten iron. Of great assistance for this purpose is the availability of slag and dross collectors or coagulants which act to form a coherent frit or skin thus enabling easy removal. In addition care should be taken to avoid low pouring temperatures which favour the formation of manganese sulphide.

At temperatures below 1370°C the free energy favours chemical reactions leading to the formation of oxide slags. The bath of molten cast iron is covered by a layer of slag which, if removed, re-forms in the presence of air. When a temperature of 1430°C is exceeded the thermochemical driving force increases in favour of carbon oxidation and suppresses the oxidation of Si, Mn and Fe. Under these conditions no slag will form on the surface of the molten iron. This state of reduction predominates to a greater degree as the temperature rises.

On cooling from superheating temperatures a stage will be reached where conditions for the formation of metal oxide slags are re-established, depending on the chemical composition of the molten iron.

Normal, cupola-melted cast iron contains particles of SiO_2. When superheated above the inversion temperature (about $1430°C$) the oxygen is evolved as CO thereby reducing or eliminating the SiO_2 particles. Graphite nucleation potential depends on the presence of silica particles in the melt and so does eutectic cell count and sensitivity to chilling or undercooling.

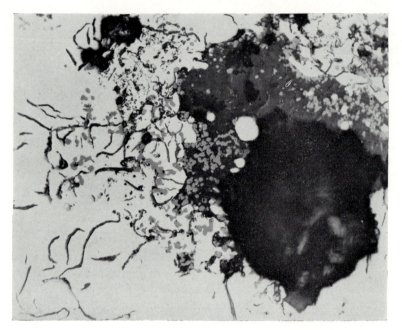

FIG. 8.

Addition of Alloy Elements to Cast Iron

A very wide range of composition and physical properties can be obtained with cast irons containing the five basic elements: carbon, silicon, manganese, sulphur and phosphorus. However, in many cases special characteristics are required such as added resistance to corrosion, heat and wear. These, and other particular attributes, can be developed in suitable base iron compositions by the addition of alloying elements. About twenty alloying elements are in everyday use, either singly or in combination. They range through the entire list of metals and metalloids from aluminium to zirconium and are added in quantities of less than 1% up to 30% or more of the composition.

It has to be decided how best to make these alloying additions as some of them are in the rare metal category and relatively expensive. Thus, it is important that their assimilation into the melt should be as high as possible. Indeed, it is desirable in all cases that losses be kept to a minimum. This is most likely to be achieved when the additions can be made direct to the molten iron at the furnace spout or in the ladle, as distinct from additions to the furnace charges. In the latter case melting losses are possible and, in the case of cupola melting, difficulty arises in keeping the addition to the charge or charges in which it is placed and in locating these charges when delivered molten. When charges of more than one grade of iron are present in the cupola at the same time the problem of segregation and isolation becomes more acute.

The decision whether to make a furnace or ladle addition is often solved by the quantity of alloy to be introduced. For direct spout or ladle additions a safe limit can be fixed at around 2%. Quantities in excess of this are liable to cause too great a decrease in metal temperature.

Spout and Ladle Additions

Fortunately, most of the rare alloying elements are needed in small ratios only and are invariably introduced at the furnace spout or in the ladle. Examples of these are boron, molybdenum, tin, titanium, tungsten, vanadium and zirconium. Crushed and suitably graded ferro-alloys are commonly employed for the purpose. Some of these have high melting points, in which case they enter the melt by becoming dissolved in it.

To overcome the temperature drop caused by cold ladle additions, exothermic or heat-generating compounds have been developed. Very often the effect of these on conserving or increasing the melt temperature is more apparent than real. At best they provide enough heat to counteract their cooling effect. In certain circumstances their heat contribution is negligible. Often they are placed on the bottom of hot ladles and the heat reaction takes place prior to the tapping of the iron. There are occasions on which they perform a useful function, but where nitrates form the heat-generating chemical, fluid slags are formed and fumes are generated.

Packeted and briquetted ladle additions have the advantage of convenience and simplicity. They are in units of predetermined weight and are easily added to the melt in the required number. The packeted granular materials allow of introduction by way of the metal stream at the furnace spout. Distribution in this way can be over a period during tapping and tends to afford greater homogeneity. The preferred technique is to drop on to the molten metal stream from a height of 3 ft or 1 m, as near to the tap-hole as freedom from turbulence will allow. This ensures sufficient kinetic energy so that the granules can penetrate the stream rather than bounce off or float along on its surface. The introductory flow should not commence until the molten iron has attained a depth 2 in. in the ladle. It should continue until about three-quarters of the total quantity of metal has entered the ladle. Ladle linings should always be clean and free from slag build up. Also the metal delivered from the furnace should be clean and of good temperature. The great advantage of ladle or spout additions is the assurance that they are present exactly where required.

Cupola Additions

Alloys commonly added to cupola charges are silicon and manganese. They can be in the form of siliceous and high manganese pig irons, spiegeleisen or the appropriate ferro-alloys. It is more convenient to store silicon and manganese in concentrated ferro-alloy form containing 75% to 80% of the element concerned, than in the form of pig iron or spiegel which contains 70% to 80% of iron. The normal melting loss or burn out of silicon and manganese in cold blast acid cupola melting is 10% and 25% respectively.

So as to minimize this loss and to avoid crushing and weighing of the ferro-alloy, the additions can be made in the form of briquettes. The bonding agent of the briquettes protects the alloy from oxidation and delivers it intact into the melting zone of the furnace. Briquettes are made to contain definite weights of alloy, in lb or kilos, and it is necessary only to place the required number into the charges.

Chromium additions can be made to cupola melted charges, either as loose ferro-alloy or in briquette form. Where high chromium ratios are required, however, the alloyed cast iron is best made in an arc or induction furnace. Copper and nickel additions when desired in amounts beyond about 2% can be added in metallic form, as ingot or scrap, to the cupola charges. Neither of these metals undergo any oxidation losses during their passage through the cupola. Moreover, they enter the melt readily and are taken up into solution, forming no separate carbides or other separated metallic compounds. "Ni-resist" and austenitic heat- and corrosion-resisting cast irons require copper and nickel additions of 7% and 14% respectively. Such high alloy additions must be added to the cupola charges unless they are pre-melted and poured into the base iron in the ladle.

Acicular cast irons contain nickel and molybdenum. The former up to 4·5% in heavy section castings and the latter up to 1%. Also some chromium is present up to 0·5% in thick sections, in which copper may replace some of the nickel. Copper and nickel can form part of the furnace charges but the molybdenum and chromium may be added at the spout or in the ladle. Phosphorus content should be kept at a low level in these high-duty cast irons. A maximum of 0·12% is recommended[11] as greater amounts will form a complex quaternary eutectic with carbon, iron and molybdenum, which removes 1·3 units of molybdenum from the matrix for each unit of phosphorus present at beyond 0·12%.

The tendency of phosphorus to combine with carbon, molybdenum and iron to form the quaternary eutectic applies equally to any composition of cast iron which contains these four elements; especially so where thick sections, which cool slowly, are concerned. Under such conditions, structural equilibrium is approached and favours development of the phosphorus–molybdenum complex. Irrespective of whether molybdenum is present or not, phosphorus is a disadvantage in any grade of high-duty cast iron and should be kept low. Above 0·2% it forms increasing quantities of iron phosphide which solidifies at 950°C and creates liquid shrinkage problems in castings of varied section thus leading to unsoundness.

On the other hand, it facilitates the production of thin-sectioned castings such as are required in the textile, building and electrical industries. Here, the low melting-point iron phosphide imparts a greater fluidity to the molten iron thereby assisting in the filling of moulds for castings having small thicknesses and large surface area.

High phosphorus pig irons and scrap cast iron are widely available as raw material for furnace charges. Therefore, additions of phosphorus are rarely needed in the form of ferro-alloy. Even so they are available as ferro-alloy briquettes for cupola additions and as packeted units for introduction to the molten iron in the ladle. Phosphorus unites readily with cast iron by either method and if it is required to be restricted to special castings, it is better to add it at the spout or in the ladle.

Effects of Alloy Additions

The technical literature contains a mass of information and data under this heading. Indeed it forms subject-matter for a volume of its own. Here, the space available allows of brief references only.

Aluminium

Small quantities are made as ladle additions to reduce chill and effect de-oxidation. Percentages between 0·01 and 0·2 can cause pinhole porosity due to pick-up of hydrogen.[12] There are four types of aluminium cast irons, according to structure.[13] Pearlitic with 9% Al, white with 11–19% Al. At 21–26% Al the irons are ferritic, changing to ledeburitic at 28–32% Al. These cast irons have high heat resistance and high electrical resistance. Their magnetic properties are of interest. The high ratios of aluminium are best added by tapping molten iron on to molten aluminium in a mixing ladle.

Bismuth

Introduced as a ladle addition at a rate of between 0·002% and 0·01% to malleable iron melts. It enables a higher silicon content in the base metal without the formation of primary graphite and/or the use of heavier sections in the white iron castings.

Boron

Also used as a ladle addition to white iron for the production of malleable iron castings. It neutralizes the graphite-coarsening effect of bismuth and forms boride of chromium, thus suppressing the inhibiting effect of free chromium on the annealing cycle. Boron and bismuth are added together to malleable iron in proportions of 0·002% boron and 0·008% bismuth. Boron returned in the internal scrap does not interfere with the mechanical properties of subsequent melts but is beneficial in neutralizing the effects of nitrogen.

Chromium

Acts as a powerful carbide stabilizer thus reducing the formation of free graphite and encouraging chill formation. Relatively small amounts 0·5–1·5% are made as ladle additions, often in conjunction with nickel. Wear- and heat-resistance are both increased by chromium. Where silicon is given a graphitizing value of 1, chromium has a value of minus 1·2. The carbon content of the eutectic is reduced by about 0·5% for each 1% addition of chromium. At about 3% chromium the formation of free graphite is suppressed and a white iron structure forms.

Chromium decreases the extent of the gamma region and raises the critical transformation temperature. The gamma field is almost completely eliminated at 20% chromium. For grey cast irons the general effect of chromium is to increase the tensile strength by approximately 3% for each 0·1% addition. Hardness is raised by 6–10 points Brinell for each 0·1% chromium increment.

Copper

Ladle or furnace additions can be made. Copper is soluble in cast iron up to about 3% and alloying is usually restricted to this value or below. It has a chill-restraining effect equivalent to about one-third that of silicon. The influence of copper upon the structure and mechanical properties of grey cast iron has been investigated extensively by Dilewijns and De Sy.[14] Their work covered additions of 0·1–2% copper to cast irons with silicon contents of 1·5% and 2%, related to degrees of eutectic saturation between 0·75% and 0·95%.

The test bars made for evaluation of the mechanical properties and hardness values had diameters ranging from $\frac{1}{2}$ to 6 in. Test results show that the influence of copper at constant eutectic saturation varies according to composition and wall thickness, between the following limits: at 1% copper the hardness increase is 15 to 35 points Brinell and increase in tensile strength is 2–4 kg/mm². At 2% copper the increase in hardness is 30–45 points Brinell and the rise in tensile strength 4·5–7 kg/mm².

Copper additions reduce the free ferrite content and increase the hardness of the pearlite. There is no practical use in making copper additions above a level of 2·0–2·5% for the sake of improvement in mechanical properties. Copper added to furnace charges suffers no oxidation loss. For copper additions only scrap from pure or electrolytic copper is suitable. Great care is needed to avoid copper scrap of unknown origin as it may introduce harmful trace elements such as lead, antimony and tellurium.

Manganese

Is one of the five basic elements normally present in cast iron. To be effective as an alloy it must be present in a ratio in excess of that required to neutralize the sulphur. For grey iron compositions this ratio is given as 1·7 times the sulphur, plus 0·3%. It combines with the sulphur to form manganese sulphide (see p. 175) and, having a specific gravity much below that of molten cast iron, tends to separate out from the melt by flotation. By so doing it takes part in the formation of ladle slags, as discussed previously.

As an alloying addition to cast iron, manganese is perhaps best known for its austenite stabilizing action, in conjunction with nickel. Non-magnetic cast irons of high electrical resistance contain from 6% to 15% of manganese.

Molybdenum

As one of the rarer elements it is seldom added to cast irons in excess of 1·0%. Its main function is to improve the mechanical properties of high-duty cast irons and for special compositions such as acicular cast iron. The need for low phosphorus content in the presence of molybdenum has already been mentioned. Otherwise a quaternary eutectic is formed containing 4·4% P and 5·8% Mo which robs the matrix of the Mo.

It forms stable carbides and offsets the graphitizing influence of about 0·35% Si. During solidification it favours formation of the carbide/austenite eutectic and is a powerful promotor of pearlite. At about 0·5% Mo, free ferrite is eliminated from the matrix. Molybdenum raises tensile strength about 3·5–4·0% for each 0·1% added. It also increases hardness in the order of 40 points Brinell for each 1·0% addition of Mo.

Nickel

This is perhaps the best known and most widely used alloying element for cast iron, being employed from a ratio of 1% in low alloy irons up to about 20% for austenitic compositions. Small amounts up to 2·0% are readily made as ladle- or spout-additions but the higher quantities are introduced by way of the furnace charges. There is no oxidation loss on melting.

The carbon content of the eutectic is lowered by 0·4% for each 5% of nickel alloyed. Also the eutectic temperature is raised by 6°C for each 1% of nickel added. During the solidification of the eutectic the presence of nickel encourages graphite formation to approximately one-third that of silicon. Thus for a 1% nickel addition to low alloy grey cast irons the silicon content may be lowered by 0·3% without increasing the risk of free carbide or chill appearing in the structure. Progressive additions of nickel lower the critical temperature and thus extend the gamma or austenitic field. At 12–14% Ni the austenite phase is stable at room temperature.

Nickel also lowers the carbon content of the eutectoid pearlite by about 0·04% for each 1% Ni present. However, the pearlite stabilizing power of nickel promotes the formation of pearlite rather than ferrite in low- and medium-strength irons, which results in an increase in combined carbon content. With an unalloyed cast iron containing appreciable ferrite, increasing additions of nickel will result in pearlite formation. The pearlite becomes increasingly finely divided until at about 6–8% nickel the eutectoid transformation is suppressed and replaced by martensite to form the series of air-hardening cast-iron compositions. A well-known example is "Ni-Hard" which has the composition TC 2·8–3·6, Si 0·4–0·7, Mn 0·3–1·2, S 0·15 max, P 0·3 max, Ni 3·3–5·0, Cr 1·2–1·7 and Mo 0–0·4%.

Additions of 1% or 2% nickel to grey iron compositions have the effect of raising the level of the mechanical properties a little. This effect is mainly due to a reduction in section sensitivity imparted by the nickel. That is to say the hardness of thin sections is reduced and that of heavy sections increased as a result of grain refinement. The uplift of mechanical properties derived from nickel is not so great as achieved with Cu, Cr, Mo or Va additions.

Phosphorus

Although this element is one of the five basic constituents of cast irons and, therefore, not usually classed as an alloying material, it does confer some useful properties, as indicated in a previous paragraph. These can be referred to more fully here. G. Henon[15] has provided an excellent review from a study of a voluminous bibliography which indicated that phosphorus has useful attributes for the improvement of quality in certain cases.

One of these is its effect on the flowability of molten cast iron. In hypereutectic irons an increase of 0·2% improves flowability equal to an addition of 0·1% C or 15°C rise in temperature. Also, it has been shown that phosphorus can reduce the tendency of an iron to produce chill. Because of its influence in improving wear resistance, a high level of phosphorus is recommended for components such as brake shoes for railway wagons.

Maximum mechanical properties exist at around 0·3–0·4% but decrease beyond this value. Phosphoric cast irons have a good resistance to heavy forms of corrosive attack, notably from hydrochloric, sulphuric and acetic acids. Tests indicate that heat conductivity is improved by an increase in phosphorus. A combination of high silicon and high phosphorus provides an iron composition well suited to the production of electric resistance grids. The presence of phosphorus reduces magnetic permeability and at high levels about doubles hysteresis losses. Large-scale trials have shown that phosphorus contents between 0·35% and 0·45% have improved the life of ingot moulds in steel works and have reduced the tendency towards "corner shelling". Commercial advantages accrue from the utilization of higher phosphorus irons where they are suitable and justified as they are invariably less costly than low phosphorus irons and hematites.

Silicon

Again, as one of the basic components of cast iron, it is not normally regarded as an alloying agent. Together with carbon it is of great importance as the ratio of these two elements forms the foundation upon which the entire edifice of cast iron specifications is founded. The two main factors which determine whether or not a cast iron remains grey and machinable are chemical composition and speed of cooling. The latter is determined by the mould and section thickness of the casting and little can be done to influence it. Silicon exerts a powerful action on the carbon combined as iron carbide, a very hard material, transforming it into iron and graphite. Where cupola melting is concerned, silicon can be controlled more readily than carbon and so becomes the dominating constituent of all general engineering grey iron compositions.

Apart from this, silicon is used as an alloying element in the production of heat- and corrosion-resisting cast iron compositions. A silicon content of between 5–6% confers heat-resisting properties up to a temperature of about 900°C and is known as "Silal". This composition can be rendered austenitic by adding to it 18% nickel and 2–5% chromium. It then becomes less brittle and resists temperatures up to 1000°C.

In order to resist acid attack the silicon content has to be at a level of between 15–18%. For maximum resistance to 10% solutions of sulphuric, nitric and hydrochloric acids the silicon contents are 15, 16 and 18% respectively. These high silicon irons usually contain less than 1·0% carbon. They are known under a variety of trade names and are best melted in an electric arc furnace.

Tin

In grey cast irons, small additions of tin can be used, with advantage, as an alternative for up to 0·5% of chromium. This practice has been applied to the production of automobile cylinder-block castings. Sections of 1½ in. or less, develop a completely pearlitic matrix with 0·1% Sn, irrespective of phosphorus content up to 0·6%.[16] In heavy sections 0·1% Sn raises the hardness level and renders it more uniform. Tin retards the tendency for the cementite of the pearlite to spheroidize and decompose at elevated temperatures, thereby causing a drop in hardness. The pearlite stabilizing effect of tin applies to both flake and nodular graphite grey cast irons even when they are of hypereutectic composition. Even so, the higher and more uniform level of hardness is coupled with easier machinability and less sensitivity to self annealing if the castings are left to cool in the moulds.

Tin bearing castings show less tendency to chill at corners and free edges. For although tin additions increase hardness values progressively, they do not produce free carbide in the structure, as does chromium. The growth and scaling resistance of flake and nodular graphite cast irons are improved by small amounts of tin and the reduction in the number of variables in the melting technique is regarded as a most important advantage. The small tin additions are made to the molten iron in the ladle. It has been found, in practice, that foundry back scrap from tin-treated melts, when remelted in the cupola, gives an almost 100% recovery of tin.

Titanium

This element plays a multiple role in cast iron because it has a strong affinity for nitrogen, oxygen and carbon. Divergencies of results as between different

G

investigators can be explained on this score. For example, its graphite-refining action may be masked if the molten iron contains a high level of nitrogen or oxygen because an addition of titanium at the usual levels of 0·1–0·3% could be expended on fixation of the two gases. When not exhausted in this way, titanium refines the flakes of graphite progressively until the interdendritic or undercooled D form of graphite results. The transformation of graphite in this way is augmented by gassing the titanium-containing melt with carbon dioxide. Reversion to normal flake graphite occurs if the melt is flushed with hydrogen. A patent covering this procedure was granted to the B.C.I.R.A.

Chill reduction has been claimed for titanium additions at low levels but this could be due to the fortuitous presence of elements such as aluminium and calcium in the ferro-titanium alloy. Or it could result indirectly from the action of titanium on the nitrogen or sulphur. Again the temperature of superheating and period over which the iron is held molten can influence depth of chill more profoundly than composition of the metal. On the other hand, small percentages of titanium facilitate first-stage graphitization during the anneal of malleable castings. They also assist in preventing cracking of the white iron castings.

Titanium raises the temperature of formation of the iron–carbon eutectic and slightly reduces its carbon content. The affinity of titanium for oxygen is greater than that of Mn, Si and Va but less than that of Al, Mg and Ca. Only zirconium exceeds titanium in its affinity for nitrogen. Nitride of titanium is very stable and remains in the structure as solid inclusions of rose, orange or yellow colour. Titanium also forms stable carbides and combines with sulphur. It reduces the growth of cast-iron components subjected to repeated heating and cooling. It is not regarded as an inoculant and when ferro-titanium tends to act as such it is due to the presence of silicon and carbon in the alloy. Conversely the known action of titanium to produce undercooled graphite is possibly due to its influence in destroying nuclei.

Production of spheroidal graphite cast iron is affected adversely by the presence of titanium. It should not exceed a level of 0·04%, especially where traces of lead are also present. Where it is desired to make additions of titanium to cast iron it is best carried out using ferro-titanium of 67% titanium content. Ferro-titanium of 30–40% titanium content gives a poor titanium pick-up which is both irregular and uncertain.

Tungsten

The British Cast Iron Research Association[17] made an investigation into the possibility of using tungsten as an alternative to molybdenum for acicular† cast-iron production. They found that the phosphorus content of the iron must be even less than when molybdenum is used; it should not exceed 0·1%. Tungsten additions of the order of 2% are required. These may be added

---

† In acicular cast iron the pearlite is replaced by an austenite transformation product intermediate between martensite and pearlite. The action of nickel and molybdenum or tungsten retards transformation so that cooling extends below the pearlite range before transformation can occur.

directly to the ladle as ferrotungsten. Recoveries of 80% can be achieved by making the addition to the metal stream as it enters the ladle. Manganese has a beneficial influence up to 0·8% but in excess of this it encourages the formation of martensite. Copper cannot be used as a complete or partial replacement of nickel as for acicular structures developed by molybdenum.

The hardness of acicular, nodular graphite irons is higher than for irons of similar analysis having a flake graphite structure. It appears that tungsten is only about half as effective in forming acicular structures as molybdenum. Difficulties are encountered in producing fully acicular structures in castings with a wall thickness of less than 1 in.

Vanadium

For a long time vanadium has been a favoured alloying element in cast irons for large diesel engine liners. It is a carbide stabilizer and is seldom added in amounts beyond 1·0%. Provided that mottle is absent 0·1% V increases tensile strength by about 0·8 tons/in². Hardness is increased by 8–10 points for each 0·1% V addition.

Vanadium produces refinement of the graphite structure and reduces segregation and tendency towards the formation of graphite rosettes. Section sensitivity also is reduced by the evening out of the graphite size from the edge to the centre of thick sections. As a carbide stabilizer, vanadium will offset the graphitizing effect of about 1·75% Si. In amounts from 0·15% to 0·35% it confers wear-resisting properties, increases the stability of pearlite and gives added resistance to growth and creep at temperatures between 450° and 650°C. Special pig irons containing vanadium are available but they are often of hypereutectic composition. The alternative is to introduce the element by means of spout or ladle additions of ferro-vanadium (35–40% V). From these the recovery is about 85%.

Zirconium

Acts as a powerful deoxidizer and is added as a ladle addition in the form of zirconium–silicon alloy containing 37% Zr and 50% Si. When added in quantities above 0·25% the recovery is about 25%. Its effect is to lessen the tendency towards chill formation and to encourage production of a ferrite structure.

Small additions of zirconium up to 0·25% are unlikely to leave any residual in the melt. This is because of its action in combining with oxygen and sulphur. It may well be that the chill-reducing effect of zirconium is conferred by its neutralization of sulphur. In this respect it functions in a similar manner to manganese.

INOCULATION OF CAST IRON

Much has been written on this subject since Meehan treated molten cast iron with calcium silicide in the early 1920's. The process is distinct from alloy additions made to the molten iron for the sole purpose of affecting the chemical composition of the melt. In fact normal inoculation treatment is concerned with the introduction of quantities of material, usually of about $0.1–0.3\%$ which exert little or no influence on the chemical analysis. The addition is made to the molten metal for the purpose of modifying its microstructure and, as a result, improving the mechanical and physical properties to a degree not explainable in terms of compositional change. Apart from chemical composition, there are a number of factors which influence the properties of cast iron.

Factors of Influence

Even if chemical composition remains steady within fixed limits, the following factors will affect the strength, machinability and other characteristics:

1. Amount of superheat.
2. Degree of oxidation.
3. Moisture in the air blast.
4. Charge make-up.
5. Weight of charge.
6. Speed of cooling in moulds—section thickness.
7. Pouring temperature.

All of these, except item 6, depend on furnace and melting practice and have particular reference to cupola melting. They can exercise their effect either singly or in combination.

Much of the required control is achieved by standardization of melting practice in accordance with correct technique. However, atmospheric moisture is usually variable, and components of charges vary in proportion, composition and properties. Cooling speed depends on casting design, section thickness and the effect of mould and cores. Thus there are a number of factors that can exert an important influence on the properties of castings which cannot be brought under full control. Fortunately the technique known as inoculation provides a means of counteracting and minimizing the effects of these variable factors.

Importance of Graphite

The quantity and form of carbon in cast iron exerts a major influence on its properties. In all grey cast irons most of the carbon is present as graphite and the remainder is combined as iron carbide, usually as a constituent of pearlite. Normally the graphite is present in flake formation. These flakes may be large or small, abundant or sparse, straight or curled, and they exert a big influence on strength and machinability. Two of the factors already listed, chemical composition and speed of cooling, play a major part in determining the quantity, shape and size of the graphite flakes.

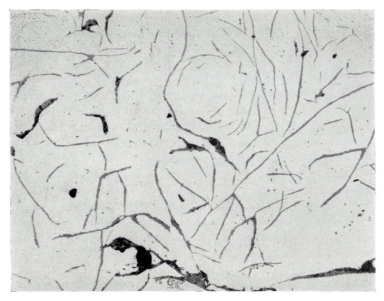

FIG. 9.

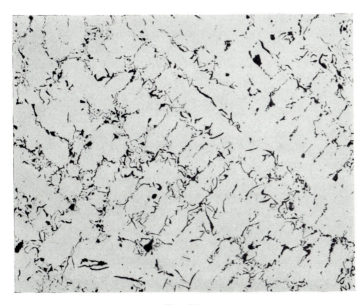

FIG. 10.

Chemical composition can be controlled more readily than cooling speed. If the rate of cooling is too fast, the formation of graphite is stopped or retarded. It takes time for the graphite to form. Very slow cooling encourages the precipitation and growth of graphite with large, coarse flakes as shown in Fig. 9. Both extremes are undesirable, the first causing hard castings, difficult to machine and the latter conferring low strength and brittleness.

Undercooling

Between these two conditions of graphitization is a third known as under-cooled graphite, appearing as in Fig. 10. As the molten cast iron cools, graphite should form from the melt at the eutectic temperature of about 1125°C for grey cast irons. Under certain conditions the molten iron falls below this temperature and solid crystals may have formed before the graphite separates. When this happens the melt is said to have undercooled and the graphite, when it does form, is confined to positions at the crystal grain boundaries where it remains in very small flake size due to restriction to growth. According to standard grading this is known as type D graphite. Castings which undercool in this way are likely to be too hard to machine at corners and free edges. They also have reduced resistance to shock loading.

STEEL SCRAP IN CHARGES

Charges for cast iron ordinarily contain the raw materials pig iron and cast scrap. The latter includes foundry returns and the ratio of pig iron and returned scrap will determine how much, if any, additional scrap, from other sources, has to be included. Purchased cast iron scrap is seldom of controlled composition and as such is not a desirable constituent for high-duty iron charges. Mild-steel scrap is of known specification and is thus a suitable component of furnace charges. It facilitates the production of the grades of cast iron with high mechanical properties and is an economical alternative to the use of refined pig irons provided that the requisite degree of metallurgical knowledge and control is applied to its use.

Because steel scrap is of low carbon and silicon content it reduces the level of these two elements in the resulting cast iron according to the amount used in the charges. At the same time the tendency for the iron to undercool increases as the ratio of steel scrap increases. In order to overcome this disadvantage the method of control already referred to as inoculation is applied.

CARBON EQUIVALENT VALUE

When carbon is dissolved in molten iron the temperature at which solidification commences is progressively lowered as the carbon increases until the limit is reached at 4·3%. Iron with 4·3% carbon is called the eutectic alloy and it solidifies at 1140°C. Grey cast iron also contains silicon and phosphorus which lower the carbon content of the eutectic by 0·33% for every 1% of each present. Therefore the carbon equivalent value of any cast iron composition is found from the expression:

$$\text{Carbon equivalent} = \text{Total carbon} + \frac{\text{Silicon}\ \% + \text{Phosphorus}\ \%}{3}$$

The lower the carbon equivalent value falls below 4·3 the more likely it is that undercooling will take place and graphitization occurs with increasing reluctance and will become completely suppressed if the undercooling attains the temperature at which the carbide eutectic forms.

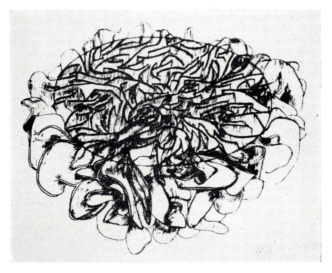

Fig. 11.

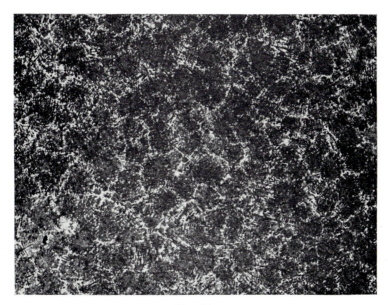

Fig. 12.

Solidification Process

When the carbon equivalent level is below the eutectic ratio of 4·3 the first crystals to form from the molten cast iron are those of austenite, containing 2% of carbon. Thus the concentration of carbon in the remaining liquid is raised progressively until it reaches the 4·3 eutectic level. At this stage the eutectic begins to grow at numerous points called nuclei. From these nuclei the graphite flakes grow in clusters known as eutectic cells, until they touch one another and all the liquid has solidified. Figure 11 represents a cluster of graphite flakes in a single eutectic cell and Fig. 12 a collection of eutectic cells as they appear in the cast iron structure.

INOCULATION TREATMENT

Since calcium silicide was developed as the first inoculating agent many other materials are known to function in a similar way. Those commonly employed are ferro-silicon, aluminium, graphite, zirconium and combinations of these. It is known now that ferro-silicon acts as a good inoculant only when it contains 1–2% aluminium and up to 1% calcium. There is a risk with aluminium-containing inoculants that the molten cast iron may pick up hydrogen from moisture in the mould, thereby encouraging the development of subsurface pinholes in the castings.

A good inoculant should possess the following characteristics:

1. Slow fading or reversion with constant results from a standard addition.
2. Enter the molten iron readily with no adverse effects from an over-addition.
3. Produce maximum inoculation and chill reduction from minimum addition.
4. Leave little residue and that easy of removal.
5. Reduce section sensitivity to a minimum.

The last point above has reference to the extreme change in properties and structure which results between thin and thick sections in uninoculated grey cast iron. Very thin sections may be chilled, brittle and unmachinable while the same iron in thick sections may be of open grain, soft and of very low strength. A properly inoculated grey cast iron will not show these extremes.

How Inoculation Works

It has been explained how graphite grows from eutectic cells as grey iron solidifies, and that each eutectic cell grows from a nucleus. Inoculation treatment increases the number of nuclei in the molten cast iron and, therefore, creates a larger number of eutectic cells with their clusters of graphite flakes. In any given volume of cast iron, the larger the number of eutectic cells, the smaller their size must be. Thus inoculation treatment is also a means of controlling grain size and density of structure. As eutectic cell count increases and size diminishes so does the flake size of the A-type graphite become smaller. These changes are accompanied by an increase in mechanical properties coupled with improved machinability.

Adding the Inoculant

Grading or particle size must be suitable according to the means of introduction and the volume of iron to be treated. Crushing between $\frac{1}{8}$ and $\frac{1}{2}$ in. mesh, free from fines and dust, covers most requirements. Maximum inoculation effect occurs immediately after the addition is made and becomes less with the passage of time. So, the treatment should be applied to the molten iron at as late a stage as possible prior to casting. This requirement is fulfilled by adding the inoculant at the furnace spout as the iron runs into the pouring ladle.

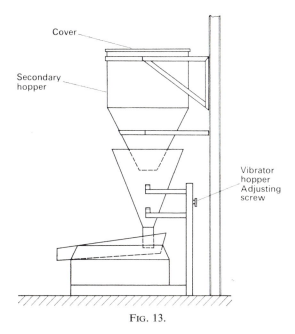

Fig. 13.

Contact with the molten stream should be made at a point as near the tap hole as possible where the flow becomes smooth. Delivery should be at a controlled rate through a steel pipe allowing a minimum vertical fall of 4 ft or 1 m. This gives the inoculant sufficient energy to penetrate the surface of the metal stream rather than float on its surface. An inch or two of molten iron should be allowed to enter the ladle before inoculation commences and treatment should finish by the time three-quarters of the volume of iron to be inoculated has entered the ladle. The molten iron must be delivered from the furnace free from slag

G*

and of good temperature. Also the ladles must be clean and free from slag build-up on the linings.

A device in the form of a hopper with vibratory control is illustrated in Fig. 13. This dispenses the inoculant to the spout delivery tube pictured in Fig. 14. The fading of the inoculating effect has been mentioned. Its effect on chill reduction is strongest at the moment the inoculant has dissolved in the melt. Fading times or reversion periods vary with the different inoculating agents and foundry conditions, from between 15 and 30 minutes. With a view to avoiding this fading effect, mould inoculation has been applied.[18]

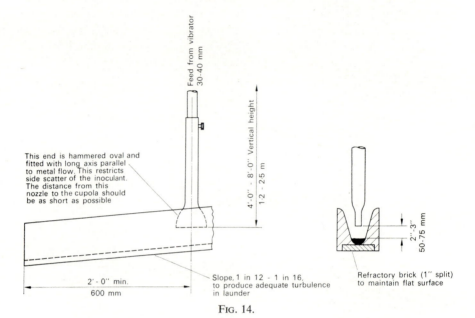

Fig. 14.

Mould inoculation involves the placing of an appropriate small quantity of inoculant inside the mould, usually at the base of the down runner. It can be used independently or as a last second boost to ladle inoculation. Some means must be found of getting the inoculating effect distributed uniformly throughout the iron entering the mould. An attempt towards this end which has achieved some success combines the inoculating agent with a fluxing material into a small pellet.[19] These pellets containing a controlled weight of inoculant are inserted into the running system.

Improved Inoculants

From time to time attempts are made to increase the potency of inoculants and to lengthen the period of time over which the treated iron can be held molten without reversion taking place. Barium-containing inoculants have been introduced for which less rapid fading is claimed.[20] On the other hand, it has been shown that the inoculating power of aluminium free ferro-silicon can be increased by a small percentage addition of strontium.[21] The simultaneous presence of a significant amount of calcium nullifies the effect of the strontium and it is important that the calcium content of any strontium inoculant should be as low as possible. Maximum inoculation is achieved with a 2% addition of a 75% Si ferro-silicon containing between 1% and 4% strontium. Practical tests have shown the superiority of the strontium-containing inoculant in reducing chill and increasing eutectic cell number, particularly in thin sections.

Dawson's work shows that for nodular iron the strontium must be alloyed with the ferro-silicon although for flake graphite irons it is effective when added as a mixture with ferro-silicon. Strontium-containing ferro-silicon is very effective in preventing chill in nodular iron castings of thin section but only when cerium is absent. Very little residue is formed from the use of strontium ferro-silicon which dissolves readily in the melt. The reduced slag formation is attributed to the low content of aluminium and calcium in the alloy.

Inoculation Control Test

The quantity of inoculant to be used is determined by the conditions of operation in each individual case. It depends on the melting unit, the charge composition, the casting section and the inoculating agent. For a given casting section a higher degree of inoculation will be needed as the proportion of steel scrap in the charge increases. A high superheating temperature also makes it necessary to add more inoculant. Thus it becomes apparent that some means of assessing the success of the inoculating treatment is required. Any such test needs to be simple, quick and capable of being conducted on the spot, in the foundry.

The wedge test fulfils these requirements and has become universal in foundries where controlled inoculation technique is applied. It must be performed under strictly standardized conditions which apply particularly to the use of dry, clean sampling ladles, the taking of samples, pouring and quenching procedure. The wedges illustrated in Fig. 15 have been designed to provide a scale of values best suited to the requirements of most foundries. Their dimensions and angles have been so arranged as to give the same chill readings for an iron of constant composition and treatment irrespective of which one is used. Normally, two tests are made, one before and the other after inoculation. The chill is measured on the fractured wedge bar as indicated in Fig. 15. The dimensions may be recorded in millimetres or $\frac{1}{32}$ in.

For maximum strength and toughness in the castings the chill value multiplied by three gives the smallest section thickness which should be poured from that iron. With sections below this thickness the risk of chilling and undue hardness will increase rapidly. A good condition of strength, toughness and hardness with increased machinability is obtained in minimum section thicknesses based on a dimension derived from the inoculated chill value multiplied by four. If the chill shown by a fractured wedge bar is greater in value than half the dimension of the base of the wedge, the next largest wedge size should be used.

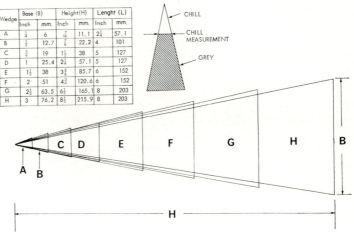

Wedge	Base (B) Inch	mm.	Height(H) Inch	mm.	Lenght (L) Inch	mm.
A	1/4	6	7/16	11.1	2 1/4	57.1
B	1/2	12.7	7/8	22.2	4	101
C	3/4	19	1 1/2	38	5	127
D	1	25.4	2 1/4	57.1	5	127
E	1 1/2	38	3 3/8	85.7	6	152
F	2	51	4 3/4	120.6	6	152
G	2 1/2	63.5	6 1/2	165.1	8	203
H	3	76.2	8 1/2	215.9	8	203

FIG. 15.

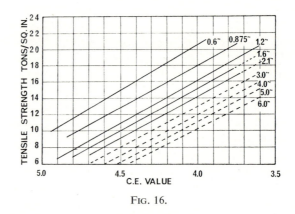

FIG. 16.

STANDARD SPECIFICATIONS FOR CAST IRON

These are based on carbon equivalent value calculated by the formula given on p. 188. B.S. 1452:1961 covers seven grades 10 to 26. The grade number refers to the minimum tensile strength required on a test piece machined from a separately cast 1·2 in. (30 mm) diameter bar. Test bar diameter should vary in accordance with the dominating cross-sectional thickness of the casting represented. The standard diameters range from 0·6 to 2·1 in. Figure 16 shows a chart in which test bars of these diameters have their tensile strength related to the carbon equivalent value.

Response to inoculation treatment is greatest for the low carbon equivalent cast irons with values from 3·5 to 3·9. That is for the high-tensile irons grades 17, 20, 23 and 26. Inoculation technique plays an important part in making possible the production of these high-duty cast irons with constant and reproducible mechanical specifications. It has contributed greatly in enhancing the reputation of cast iron as a reliable material of engineering construction and raising it from its former low rating of from 8 to 10 tons in^2 tensile strength, in the strength of material tables.

SPHEROIDAL GRAPHITE CAST IRON

Some twenty years after the development of the inoculation technique another outstanding discovery helped to extend greatly the sphere of usefulness of cast iron in engineering. This is the process for the production of spheroidal graphite in iron as cast, first described by Morrogh and Williams[22] in 1948. It was accomplished by introducing cerium into the melt. Soon after this announcement was made, it was followed by another giving particulars of the magnesium process developed by International Nickel Limited. Up to this time only one method of producing the spheroidal graphite structure in cast iron was known, namely the long annealing of hard, white-iron castings as practised in the production of white-heart malleable cast iron.

The possibility of being able to obtain spheroidal graphite in the structure of iron as cast enabled it to possess a measurable degree of ductility which has extended its usefulness enormously. For a number of applications it can replace steel castings since it has better machinability, greater damping capacity and a wider range of tensile strength. The forecast that it would replace malleable iron castings has not been fulfilled, but it has created its own sphere of usefulness for large and heavy castings which it would be impractical to produce by the traditional malleable annealing technique.

Both cerium and magnesium, the elements used to impart nodulization to the graphite, combine readily with sulphur. Thus, it is advisable on the grounds of economy in the use of these alloys that the sulphur should be at 0·02% or less in the melt prior to their introduction.

Spheroidizing Elements

Although a number of elements other than cerium and magnesium are known to act as graphite spheroidizers in various degrees of potency it seems unlikely that any of them will attain general use in practice. Misch metal was used as a carrier for cerium in the original process and this also contains the rare-earth metals lanthanum, yttrium, samarium, neodymium and gadolinium. Claims have also been made for lithium, calcium and sodium together with salts of the alkali metals. Despite all this it is the magnesium process which has gained the most commercial use. It has predominated over the cerium process mainly because it can be applied to a wider composition of base iron, the cerium addition being applicable to irons of hypereutectic analysis.

INTRODUCTION OF MAGNESIUM

Since the advent of the use of magnesium as a spheroidizing agent much effort has been expanded on devising an efficient and safe way of adding it to the melt. Magnesium vaporizes at 1120°C which is well below the temperature of molten cast iron. Early attempts to plunge magnesium metal were fraught with danger and some ended disastrously. Nickel was one of the first and most successful carriers for magnesium which is only sparingly soluble in molten iron.

Another widely used vehicle is ferro-silicon containing 45–50% silicon and from 5–20% magnesium. Clark and McClahan[23] have shown that the magnesium recovery of these alloys increases in efficiency as the magnesium content is reduced from 20% to 5% and attains up to 70% with a 2% magnesium ratio. However, the ferro-silicon alloy most used contains about 7% magnesium.

The American Cast Iron Pipe Company[24] has developed a novel method for the introduction of metallic magnesium into molten cast iron. Metallurgical coke is impregnated with 43% by weight of molten magnesium. The coke lumps are of two sizes, 3 in. down and 6 in. down. When plunged into cast iron at 1425°C magnesium recovery was 35% and 45%, respectively, over treatment times of $1\frac{1}{2}$ minutes for the smaller and 3 minutes for the larger lumps.

Another reference[25] to the use of magnesium-impregnated coke gives further details. The metal charge of pig iron, 17% steel scrap and ductile returned scrap is melted in a cold-blast acid-lined cupola. The molten base iron is collected in a 6000 lb receiving ladle from which 1400 lb lots are treated in a preheated pouring ladle. This ladle has a height of twice its diameter. A preheated plunging bell containing $7\frac{3}{4}$ lb of magnesium impregnated coke (43% Mg) and 4 oz misch metal is immersed in the melt. The treatment extends over 90 seconds and the treated iron is transferred to a second ladle and 1% ferro-silicon (85% Si) is added. All treated iron must be poured into the moulds within 8 minutes. Part of the $7\frac{3}{4}$ lb of impregnated coke is used to desulphurize the melt from 0·06% to 0·007% S thereby eliminating treatment with calcium carbide.

In the majority of cases where magnesium metal is added to the melt, pressurized ladles or autoclaves are used. Description of the construction of a pressure vessel is given in *Russian Casting Production*.[26] For ordinary cupola-melted cast iron the pure magnesium addition is 0·15–0·20% by weight of molten metal. Due to the small weight of magnesium and the relatively short time of treatment the temperature drop is 30–50°C. Pressure in the autoclave is important for maximum pick-up of magnesium and should be as follows (Table 1):

TABLE 1.

Metal temperature (°C)	Autoclave pressure atmospheres	Immersion time of Mg in seconds
1300 ± 20	2·0–2·5	120–150
1350 ± 20	3·0–3·5	90–120
1400 ± 20	3·5–4·0	60–90

Volatilization of magnesium ceases at above 3 atmospheres. From a survey made by the American Foundrymen's Society, Ductile Iron Division[27] it appears that what has come to be known as the "sandwich method" of magnesium treatment was employed for 60% of the total tonnage treated (820,000 tons) in the U.S.A. during the period under review. This method employs a ladle with a base lining of sufficient thickness to allow the formation of a recess or pocket as shown in Fig. 17. Into this pocket the requisite quantity of magnesium-bearing alloy is placed and covered with steel punchings. In this way the reaction is confined to the bottom of the ladle where it occurs at a reduced rate and temperature. The molten iron entering the ladle is directed

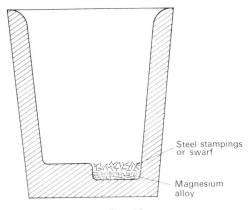

Steel stampings or swarf

Magnesium alloy

FIG. 17.

to the side away from the depression containing the alloy. Recoveries in treating molten iron at 1525–1550°C with magnesium ferro-silicon alloy, as given by R. W. White,[28] are:

Plunging method 25% magnesium recovery.

Sandwich method 51% magnesium recovery.

With the latter method there is an added bonus in converting steel punchings at 1 cent per lb to ductile iron at 6 cent per lb.

Also, in the same publication as the last reference, W. H. Dawson gives results obtained by the injection method of blowing finely divided magnesium through a 2 in. o.d. × $\frac{1}{2}$ in. i.d. graphite tube with a stream of nitrogen. Using a grading of magnesium particles 95%—16 mesh screen and retained on the 40 mesh, magnesium recoveries were 40–50% of that added. A minimum depth of iron in the ladle (12–15 in.) is also a factor and the surface of the molten metal should be 9–12 in. below the ladle rim. With low sulphur iron a suitable injection time is 45 sec.

Again, in the same reference R. Carlson gives details of the plunging method using a magnesium-bearing alloy, for centrifugally cast spheroidal graphite iron pipe in the foundry of the American Cast Iron Pipe Company. By plunging, 31% less Mg alloy is required than by the open-ladle method. The temperature drop for plunging is the same as for open-ladle treatment, namely 40°C. With injection an average temperature drop is 80°C and for the sandwich method using 2% steel punchings, about 68°C.

Returning to reference 27 and the sandwich method. Steel punchings or chippings predominated as a cover material with ferro-silicon next in preference. The most widely used nodularizing agent was magnesium-ferro-silicon with the 9% Mg grade predominant over the 5% Mg grade. Nickel-magnesium alloys were next in order of usage. Pure magnesium was used very little.

Magnesium Introduction by Fluxes

This Company and many other workers have investigated flux compositions as a means of adding magnesium to molten cast iron. The halide salts of magnesium and rare-earth metals are less costly than the metallic elements and the reaction with molten iron is much less violent than from metallic magnesium or the magnesium alloys in general use. Essential requirements for a successful flux technique are efficient operation at temperatures between 1350° and 1450°C and for efficient assimilation, turbulent mixing into the melt. The latter implies injection by pressurized gas. Simply tapping on to flux placed at the bottom of a ladle does not give the degree of metal-flux contact required and flux ratios have to be double those used for injection.

Flux ingredients which have been tried in different combinations and ratios are: halide salts of magnesium, calcium and the rare-earth metals, in some cases aluminium and cryolite are included. Reduction of the halide salts of magnesium is accomplished by the presence of calcium silicide. Aluminium is better absent as it encourages pinhole formation when remaining as a residual in the melt. Rabkin[29] describes a method of reducing magnesium oxide with aluminium at temperatures of 1360–1400°C. A powdered mixture of MgO/CaO/Al is briquetted under a pressure of 1000 kg/cm^2 and held in the bottom of the ladle. Calcined dolomite can be used as a source of MgO. The briquettes are used in a ratio of 2% of the weight of melt to be treated. As a result of the reaction atomic Mg is released.

Another Russian[30] development injects 1·3–1·5 wt.% of a 50–50 mixture of fused $MgCl_2$ and calcium silicide. The process can be operated on a continuous-flow basis, out of contact with the atmosphere. Temperature of the iron at the cupola spout is 1320–1420°C with sulphur at 0·05–0·07%. Optimum assimilations of the Mg and Ca are achieved from flow rates of 2·5–3·0 kg/sec, contact time of flux with the metal stream 2·5–3·0 sec and grading of the powder 4 mm. The amount of flux required depends, other things being equal, on the sulphur and oxygen contents of the base iron. With sulphur content as quoted satisfactory nodularization was obtained by 1·3–1·5 wt.% of the flux. At this ratio the sulphur was reduced to 0·01–0·005% and oxygen content to 0·0015–0·001%. Stable nodular graphite is ensured when the Mg and Ca residuals are at 0·02–0·04 and 0·01–0·03% respectively.

The process carried out as described results in a temperature drop of only 10–20°C. The chilling tendency is low and many castings need no high-temperature annealing even when high ductility is specified. A further advantage claimed is that in the presence of the calcium the magnesium level in the melt can be considerably lower than when magnesium metal or its alloys are used. There is, therefore, less dross formation and a reduced tendency for the occurrence of dross and cope defects in the castings.

Kusaka (Brit. Pat. 1,004,157) uses a flux mixture of MgF_2 10–30%, rare-earth fluorides 10–30% and calcium silicide 40–80%. It is injected in a ratio of 2 wt.% at a melt temperature of 1400–1500°C. The sulphur content of the base iron being 0·04–0·06%. The rare-earth metal and the magnesium fluorides react with the calcium silicide to leave in the melt a weight ratio of Mg:R.E. = 1 : 0·9–4·0.

Electrolysis Method

Allam and Bray[31] have published details of experimental work which they have carried out on graphite nodularization produced by electrolysis of molten cast iron and fused flux. They apply first a basic slag containing CaF_2 to obtain the requisite degree of desulphurization. The sulphur-containing slag is removed and replaced by one containing $MgCl_2$. Residual magnesium content is controlled by the initial sulphur content and the process is applicable to cast irons of a wide range of chemical analysis at a temperature of 1360–1370°C.

For industrial application a basic-lined electric induction furnace is recommended with a fume extractor to remove chlorine fumes evolved from the $MgCl_2$. The process works at a current density of about 1 A/cm² and a voltage range of 7·7–7·8 V for a period of 10–12 minutes.

THEORY OF NODULAR GRAPHITE FORMATION

Figure 18 shows the appearance and structure of a graphite spheroid as seen under the microscope in a sample of spheroidal graphite iron produced by the magnesium process. A more general view of graphite spheroids in a ferritic matrix is shown in Fig. 19. When grey cast iron commences to solidify it consists of a mixture of austenite and iron–carbon eutectic. Graphite forms from the

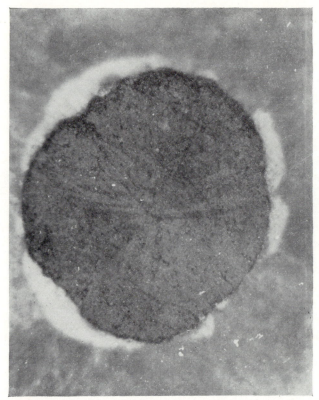

Fig. 18.

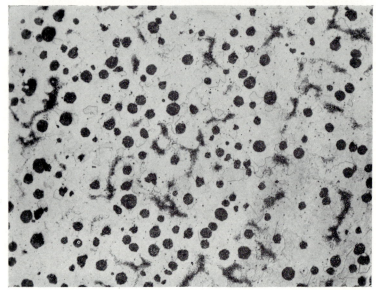

Fig. 19.

eutectic at a number of points called nuclei. Clusters of graphite flakes grow in the liquid austenite until they contact each other. They form what are known as eutectic cells and the ends of the graphite flakes are in direct contact with liquid austenite.

In spheroidal graphite cast iron eutectic cells also form from nuclei present in the austenite matrix. The graphite does not grow in flake clusters with the edges of the flakes in direct contact with liquid austenite but as spheroids surrounded by an envelope of austenite through which carbon must diffuse to feed the growth of the graphite spheroid. Each spheroid represents a single eutectic cell and the cells appear in much greater numbers in spheroidal graphite cast iron than they do in flake graphite irons. Why magnesium or cerium additions to suitable molten base iron compositions can cause these changes is not known, but a number of explanations have been suggested. Some of these deal with grain boundary concentrations and surface tension. Others with heat energy differentials. A review of them is given by S. Banerjee.[32]

A theory put forward by A. A. Gorshkov[33] is that the graphite spheroids are nucleated by tiny bubbles of hydrogen. Pure Ni–C and Co–C alloys have high solubility for hydrogen and give rise to numerous tiny bubbles during solidification. The carbon diffuses into these bubbles and forms graphite nodules. Graphite nodules can nucleate from the molten condition because no metastable phases form such as cementite or iron carbide as in Fe–C alloys. According to Stokes' formula bubbles of 0·001–0·01 mm diameter can survive for hours, since they float upwards at low speeds of 0·4–40 cm/hr. The carbon can penetrate inside these bubbles and completely fill them. Another necessary condition is absence of oxidizing gases or oxygen since graphite burns to form CO or CO_2 in either oxygen or air at 700°C. Molten cast iron contains ample hydrogen to form very large numbers of microscopic bubbles which form nuclei for the formation of graphite spheroids since they already have their interior lined with carbon from the reaction with added magnesium.

$$\text{Mg vapour} + \text{CO gas} = \text{MgO solid} + \text{C solid}$$

Below 1814°C the reaction proceeds from left to right and above it from right to left.

Furthermore, the alkali earth metals employed successfully as nodularizers, all have the property of being able to absorb large quantities of hydrogen at room temperature and to liberate it when the temperature is raised. When immersed in the molten cast iron microscopic bubbles are evolved which form nuclei for the rapidly diffusing carbon which remains as spheroids after the iron solidifies. To prove this theory cerium, lanthanum and other alkali earth spheroidizers were ignited at 1100°C, under helium, to remove the hydrogen. Then when added to cast iron melts previously despheroidized with lead and bismuth, they failed to exert a respheroidizing effect such as could be achieved with the same elements from which the hydrogen had not been removed.

REQUIREMENTS OF THE MOLTEN BASE IRON

It has already been stated that the magnesium process became the most used commercially for the production of spheroidal graphite cast iron because it allows greater latitude of composition of the base irons. Nodular graphite iron and ductile iron are alternative terms for the material, spheroidal graphite iron being contracted to S. G. iron. C. K. Donoho[34] gives the requirements for the molten cast iron that is to be rendered ductile by magnesium additions. The range of total carbon content is wide, from 3·0 to 4·0%. Final silicon levels range from 2·0 to 3·0%. Magnesium has a greater degree of solubility in irons of higher silicon content. Silicon contents of up to 6% are used for increasing heat resistance up to 760°C. Manganese can range from 0·15 to 0·80% but for maximum ductility and fast ferritization by heat treatment the manganese should be low. Manganese on the high side of the range is desirable for highest as cast strength and best response to hardening heat treatments. Phosphorus is usually limited to 0·1% as it reduces elongation and impact strength. The phosphide constituent begins to appear at about 0·11% phosphorus. Sulphur is held as low as possible since magnesium and rare-earth metals react first with sulphur.

The carbon equivalent value is limited to about 4·5% for most purposes as graphite flotation may occur above this level, especially where thick sections cool slowly. It is not a rigid requirement and may be exceeded for very thin sections which freeze quickly. Alloying can be done by means of ladle additions. Nickel increases the strength properties of the as cast material but tends to retard ferritization as it is a pearlite stabilizer. At 3% nickel acicular structures are obtained and at 20% the matrix becomes austenitic. Chromium is not a usual addition to S.G. iron as at about 0·15% it makes castings difficult to ferritize. Molybdenum increases the strength and hardness of heavy sections and improves strength at elevated temperature. It is added up to 1·0% and together with nickel produces a high strength acicular matrix. Copper may be used up to 1·0% as it stabilizes pearlite and gives added strength and hardness but inhibits ferritization by annealing. It has the adverse effect of increasing the effect of trace elements which encourage the formation of degenerate graphite shapes.

INHIBITING TRACE ELEMENTS

During the early stages of the development of S.G. iron serious inconsistencies were encountered in the development of an all-spheroidal graphite structure. Although production technique remained constant the graphite sometimes appeared in modified flake form and not as the required spheres. Eventually this apparently anomalous behaviour was found to be caused by the presence of trace elements in the raw materials. This was a serious drawback and limited the choice of charge materials to a few selected brands and grades which, from experience, were known to be free from the undesirable trace elements. Fortunately it was discovered that small percentage additions of cerium along with the magnesium had the effect of neutralizing the inhibiting effect of the adverse trace elements. To this extent a marriage of the original cerium process with the magnesium process becomes possible.

Table 2, after Donoho, gives details of inhibiting trace elements which encourage the formation of degenerate graphite shapes and indicates maximum levels, harmful levels and levels which can be neutralized by addition of about 0·02% cerium.

TABLE 2.

Inhibiting elements	Max. level %	Harmful level %	Level for neutralization with 0·02% Ce
Lead	0·002	0·01	0·014
Bismuth	0·002	0·005	0·006
Antimony	0·002	0·01	0·015
Aluminium	0·05	0·15	0·50
Titanium	0·05	0·08	0·15
Selenium	0·03	0·05	—
Tellurium	0·03	0·05	—

Cope or Dross Defect

The addition of magnesium to molten cast iron causes the formation of magnesium oxide and magnesium sulphide, both of which float on the surface of the liquid metal. In addition, where heavy sections and slow cooling occurs there is the tendency for the graphite spheroids to float upwards to the top of the mould. They become trapped in the solidifying skin of the casting and remain as a blemish, often associated with gas holes, and commonly known as cope defect. The introduction to the melt of magnesium in excess of the amount needed to spheroidize the graphite aggravates cope defect. According to R. Barton[35] the analysed magnesium content at which 100% spheroidized graphite structure is obtained is generally at 0·035–0·40% for irons with a pretreatment sulphur content in excess of about 0·015%. At or below this sulphur level a reduced magnesium content is sufficient. If the sulphur, after the addition of magnesium is present as magnesium sulphide, this contributes to the magnesium content as analysed. By deducting the stoichiometric magnesium equivalent of the sulphide the calculated magnesium content is obtained. This, to give a fully spheroidized structure, is not markedly influenced by the initial sulphur content. Also the use of hypo- or hypereutectic base irons has little effect on the net magnesium content needed to produce 100% spheroidization.

There is a need, therefore, to standardize and control the levels of sulphur and magnesium in order to minimize the occurrence of cope defect. Similar comments apply to pouring temperatures and running systems. Even so, atmospheric oxidation will take place to form dross. To curtail this a fluxing treatment has been applied to the molten iron in the ladle and volatile additions to the mould cavity with the object of providing a non-oxidizing atmosphere while the casting is being formed. The molten flux is of a composition such as to rapidly absorb the dross in the ladle as it forms during addition of the magnesium.

Reversion to Flake Graphite

Russian workers[36] have investigated the behaviour of carbon, magnesium and sulphur when magnesium-treated iron is held in the molten state. They added 0·3–0·4% by weight, of lump magnesium to a pressurized ladle. In one set of tests the magnesium was added alone and in a second set the magnesium addition was followed by one of 0·5% of 75% ferro-silicon. Carbon decreased with holding time in the first set of tests depending on the amount of magnesium volatilized. Sulphur reduction was 74–92% during the holding period and fell most rapidly over the first 10–20 min. About 45% loss of magnesium occurred during the first 20 min mainly due to desulphurization. Loss of magnesium as vapour ceases when the temperature falls to 1107°C.

Carbon decreased with holding time in the second series of tests and 80% of the sulphur was removed during the first 20 min of holding time. Magnesium losses were 44·2% in the same initial period. Over the entire holding period magnesium loss was less for the second treatment indicating that ferro-silicon additions promote stabilization of the residual magnesium. Inoculation with about 0·5% ferro-silicon has become part of the recognized technique for production of spheroidal graphite cast iron, subsequent to the magnesium addition. Data obtained from these tests indicated that reversion to flake graphite occurs when the residual magnesium content falls below 0·03%.

R. W. White[37] says that 0·01% magnesium is lost every 10 min during which a ladle of treated metal stands at 1480°C. Therefore, the time between furnace and mould should be as short as possible. Recommended gating ratios are 2:4:1·5 with offset pouring basin, sprue well, bottom gating, tangential ingates and short metal flow. Pouring speed of 16 lb/sec/in^2 area of sprue base is suggested. Feeding heads should have a diameter D = casting section plus 2 in., height = $2D$. The connecting neck should be 0·8 casting section size and of maximum length $D/3$, angled to the feeder diameter at 45 degrees.

PRODUCTION TECHNIQUES

A low sulphur content of 0·02% is desirable in the base iron prior to the introduction of the magnesium. Therefore, acid cold-blast cupola-melted irons are unsuitable without a subsequent desulphurization treatment which causes a temperature drop in addition to that occasioned by the treatment with magnesium and the final ferro-silicon inoculation. The basic lined cupola can produce molten iron with the requisite low sulphur but this is usually accompanied by high total carbon ratios and high silicon losses. Thus, if the cupola furnace is used as the melting unit it is better if it acts as part of a duplexing system in which the final adjustments to composition and temperature are made in an auxiliary furnace such as an electric induction or arc furnace.

Cupola-induction furnace duplexing is beneficial in that it makes best use of the advantages of both types of furnace. The cupola is an efficient melting unit but not proficient for superheating. Whereas, the induction furnace is an economical means of imparting superheat, but relatively expensive in producing molten iron from solid charges. Since S.G. iron was first developed, many accounts of production techniques, duplex and otherwise, have been published in the technical literature. The references appended indicate where some of these may be located and studied: acid cupola melting, porous-plug ladle desulphurization and adjustment to carbon, silicon and temperature in an induction furnace, magnesium addition and final inoculation performed in the pouring ladle.[38] Manufacture of S.G. iron castings in a jobbing foundry deals with the making of propellers, turbine casings and liner castings, employing cupola, rocking arc and oil-fired reverberatory furnaces.[39] An account is given of desulphurization of iron melted in an acid-lined, water-cooled, hot-blast cupola, by means of the shaking ladle technique using calcium carbide and adding magnesium by the sandwich method.[40] Both pressure and open ladle techniques are described in an instance where S.G. iron castings have taken the place of steel forgings. The melting unit here being a 5 ton mains frequency electric furnace.[41]

SPECIFICATION AND HEAT TREATMENT

British Standard 2789: 1961 lists six grades of S.G. iron with tensile strengths ranging from 24 to 47 tons/in.^{2} Elongation falls from 17% minimum required for the 24 ton iron to 2% for the highest tensile grade. Average impact values for grades 24 and 27 are given as 10 and 6 ft/lb minimum, respectively. Although in many cases castings will conform to specification without heat treatment, the range covered is a wide one. For the highest elongations and impact values heat treatment is not unusual. Standardized treatments given by Handley and Jones[42] are: ferritization or full anneal, heat up to 920°C in 6 hr and hold at this temperature for 4 hr, cool to 720°C in 4 hr and hold for 16 hr, cool to 500°C in 4 hr. For normalizing heat to 920°C in 6 hr and hold for 4 hr followed by air cooling. With full anneal, the pearlite retained should not exceed 5% of the structure.

REFERENCES

1. PERRY, W. R., *The British Foundryman*, **59**, 239–44 (1967).
2. EVANS, N. L., *Foundry Trade Journal*, 1–22 Dec. and 5 Jan. 1932–3.
3. LEYSHON, H. J., and SELBY, M. J., *BCIRA Journal* **15**, 108–14 (1967).
4. VOLIANIK, M. N., *Modern Castings* **51**, 52–55 (1967).
5. TONKS, W. G., *BCIRA Journal of Research and Development* **6**, 214–25 (1956).
6. COATES, R. B., and LEYSHON, H. J., *BCIRA Journal* **12**, 479–94 (1964).

7. EKETORP, S., *Journal of Metals* **12**, 44–48 (1960).
8. COATES, R. B., and LEYSHON, H. J., *BCIRA Journal* **11**, 635–40 (1963).
9. MORGAN, A. D., *BCIRA Journal* **10**, 438–45 (1962).
10. TONKS, W. G., *BCIRA Journal of Research & Development* **6**, 214–25 (1956).
11. ROTE F. B., and WOOD, W. P., *Trans. American Soc. for Metals* 35 (1945).
12. DAWSON, J. V., and SMITH, L. W. L., *BCIRA Journal of Research and Development* **6**, 226–48 (1956).
13. HUGONY, E., *Fonderie* **239**, 38 (1966).
14. DILEWIJNS, J. A., and DE SY, A. L., *Foundry Trade Journal*, **123**, 37–44 and 153–61 (1967).
15. HENON, G., *Fonderie* **257**, 255–71 (1967).
16. GILBERT, G. N., *BCIRA Journal* **12**, 298–312 (1964).
17. GLOVER, D. J., *BCIRA Journal* **14**, 718–31 (1966).
18. DAWSON, J. V., *BCIRA Journal* **14**, 546–54 (1966).
19. DAWSON, J. V., and MAITRA, S., *The British Foundryman* **59**, 117–27 (1967).
20. SCHAUM, J. H., *Modern Castings* **41**, 134 (1962).
21. DAWSON, J. V., *Modern Castings* **49**, 171–7 (1966).
22. MORROGH, H., and WILLIAMS, W. J., *Journal of the Iron & Steel Institute* **158**, 306–22 (1948).
23. CLARK, R., and McCLAHAN, I., *Modern Castings* **48**, 78–81 (1965).
24. SNOW, W. E., *Foundry* **92**, 99–103 (1964).
25. DALTON, R. F., *Modern Castings* **48**, 50–51 (1965).
26. SHABLINSKII, V. B., and PETROVA, E. V., *Russian Casting Production* 259–61 (1961).
27. A.F.S. Committee 12H, *Modern Castings* **52**, 110–11 (1967.)
28. WHITE, R. W., *Modern Castings* **44**, 628–31 (1963).
29. RABKIN, A., *Russian Castings Production* 54–55 (1964).
30. MIL'MAN, B. S., *et al. Russian Castings Production* 55–56 (1967).
31. ALLAM, A., and BRAY, H. J., *Iron and Steel* **40**, 516–20 (1967) and **41**, 12–16 (1968).
32. BANERJEE, S., *The British Foundryman* **58**, 334–43 (1965).
33. GORSHKOV, A. A., *Russian Castings Production* 201–3 (1965).
34. DONOHO, C. K., *Modern Castings* **46**, 608–10 (1964).
35. BARTON, R., *BCIRA Journal* **11**, 741, 66 (1963).
36. KHROPOV, A., and BEDAREV, V. J., *Russian Castings Production* 164–7 (1963).
37. WHITE, R. W., *Foundry Trade Journal* **116**, 601 (1964).
38. COATES, R. B., and LEYSHON, H. J., *Foundry Trade Journal* **121**, 273–7 (1966).
39. NICHOLLS, C., *The British Foundryman* **60**, 1–10 (1967).
40. CURRY, T. W., *Foundry* **93**, 82–86 (1965).
41. TURNER, H., *The British Foundryman* **56**, 510–20 (1963).
42. HANDLEY, R. L., and JONES, T. I., *Foundry Trade Journal* **116**, 595–7 (1964).

C. TREATMENT OF NON-FERROUS HEAVY METALS

CHAPTER 6

COPPER AND HEAVY NON-FERROUS CASTING ALLOYS

J. L. Francis

High Conductivity Copper Castings

Relatively few castings are produced from elementary metals of purity greater than 99%. However, for the electrical industry castings of high conductivity are required and these must be poured from copper containing a minimum of impurities. It is not easy to obtain sound castings from pure copper unless the correct melting and fluxing technique is known and applied.

The assessment of electrical conductivity is based on the resistance of pure annealed copper at 20°C which has a value of 1·724 microhms/cm cube and is taken as 100%. It is known as the International Annealed Copper Standard (I.A.C.S.). Any less conductivity value has a higher resistance and is expressed in reciprocal percentage form, i.e. double the standard resistance would give 50% conductivity.

Thermal conductivity is also a requirement of copper castings. In most cases where maximum electrical conductivity has been achieved the optimum thermal conductivity will have been obtained as well.

Porosity Due to Gases

Any cavities or unsoundness present in the metal will reduce conductivity very greatly. Such defects can result from liquid shrinkage and gas evolution during solidification. The former is a natural phenomenon of cooling liquids and can be overcome by the provision of adequate feed metal, but gas porosity is more insidious.

Molten copper of 98–99% purity, at 100°C above its melting point (M.P. 1083°C) can hold in solution 10 cm³/100 g of hydrogen gas. About half of this quantity is expelled from solution as the molten copper cools to its solidifying temperature. During the phase change from liquid to solid a further 3 cm³/100 g of hydrogen is evolved leaving approximately 2 cm³/100 g which is maintained in solid solution. Porosity due to hydrogen is most likely to result from that which comes out of solution during the change from liquid to solid metal because at this point crystals are forming and the escape of gas is thus impeded.

Steam Reaction

Another gas which must receive consideration is oxygen which combines with molten copper to form cuprous oxide (Cu_2O) which is soluble in the melt and forms a eutectic containing 0.39% oxygen when the copper solidifies. No oxygen is liberated from the cuprous oxide dissolved in pure molten copper, provided that hydrogen is absent. If hydrogen comes out of solution in the presence of cuprous oxide it reacts with it to form water vapour or steam according to the equation:

$$2\,H + Cu_2O = 2\,Cu + H_2O$$

This is known as the "steam reaction" and causes gross porosity in high conductivity copper castings. Hydrogen released from solution in molten metal is in the nascent, atomic state, in which its activity and ability to diffuse is very great. Although steam is not soluble in molten copper it causes porosity when not able to escape from the castings on which a solid skin will have formed at this stage.

Little harm results from the presence alone of either oxygen or hydrogen. It is the simultaneous presence of both cuprous oxide and atomic hydrogen which causes the trouble. Therefore, to avoid steam reaction it is sufficient to ensure freedom from one of the gases only.

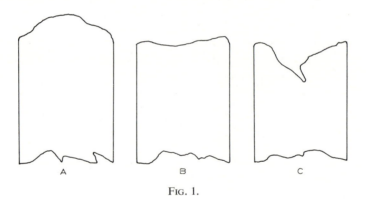

Fig. 1.

Fluxing and Deoxidation Technique

The standard procedure developed by this company more than 35 years ago aims at keeping the cuprous oxide formation to the lowest level possible. Thus, melting is performed under cover of a reducing flux which inhibits pick-up of oxygen. Immediately prior to pouring, freedom from oxide is ensured by plunging a standard quantity of special deoxidant contained in a copper tube. Importance is attached to having the deoxidant available in a form which can be inserted well below the surface of the melt so as to achieve maximum efficiency from the treatment.

A means of measuring the success of the technique is available in the form of a simple test which can be performed easily and quickly in the foundry before any castings are poured. It takes the form of a mould for a small test-piece 2 in. diameter by 6 in. long (50 mm × 150 mm) cast vertically and shown in the diagrams Figs. 1A, B, and C. A gassy melt, indicative of steam reaction, will solidify with a convex surface as Fig. 1A. If the surface is level or slightly sunken as in Fig. 1B, the oxygen content has been reduced to the low level of 0·008%. Complete deoxidation gives a pronounced sink as shown by Fig. 1C. Having obtained a deoxidized melt, it should be transferred into the moulds with a minimum of turbulence and contact with the oxygen of the atmosphere. Transference from one vessel to another should be avoided and as short a stream as possible used during the casting operation. Otherwise oxygen may re-enter the metal to form more cuprous oxide.

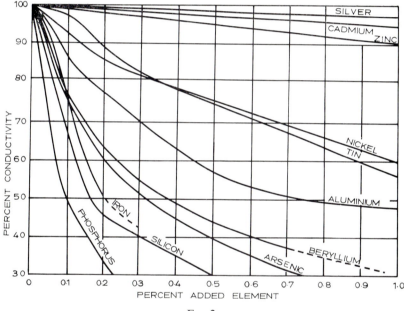

FIG. 2.

Effect of Residual Metals on Electrical Conductivity

Phosphorus as phosphor-tin or phosphor-copper is a commonly used deoxidant for copper and its alloys, and is very effective. Unfortunately small amounts of residual phosphorus reduce the electrical conductivity of copper to a low level. Of the eleven elements given in Fig. 2 phosphorus exerts the greatest deleterious action and 0·05% of it lowers the conductivity by about 30%. Thus, if phosphorus is employed as a deoxidant the treatment must be carefully controlled to the stage represented by the test-piece shown in Fig. 1B and subsequently carried to completion, as Fig. 1C, by plunging a deoxidant such as calcium which, in small residual quantities, decreases the electrical conductivity only slightly.

Alternative Degassing Techniques

Lithium acts as a deoxidant and a degasser by combining with both oxygen and hydrogen. It is available in a form convenient for plunging into the melt and is used in a ratio of about 0·02% by weight. It has a negligible effect on conductivity but arrests the tendency of complex shaped castings to form cracks at intersections.

Hydrogen can also be removed from molten copper by gas flushing. Nitrogen is bubbled through the melt and the hydrogen diffuses into the bubbles and so passes out of the melt in the bubble spaces. Unless the nitrogen is passed through a moisture absorber it may carry water with it and so nullify its effectiveness. Commercial nitrogen is not always completely dry.

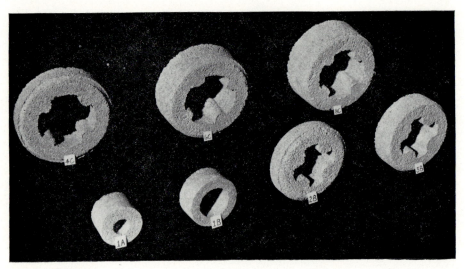

FIG. 3.

To overcome this risk and to avoid the inconvenience of gas cylinders, pipes, valves and injection lances, in a foundry, a proprietory degassing briquette has been developed as shown in Fig. 3. It is applied by simply plunging it into the melt where it evolves scavenging gases. Figure 3A shows the construction of suitable plumbago plungers. This operation can be performed by the furnace-man and needs no technician as for flushing with gas from cylinders. Where gas flushing is used deoxidation treatment is applied subsequently so as to remove any oxidation resulting from the turbulence caused by gas bubbling.

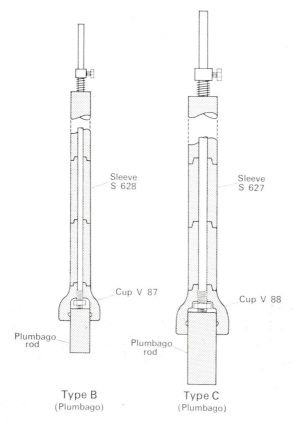

Sleeve
S 628

Sleeve
S 627

Cup V 87

Cup V 88

Plumbago
rod

Plumbago
rod

Type B
(Plumbago)

Type C
(Plumbago)

FIG. 3a.

High Conductivity Copper Alloys

Some conditions of use require greater hardness, strength and wear properties than possessed by pure copper. To meet these the elements cadmium, chromium and tellurium are added; the principal high conductivity alloys are:

Cadmium copper containing 0·7 to 1% Cd
Chromium copper containing 0·4 to 0·8% Cr
Tellurium copper containing 0·3 to 0·7% Te

All these have I.A.C.S. conductivity values above 80%. For cadmium and tellurium copper the melting technique is as for high conductivity copper. In the case of chromium copper a special flux is available for suppressing the formation of chromium oxide.

Commercial Grade Copper Castings

Copper castings of commercial quality are those for which the maximum in electrical or thermal conductivity is not required. They may contain about

2 % of other elements of which tin and zinc are commonly present. These increase the hardness and improve machinability. Also they make it much less difficult to produce sound castings. In fact the melting, fluxing and deoxidizing practice for commercial quality copper castings is exactly the same as that for copper-tin bronzes and gun-metals. Pouring temperatures suggested for copper castings range from 1250°C for light sections up to $\frac{1}{2}$ in., down to 1150°C for heavy sections of $1\frac{1}{2}$ in. and above.

Copper-Tin Bronzes and Gun-metals

Listed in British Standard Specification BS 1400 are three copper-tin bronzes containing lead, two containing zinc and three alloys containing tin, lead and zinc. The first three are known as leaded bronzes, the next two as gun-metals and the final three as leaded gun-metals. All eight of these alloys can be given the same fluxing and deoxidizing technique.

Gases which cause Porosity

In the *Proceedings of the Institute of Metals*, 1944, April and August, two papers were published on the degassing of bronze, one by W. T. Pell-Walpole and the other by W. A. Baker and F. C. Child. The gases commonly present in melting atmospheres are water vapour, hydrogen, carbon monoxide, carbon dioxide, sulphur dioxide and nitrogen. Carbon dioxide and nitrogen are inert but the other four gases can cause unsoundness.

Hydrogen Porosity

The main cause of gas porosity is hydrogen. As we have already seen in the case of copper, hydrogen dissolves in the melt increasingly as temperature rises and is expelled from solution as the metal solidifies. Hydrogen is derived mainly from the dissociation of water vapour according to the reversible reaction

$$H_2O \rightleftharpoons HOH$$

Water vapour is present in variable amounts in atmospheric air and is evolved from damp refractories, moist fluxes and undried furnace tools. It forms in considerable quantities from the combustion of fuels such as coke, gas and oil. Also it may emanate from corrosion products, grease and dirt on scrap used in the furnace charges.

Control of Hydrogen by Oxidation

The authors already mentioned have shown that the presence of oxygen in bronze limits the amount of hydrogen that can exist in equilibrium with it (Fig. 4). However, the presence of zinc lowers the solubility of oxygen and if it is present above 0·5% degassification by oxidation becomes less effective. The presence of phosphorus acts similarly. Even so, oxidizing fluxing influences the hydrogen content in other ways.

Consider again the equation representing dissociation of water vapour. It is reversible and can work from left to right or vice versa. Its state of equilibrium can be upset by altering the balance on either side of the equation. Thus, if oxygen is added to the right-hand side it tends to drive the reaction from right to left. In other words, it helps to keep water vapour from splitting up into hydrogen and oxygen. Water vapour is insoluble in bronze and can escape from the melt taking along with it the hydrogen which might have been evolved by dissociation and dissolved in the metal.

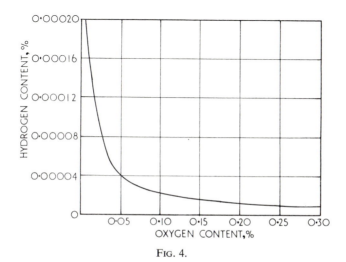

Fig. 4.

Furthermore, oxygen generating fluxes for bronze are available in block form. These blocks can be placed on the bottom, inside crucibles preheated to redness. Ingots and scrap are charged on top and the briquetted flux remains adhering to the base of the crucible during the melting-down period.

Refractory crucibles for bronze melting are of clay-graphite or silicon carbide construction. In either case the flux in direct contact with the refractory at furnace temperature will evolve oxygen which will react with the carbon of the refractory to form carbon monoxide gas. Bubbles of this will rise through the molten charge to produce a gas scavenging action. Hydrogen dissolved in the molten bronze will diffuse into the bubble space in the atomic state. Once inside the bubble it becomes molecular and loses its power of diffusion. Thus confined it passes to atmosphere with the passage upwards of the gas bubble.

Degassing by this method relies for its efficiency on a steady supply of small gas bubbles rather than fast flowing large ones. The former condition is more readily achieved from a slowly disintegrating submerged flux block than from

H

gas under pressure bubbled into the melt through graphite lances. There exists also the possibility that some dissolved hydrogen may react directly with the evolved oxygen to form water vapour. Being insoluble it too will form bubbles which can enclose further hydrogen and so conduct it away to atmosphere.

Oxidation–Deoxidation Flux Degassing

After melting under oxidizing conditions a deoxidation treatment will be necessary to remove excess oxygen. This is accomplished by plunging the requisite amount of phosphor copper into the melt. By reason of the use of the opposed oxidation and reduction treatments the procedure has come to be known as the oxidation–deoxidation flux degassing technique. Both aspects are standardized and under control when the recommended fluxes and deoxidants are applied in the correct ratios. As a result the bronze when poured will contain a residual quantity of about 0·02% phosphorus at which level the oxygen remaining will be negligible.

An alternative procedure for melting bronze and gun-metal can be used. It consists of the use of a neutral or reducing flux cover and subsequent degassing of the melt with nitrogen at a rate of 1 litre per kg of melt. Or an appropriate size of carbon dioxide generating block (Fig. 3) can be plunged. Following degassing the deoxidation treatment is carried out.

When oxidizing flux degassing was first advocated, it was considered that there would be a steady reduction of phosphorus content, from phosphor bronzes before hydrogen elimination could become effective. However, experience has shown that the theoretical combination of phosphorus with oxygen does not prevent removal of hydrogen and the loss of phosphorus is relatively slight. It is also known now that the decrease of zinc, tin and lead from bronzes when employing oxidizing fluxes is comparatively little.

Melting Rate

Speed of melting is a most important factor. We have seen that solution of hydrogen in the liquid metal is a function of both time and temperature. At any given temperature above the melting point there is a limit to the volume of hydrogen which can be dissolved, but this volume will not be taken up if insufficient time is allowed. With fast melting saturation will not be attained.

Oil- or gas-fired crucible furnaces should melt a 100 kg bronze in 50 min or less. A reverberatory furnace ought to melt 500 kg charge in 60 min at most. Indirect arc furnaces melt at a rate of 250 kg in 30 min. In a good oil- or gas-fired furnace it requires two-thirds of the total melting time to bring the charge to the full plastic state and the remaining one-third to superheat to pouring temperature. Once the charge has attained the required pouring temperature it should be poured into the mould without delay, it ought never to stand at temperature in the furnace waiting for the mould to be ready.

Metal–Mould Reaction

Even though completely degassed, molten bronze poured into a sand mould may still produce castings containing gas porosity. This is because it can pick up and dissolve hydrogen from the moisture in the moulding sand. Moisture is present in both green and dried sand moulds, as free water in the former and combined with the clay bond in the latter. Thus when the molten bronze enters the mould it is immediately in contact with an atmosphere containing steam.

As we have already seen, dissociation of water vapour frees hydrogen and oxygen. The latter will combine with the element in the bronze which has the greatest affinity for it, leaving the hydrogen to dissolve in the molten metal. In copper base alloys containing phosphorus from the deoxidation treatment with phosphor copper, a phosphate film or envelope may form around the casting. Such a film, with a lowered solidifying point will not be capable of stifling the water vapour reaction.[1] Any delay or failure to produce a protective skin on the alloy at the metal–sand interface will allow the reaction to proceed and hydrogen to pass into the metal only to be expelled as porosity in the solidifying casting. The presence of zinc with its relatively high affinity for oxygen increases the tendency for dissociation of water vapour to take place. Hydrocarbon gases evolved from cores are dissociated and evolve atomic hydrogen in a similar way.

Control of Phosphorus Content

A bronze or gun-metal containing 0·02 % of residual phosphorus is considered to be completely deoxidized. Phosphorus levels in excess of this ratio contribute to the incidence of metal–mould reaction. Therefore the importance of being able to determine how much phosphor copper to add as a deoxidant is obvious. The test as used for high conductivity copper (Fig. 2) is one way of obtaining an indication. Another is to use standard melting and fluxing procedure such as can be easily and accurately applied by the operator in charge of melting.

Oxidizing flux blocks of standard weight have been mentioned. So also has the method of controlled deoxidation by the plunging of correctly proportioned cartridges of granulated phosphor copper. From this procedure, the charge when ready for pouring will have an oxygen content of about 0·04%. The appropriate quantity of deoxidant for this condition can be predetermined.

Where it is the practice to rely on an oxidizing furnace atmosphere by passing excess air through the burners, the degree of oxidation of the melt is not so well under control. Consequently it may contain a level of oxygen anywhere between 0·01 and 0·1%. Deoxidation, in these circumstances, may be under or over done. Working on a factor of ignorance, the latter is the most likely and a high residual phosphorus level will encourage hydrogen porosity from metal–mould reaction, especially for castings of over $\frac{1}{2}$ in. section. That the indiscriminate addition of phosphorus to gas-free gun-metal causes hydrogen porosity, has been shown by Swinyard[2] and Rutherford,[3] particularly where thick sections and slow cooling prevail.

Effect of Silicon and Lead

Any condition which delays or prevents formation of a continuous protective film on the casting at the metal–mould interface, will contribute to metal–mould reaction. Phosphorus acts in this way and so do traces of silicon,[4] especially when lead is also present, as it is in most gun-metal alloys. It is thought that the oxidation of these two elements cause production of a film having a delayed solidification due to the formation of lead silicate. Hudson has shown that high lead bronzes are very sensitive to the presence of traces of silicon, especially where castings of thick sections are concerned. The nature of the refractories of crucibles and furnace linings should be considered from the point of view of silicon pick-up by the melt since only 0·005 % can induce metal–mould reaction.

Inhibiting Mould and Core Dressing

A mould and core dressing developed by the British Non-Ferrous Metals Research Association can be used to inhibit the onset of metal–mould reaction. It consists of a finely divided aluminium–10 % magnesium alloy suspended in a spirit carrier. The mechanism of its inhibiting action is based on the preferential reaction of magnesium with steam and hydrocarbon gas to liberate molecular instead of atomic hydrogen as a product of the dissociation. Molecular hydrogen does not dissolve or diffuse in molten metal.

Nickel in Leaded Gun-Metals

Demands that castings withstand more exacting conditions of service and are produced of higher quality are insistent. The traditional gun-metal alloys developed by empirical methods over past years, have wide freezing ranges. They are thus susceptible to metal–mould reaction, increasingly, as casting sections become greater and suffer, as a result, from interdendritic porosity, which in turn causes low mechanical properties and leakage under pressure.

To remedy this situation the British Non-Ferrous Metals Research Association[5] sponsored by the International Nickel Co. have developed a nickel containing gun-metal which has become embodied in British Standard Specifications as BS 1400 L.G.4. The idea is to close the freezing range by reducing the sum of the tin and zinc contents to not more than 10 % and to 8·5 % in really thick sections. The requisite quantity of nickel can easily be introduced by means of 50–50 nickel copper shot.

Careful selection of a gun-metal alloy of suitable composition to the work in hand has become an important factor in the production of sound castings. Feeding heads of large volume do no good on alloys of long freezing range. Feeders, where applied, should be high enough to provide head pressure and coupled with the use of denseners.

Phosphor Bronzes

Four grades of phosphor bronze are covered by British Standard Specification 1400. They are designated P.B.1, P.B.2, P.B.3 and L.P.B.1. The latter representing leaded phosphor bronze. Except in the case of the latter, the phosphor bronzes are relatively pure copper-tin alloys with less than 0·5% of zinc and rarely more than 0·5% of phosphorus. The low zinc levels make these alloys much less sensitive to hydrogen porosity from metal–mould reaction.

Melting may be performed under somewhat less oxidizing conditions than for gun-metals. If it is considered desirable to replace any phosphorus oxidized by this treatment it can readily be reintroduced by plunging standard cartridges containing predetermined weights of granular phosphor copper.

Inverse Segregation

Often called "tin sweat", this condition may occur in phosphor bronzes of high phosphorus and tin contents. It appears as beads of exudation on the outside of castings. These exudations have tin and phosphorus ratios much above the average levels of the parent alloy and are, in fact, beads of the low freezing point alpha-delta eutectoid.

The phenomenon is associated with gassy melts. Pouring temperatures, speed of solidification, high temperature gradients and a wide freezing range are all factors influencing this defect. However, well-degassed melts poured at the correct temperature are largely immune from it. Pouring temperatures are in the range 1100 to 1040°C as casting sections increase from $\frac{1}{2}$ in up to more than $1\frac{1}{2}$ in.

Aluminium Bronze

Two grades appear in BS 1400, A.B.1 and A.B.2. Both have the same specified range of aluminium (8·5 to 10·5%) but A.B.2 contains higher levels of iron and nickel. The as-cast strength of these alloys is beyond that of mild steel and the mechanical properties of the A.B.2 alloy can be further modified by heat treatment. Nowadays, charges are made up from ingot purchased to specification, plus foundry back scrap. This is the case for the majority of non-ferrous and light casting alloys and avoids the mixing and homogeneity problems involved in compounding alloys from their constituent elements.

Gas Absorption

Here again, the greatest danger is from hydrogen evolved from the dissociation of water vapour. There is, however, a progressive reduction in both liquid and solid solubility of hydrogen with increase in aluminium content. At about 8% aluminium the solubility is about one-half of that for copper.[6] Sulphur dioxide[7] is absorbed by aluminium bronze to form aluminium oxide and aluminium sulphide which compounds remain in the metal or in the slag. As with copper, much of the gas unsoundness experienced in casting is possibly due to hydrogen expelled during cooling reacting with oxides present to form water vapour. Thus, small amounts of gas can result in extensive porosity in the solidified castings.

Oxide Formation due to Turbulence

In all alloys containing aluminium, turbulence arising from pouring, stirring, or any other cause, must always be avoided. This means that gating has to be applied at the base of moulds. Thus, temperature gradients will be opposed to those needed for progressive solidification and feeding. Undue agitation of the molten alloy causes formation of aluminium oxide. Once formed this oxide tends to become trapped in the melt and forms films in the castings. These reduce mechanical properties, cause leakage under pressure and blemishes on polished surfaces.

Aluminium oxide is relatively strong and forms a resistant skin or envelope around splashes and globules of molten metal. This skin is tough enough to prevent joining up to the main stream, so that seams, folds and cold shuts will result.

Fluxing Treatment

Few elements have a stronger affinity for oxygen than aluminium and so it is not practical to apply deoxidation. Fluxes capable of a solvent action are employed and these are based on fluorides. Clean charges may, with care, be melted under a cover of charcoal provided that the charcoal is not allowed to remain for long periods on the molten surface. If it does, the iron, which is a constituent of aluminium bronze, may pick up carbon to form hard spots in the castings.

As aluminium is itself a powerful deoxidant, deoxidation in the accepted sense is not applicable to aluminium bronzes. Lithium has been mentioned in connection with high conductivity copper and this can be usefully applied to aluminium bronzes for combining with both oxygen and hydrogen.

Phosphorus is not recommended as a deoxidant as it may lead to the separation of phosphides. Although it reduces the solubility of hydrogen there is a risk that residual phosphorus, as phosphide, will lower ductility and resistance to impact and in addition give rise to hot shortness. Silicon if employed for removal of oxygen may act similarly on the mechanical properties.

Manganese and boron are useful additions. Despite their inability to reduce aluminium oxide, they increase the fluidity of the melt, thereby assisting the oxide to separate out and float to the surface where it may be skimmed off.

Unfortunately, boron can encourage the formation of hard spots as indicated in the section dealing with manganese bronze or high tensile brasses (p. 223).

This applies particularly to fluxes containing boron salts and products for fluxing these alloys are now available which are free from the elements fluorine and boron.

Alloy Additions

Iron is included in the specified compositions from 1·5 to 5·5%. Its solubility in aluminium bronze is considerably greater than in copper. It acts as a grain refiner and prevents self-annealing, a condition occurring in slowly cooled castings resulting in the resolution of the beta constituent into the alpha-delta eutectoid. By suppressing this change the iron helps to improve the mechanical properties, particularly in the case of heat-treated alloys. One per cent of iron raises the tensile strength by from 2·0 to 2·5 tons/in². This rate of increase is maintained almost proportionately for iron additions up to about 4%.

Nickel is regarded as a desirable constituent and is introduced up to 6%, especially where iron is also present. The combination of iron and nickel helps in maintaining ductility in heat treated products and in general provides a means of exercising greater control.

Lead has been added to improve bearing properties and increase machinability. It appears in the structure as a separate constituent since its solubility is practically nil. There is thus the problem of overcoming its tendency towards segregation and it is not desirable as a regular ingredient in castings. Also its effect on alloys for rolling and forging is to encourage cracking.

Degassing

To avoid the risk of gas porosity in large castings, a very low level of dissolved gas must be stipulated. Although the mean gas content of a melt may be considered to be within safe limits, account must be taken of the possibility of much higher levels at places where slow cooling occurs. Hydrogen expelled from the faster cooling thinner sections can diffuse into metal still liquid in heavier sections of the casting. Large marine propeller castings are a case in point. The relatively thin blades solidify long before the heavy boss and hydrogen may diffuse to this region imparting locally a gas content high enough to cause serious porosity when the boss finally becomes solid.

For such large castings, dissolved gas must be at a minimum. To achieve this a degassing treatment is necessary and scavenging with an inert gas is the usual technique. Either nitrogen or carbon dioxide can be used but with the former there is the possibility of interaction between nitrogen and aluminium, at high temperatures. The use of carbon dioxide does not carry this risk and the technique, described previously, of plunging a CO_2 generating briquette (Fig. 3) provides a useful alternative.

Pouring temperatures for aluminium bronze castings lie within the range 1150° to 1250°C. The higher temperature is suitable for castings having section thickness of $\frac{1}{2}$ in. or less, and for casting sections of $1\frac{1}{2}$ in. or greater the lower temperature is used.

Silicon Bronze

These alloys were developed as an alternative to tin bronzes and with a view to the conservation of tin. Alloyed with copper, silicon is about $2\frac{1}{2}$ times as effective as tin in conferring strength and hardness. Consequently the best known compositions contain about 4% of silicon. British Standard Specification BS 1400 lists only one example in which the silicon content ranges from 1·5 to 5%, along with 5% of zinc, 2·5% of iron and 1·5% of manganese.

Three proprietary alloys are available, namely P.M.G. Metal of Vickers-Armstrong Ltd., Everdur of American Brass Co. and Herculoy of Revere Copper and Brass Inc. U.S.A. Apart from an alternative to tin bronzes these alloys have distinctive characteristics of their own which show advantages over those of gun-metal. They can be readily cold worked, forged and extruded in which state their strength and ductility are similar to that of mild steel but combined with a much greater resistance to corrosion.

Strength at elevated temperature (250°C) is superior to that of Admiralty gun-metal. At sub-zero temperatures both strength and ductility increase, a valuable asset for low temperature applications. The fluidity of silicon bronze is high and its freezing range is short. Both these properties assist in the production of sound castings.

Melting Practice

Precautions against hydrogen absorption must be maintained. These involve fast melting in a slightly oxidizing atmosphere and avoidance of over-heating. Charcoal should not be used as a cover. When properly melted the alloy is relatively free from dross but about 1% of a flux containing a proportion of fluoride salts is recommended. Three-quarters of this should be added as soon as the charge begins to melt and the remaining quarter plunged prior to skimming and treating with phosphor copper. The latter is used not so much as a deoxidant but to give a final cleansing action. Large melts can be degassed by flushing with nitrogen or carbon dioxide in the way already described. An average casting temperature is 1150°C but thin castings can be poured somewhat higher and thick sectioned work at 1110°C.

Nickel Bronzes

These have been defined as alloys of copper with 10% or more of nickel and a zinc content less than that of the nickel. Bronzes of lower nickel content have already been dealt with as nickel containing gun-metals. The required nickel content is conveniently introduced in the form of 50–50 nickel copper shot.

As the nickel content increases so does the ability of the alloys to dissolve carbon and sulphur. Carbon can be picked up from charcoal covers and in the presence of oxides can be the source of carbon monoxide gas porosity in the castings. When melting with coke as fuel this must not be allowed to enter the crucible along with the charge.

Melting Procedure

The charge should be melted down in an oxidizing atmosphere derived from the addition of about 2% of its weight of an oxidizing flux. When melting in crucibles the crucible is covered with the bottom of an old pot. Superheat as rapidly as possible to an adequate temperature which depends on the nickel content. With 20% nickel heat to 1400°C and for Monel metal (60% Ni) a temperature as high as 1650°C is required.

Degassing is accomplished by gas flushing in the manner previously described or special oxygen evolving fluxes can be plunged. The melt is then deoxidized by plunging cartridges containing slugs of magnesium adjusted in weight to suit the volume of metal being treated. The magnesium plays a dual role in combining with both oxygen and sulphur.

Manganese Bronze

As these alloys contain about 55% copper and 30% zinc, they are not bronzes but brasses. Among the other elements present may be aluminium, manganese, iron, tin and nickel. Because of the high mechanical properties imparted by these added metals, the series of alloys produced is more appropriately named high tensile brass. BS 1400 lists three grades H.T.B.1, H.T.B.2, and H.T.B.3, with minimum tensile strengths of 30, 38 and 48 tons/in^2, respectively. These values are accompanied by good ductility, thereby providing an as-cast material with strength properties comparable with those possessed by mild steel.

Aluminium is one of the essential constituents, conferring on the alloy the undesirable property of dross formation. In consequence the molten metal must not be subjected to turbulence from pouring or stirring or the formation of aluminium oxide will be augmented. To reduce this hazard, ingates are applied to the bottom of moulds. Due to the relatively low melting point and high zinc content, gas absorption is not a problem.

Copper–zinc alloys form solid solutions in all proportions and develop metallographic structures composed of alpha and beta. The high tensile brasses conform to this, equivalent zinc values are given to the alloying elements as follows (Table 1).[8]

To find the zinc equivalent of a complex brass alloy, add the zinc equivalents of the extra elements to the percentage of actual zinc; if this total is X then the

zinc equivalent of the alloy $= \dfrac{100\,X}{X + \text{Cu}\%}.$ Equivalent zinc values of 46·5%

H*

and above will produce a beta structure. To obtain 15% of alpha the zinc equivalent must be below 45%.

TABLE 1.

Alloying metal	Approximate zinc equivalent
1% silicon	10% zinc
1% aluminium	6% zinc
1% tin	2% zinc
1% magnesium	2% zinc
1% lead	1% zinc
1% iron	0·9% zinc
1% manganese	0·5% zinc
1% nickel	1·2% copper

Melting Practice

Zinc vaporizes at 913°C and forms zinc oxide in the atmosphere. An efficient flux cover suppresses this nuisance associated with alloys containing appreciable zinc contents and having melting points above 913°C. There is also the problem of restricting the formation of oxide of aluminium and assisting in the removal of that which does form. This dual function is provided by fluxes containing fluoride salts. In the presence of elements such as zinc and aluminium, deoxidation is not necessary. However, small quantities of phosphor copper may be added with the object of increasing fluidity and assisting the separation of aluminium oxide from the melt.

Hard Spots

Alloys of high tensile brass and aluminium bronze are liable to defects known as hard spots (Fig. 5). These occur mostly on cope surfaces and are revealed during machining operations which are thereby rendered difficult or impossible. Hard spots damage cutting tools, cause drag marks and appear as blemishes on polished and plated surfaces. Investigations carried out by the British Non-Ferrous Metals Research Association have shown these particles to be composed mainly of iron–silicon compounds. Boron also may be a constituent. Fluxes and refractories are sources of silicon and boron, the former may also come from sand adhering to scrap returned from the foundry.

Iron is an essential constituent of the alloys themselves, about 1·0% of it being necessary to exert the requisite degree of grain refinement compatible with the development of the minimum mechanical properties specified and to prevent cracking during rolling. When silicon is present it reduces the solubility of the iron in the molten alloy, thus encouraging its precipitation from the melt. A residual silicon content of 0·04% is sufficient to initiate this action.

Purchased ingot metal may already contain hard spots which originate at or near metal/refractory and metal/flux interfaces. Here silicon and boron compounds may be reduced by aluminium contained in the alloy, to elementary form. Local concentration of these elements form nuclei for the development of hard spots. High melting temperature and fluoride containing fluxes augment the transference of silicon and boron. Prolonged holding of the molten alloy in the furnace encourages segregation and further diffusion of impurity elements to hard spot segregations and coalescence of particles, due to fusion, can occur.

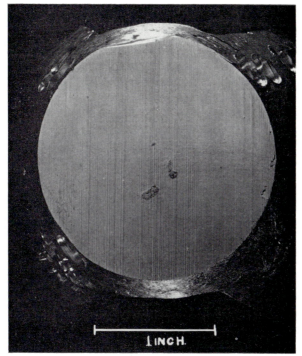

FIG. 5.

Precautions

Once hard spots have formed, no means are known for their effective removal. Therefore, precautionary measures against them should be taken. Less risk of silicon pick-up exists when clay graphite crucibles are used in place of those containing silicon carbide. Sand should be removed from foundry back scrap by shot blasting. Fluxes, when employed, should be used in minimum quantities. This Company has developed fluxes free from the elements fluorine and boron. These fluxes are suitable for crucible and reverberatory furnace melting, particularly the latter, in which melting is carried out on a large scale. In all cases melting should be fast and unnecessary high temperatures avoided.

Brass Alloys

Six brass alloys are listed in BS 1400. They are basically alloys of copper with zinc but up to about 5% of other elements such as lead, tin, iron and nickel may be present. The copper/zinc ratios mainly concerned are 75/25, 70/30, 65/35 and 60/40. Alloys containing less than 20% of zinc are known as gilding and brazing metals.

Zinc is soluble in copper in all proportions and the copper/zinc alloys are, therefore, all solid solutions. However, the alloys may possess a duplex microstructure depending on the amount of zinc present. Up to about 34% of zinc the structure consists of crystals of a solid solution known as alpha. Under cooling slow enough to allow of equilibrium conditions, a zinc content of about 39% can be introduced before a second constituent, richer in zinc, and called beta, appears. The all alpha brasses combine good strength with considerable ductility and are used for cold rolling into sheet, strip, wire and tubes.

Duplex brasses containing a mixed alpha–beta structure contain between 40 and 42% zinc. Beta crystals confer a higher degree of hardness which serves to increase tensile strength and lower ductility. Brasses in this class are widely used for general casting purposes and for hot-rolling, extruding and stamping at temperatures in the range 600° to 800°C. Naval Brass—B5 has 60% copper with about 1% tin, remainder zinc and brass for diecasting has a similar copper content without the tin but with up to 0·5% aluminium.

Melting Practice

When discussing the melting of high tensile brass, mention was made of the evolution of zinc vapour. This is common to all alloys containing substantial amounts of zinc and the brasses are certainly no exception. Thus, one of the functions of a flux for brass is to help in suppressing zinc fume. Another is to assist dirt and non-metallics to separate out from the molten alloy and pass into the slag. Flux compositions without oxidizing or reducing properties are suitable for this purpose. Besides granular materials they are also available in convenient block form of specified weight which assists greatly in applying the standard procedures so very desirable in all metal melting operations.

Brass alloys suffer little from porosity defects due to dissolved hydrogen. In this respect they differ considerably from copper and the bronzes. This is because of the pressure of the zinc vapour. Also in view of the presence of ample zinc, deoxidation as such is not required. The addition of controlled amounts of phosphor copper by plunging appropriate sized cartridges does impart increased fluidity to the molten alloy, thereby assisting the separation of dross and non-metallics into the slag which is subsequently removed by skimming. Suitable fluxes will produce a slag sufficiently fluid to be removed without dragging out valuable metal such as would occur with too viscous a slag. As already mentioned, the retention of the slag on the surface of the molten alloy up to the time of pouring will assist greatly in suppressing the escape of zinc fumes. Suggested pouring temperatures are between 1150°C for section thicknesses below $\frac{1}{2}$ in. down to 1020°C for sections of $1\frac{1}{2}$ in. or more.

Nickel Brasses

The names nickel silver and German silver are more frequently applied to these alloys. They cover a wide range of composition with nickel contents from 10 to 30% and zinc between 5 and 45%. Casting alloys contain about 20% nickel and 10 to 20% zinc. Tin and lead are often present, the balance, apart from impurities, is copper.

The British Standard compositions for nickel silver are divided into four groups with numbers running consecutively from NS 101 to NS 113. Group 1 (NS 103–113) contains all the alloys with an alpha structure (10 to 30% Ni). These alloys are suitable for cold rolling. Group 2 is a selection of NS 104 and NS 107, from Group 1 with 12 and 18% Ni, respectively. These two alloys are employed for the telecommunications industry. Group 3 (NS 101 and NS 102) list two alloys of 10 and 14% Ni, respectively with higher zinc contents so as to develop an alpha–beta structure, for extrusion purposes. Group 4 contains four alloys for sand cast components. This group has nickel contents running from 12 to 25% with copper 57 to 66%. The alpha–beta alloys are hot working materials and can be worked over a wide range of temperature (700–850°C) by all the standard hot-forming processes.

Although nickel can dissolve hydrogen up to 40 cm^3/100 g at a temperature 100°C above its melting point, the presence of zinc at 10% or more creates a vapour pressure which acts against hydrogen absorption as in the case of the plain copper-zinc brasses. The 70/20/10 copper–nickel–zinc alloy is used for both casting and wrought work. With 30% nickel the alloy has the appearance of silver and is strongly resistant to corrosion. Alloys containing 10% or less of nickel are often electroplated to improve the appearance and impart added resistance to corrosion.

Nickel, when molten, can dissolve carbon and this is encouraged by melting under reducing conditions. Therefore, charcoal covers should not be used. Carbon, if present, can cause gas porosity by reacting with oxides to form carbon monoxide. It may also engender brittleness in wrought alloys by precipitating graphite from nickel carbide, Ni_3C during annealing treatments at 700°C or above. It is recommended that the carbon content should not exceed 0·04%. Residual oxygen should be kept down to about the same level.

Effect of Other Elements

Lead is almost insoluble as in brass and bronze alloys. Its separation in metallic form increases machinability. However, for casting of thick sections it should not exceed 4%, because slow cooling encourages segregation and exudation known as "lead sweat". In wrought alloys, lead reduces strength and ductility and so is undesirable, especially where much rolling or working has to be done, as cracking is likely to develop.

Tin is added as a grain refiner and to increase the fluidity of the molten alloy. It also increases strength and hardness and for this reason is not recommended for wrought alloys. The solubility of carbon, already mentioned, is retarded by the presence of tin. A convenient way of adding tin is in the form of a shotted alloy containing equal parts of copper, nickel and tin. In casting alloys tin is present up to a level of 4%.

Impurity Elements

An undesirable element in all alloys containing nickel, is sulphur. Both silicon and aluminium are regarded as being generally detrimental for most applications and are best absent. Iron is regarded as an impurity although it has been suggested for use up to 2% in nickel brass alloys used in the food handling industry.

Melting Practice

In view of the need for excluding sulphur and carbon it is best to use a cover when crucible melting, especially if coke is used as fuel for none of this must be allowed to fall into the charge. Likewise, charcoal must not be used as a surface cover. Melting should be rapid and the furnace atmosphere oxidizing. Fluxes as for brass melting are used to remove non-metallics and suppress zinc fumes.

Despite the presence of ample zinc a deoxidation procedure is desirable. F. Hudson[9] recommends a triple treatment which comprises the addition of 0·1% manganese, 0·05% magnesium and 0·02% phosphor copper. The latter is optional, but if used should be introduced immediately prior to pouring. All the items should be plunged.

The manganese while acting as a deoxidizer also exerts a grain refining influence. Magnesium is also a powerful deoxidant and serves to remove any sulphur. No residual traces of magnesium should be left in the melt as it decreases fluidity and forms dross. Specially convenient deoxidizing cartridges are available. These contain the elements Ca, B, Mg and Mn in correctly controlled quantities compatible with the size of the melt. They help greatly in simplifying and standardizing this important final treatment.

Pouring temperatures range between 1200° and 1350°C according to the design and kind of running system applied to the castings. The higher temperatures are used for work of thin section and also for compositions higher in nickel and of lower zinc content. Chills or denseners are used with advantage at thick sections, particularly for pressure resisting work. They help to prevent shrinkage defects of the interdendritic type.

Recovery of Swarf, Turnings, Drosses, etc.

Reclamation of scrap is an important operation in the metal melting industry. Heavy scrap and returned material of appreciable section can be remelted in normal furnace charges conforming to the techniques applicable to the various alloys, as already detailed. Machine shop turnings, borings, grindings and thin sheet clippings, however, set a special problem. These materials are of low bulk density with a large surface to volume ratio. Unless they are melted under correctly controlled conditions they become oxidized and recovery yields will be at a very low level. Useful information on this matter appears in the section on Light Alloys.

Small-scale recovery may be carried out in crucible furnaces, but in the production of secondary ingot, large reverberatory or rotary furnaces are used. Alternatively electric induction and indirect-rocking-arc furnaces may be employed. The rotary and induction furnaces are especially applicable because their stirring action helps greatly in arriving at a melt of uniform composition and avoids the necessity for rabbling. In all cases it is essential to get the swarf out of the influence of the furnace atmosphere with the least possible delay.

The technique described in the Light Alloy section for the recovery of aluminium alloy swarf, makes use of a bath of prefused flux into which the charge is submerged. Charge increments must be so adjusted as not to protrude above the flux surface. This procedure is equally applicable to brass and bronze swarf, using the appropriate fluxes. There is, however, an alternative procedure which commences by forming a heel or pool of molten alloy in the furnace, from first melting a proportion of relatively heavy scrap. A suitable pool is formed by melting solid scrap in a ratio of about 20% of the total charge weight. The swarf or light scrap is mixed with flux and submerged, in suitable quantities at a time, in the molten metal bath already present. It is important that no swarf is allowed to project above the surface. The flux content of the swarf melts to form an additional protective layer on top of the metal bath.

Fluxes of low melting point are necessary as the success of recovery depends on maintaining a good blanket of liquid flux throughout all stages of the process. In fact the maintenance of this fluid surface layer forms the criterion of how much flux to use. Scrap of the nature considered here will vary greatly in quality and degree of freedom from dirt and other non-metallic materials which must be removed by the flux. Thus, if it is seen, at any stage, that the flux layer is drying up, additional flux must be charged to restore wet fluxing conditions. Throughout the process for the recovery of the swarf the temperature should be kept as little above the melting point of the alloy as possible. During the final stages the temperature can be raised to provide sufficient fluidity to run the charge off into pig moulds. A drying up of the slag at this stage is of no importance. It can be run off separately or allowed to flow out with the metal. In the latter case it will separate as a glass on top of the pig from which it will break away, when cold, to leave a clean surface.

From what has been said it will be understood that recovery fluxes, suitable for the purpose, must fuse well below the melting point of the metal or alloy and be very fluid at the working temperature. Taking account of the variability of swarfs and drosses both in quality and composition, and in the kind of furnace which may be used, a single flux composition is not practical. This Company has acquired great experience, over the years, and can, therefore, supply recovery fluxes best suited to the individual conditions and requirements. Some recovery processes blow air through the melt to remove impurities such as aluminium. Such treatment also produces large quantities of zinc oxide where gun-metal residues are concerned. In these circumstances provision must be made to prevent the slags from thickening rapidly and specially compounded fluxes are required.

Other considerations involved in the compounding of recovery fluxes must have regard to their capability to deal with oxides and to ensure that they exert no severe attack on crucibles or furnace linings. Careful selection and proportioning is made from a number of ingredients from among materials such as carbonates and chlorides of sodium, calcium fluoride and borates. Various ratios of these compounds, in suitable form, can provide flux mixtures able to perform successfully under the different conditions of operation called for by the recovery and reclaiming processes operated in practice.

Mixed Scrap

Although highly desirable, it is difficult to keep scrap segregated according to its composition especially the swarf and turnings collected from machining operations. When the scrap is collected from various supply sources the mixing together of different alloys is even more likely. Difference in appearance between ordinary brass, high tensile brass, aluminium bronze, tin bronze and gun-metal is not very great, especially where swarf is concerned. Even small proportions of high tensile brass and aluminium bronze introduce sufficient aluminium into the charge to constitute a deleterious impurity. Traces of aluminium as little as $0 \cdot 01\%$ cause loss of mechanical properties and leakage of pressure type castings when present in bronzes and gun-metals in which it forms no part of the alloy.

As aluminium oxide it forms films and "stringers" which tend to become trapped in the casting. These cause porosity by attracting gas bubbles and by impeding self-feeding properties which in any case are never very good in gun-metal alloys. Fortunately, it is not unduly difficult to remove aluminium, in quantities up to about $0 \cdot 5\%$ from molten bronze and gun-metal. Refining treatment by air blowing has been mentioned but oxidation of the aluminium can be done by fluxing.

Removal of aluminium by fluxing involves taking the melt up to a temperature a little above that normally used for casting. The flux is applied to the molten surface where it fuses. Then it must be well plunged or stirred into the bath for several minutes. Successful elimination of the aluminium depends on intimate contact between the flux and the molten alloy. Much of the labour is obviated where rotary or induction furnaces are employed.

Where aluminium is present at above 0·5% it is best tackled by diluting down to about this level with scrap known to be aluminium free. It is best to apply the flux in stages in a ratio of about ½% of the charge weight at a time. As the amount of aluminium present decreases, its removal becomes less efficient and more difficult to effect.

Fluxes for eliminating aluminium are based on chemical compounds which provide oxygen to transform the aluminium into alumina. Among these are manganese dioxide, potassium permanganate, sodium nitrate, sodium sulphate and calcium sulphate. Fluorides of calcium and sodium may also be present to help impart the necessary fluidity. Any iron or steel turnings mixed in with the swarf can be removed by magnetic separation.

High Lead Bronzes

The copper-lead bearing alloys in practical use contain between 10 and 36% of lead although in a few cases the lead may reach the 50% level. These combinations are not alloys in the true sense because in the solid state they exist as a mechanical mixture of copper with lead. The lead should be distributed uniformly throughout the copper so as to impart good anti-friction properties and a high load carrying capacity. It is the necessity for even dispersal of the lead which constitutes the main problem in producing satisfactory lead-bronze castings.

All combinations of copper and lead up to 36% of lead form a homogeneous melt at temperatures above 1083°C. On cooling below this temperature, crystals of copper solidify and separate out from the liquid. This process continues until the temperature reaches 954°C when all the copper will have solidified to form a network of primary copper dendrites. The lead, still in the molten state will occupy the spaces between the copper crystals. Finally at 326°C both metals will be solid and co-exist as separate constituents. If this cooling sequence is unduly prolonged, as it may well be when thick section castings are made in sand moulds, the lead with its higher density tends to liquate and segregate to form "lakes" or regions of concentration. Therefore, it is advantageous to cast into chill moulds.

With high lead bronzes containing more than 36% lead the conditions become somewhat more complex. When molten, at temperatures above 954°C, the melt is not a single liquid of copper and lead, but exists largely in the form of an emulsion which separates into two liquids on cooling between 1250° and 960°C. Thin section castings may not show much lead segregation, but those of thick section, which cool slowly, and especially those cast centrifugally are very apt to do so. Segregation becomes more marked at the lower temperatures and appreciable separation can occur at 1000°C and below.

Conditions governing the behaviour of lead bronzes of more than 40% lead content are not fully established. At the 40% lead ratio a single liquid exists provided that the temperature of the melt is raised to beyond 1325°C. Then in order to retain the lead in a uniformly fine state of dispersal in the solid castings, cooling from this temperature must be rapid. Because of this behaviour, lead bronzes with lead contents between 36 and 90% are said to fall within the miscibility gap.

Effect of Added Elements

Tin is sometimes added to increase the hardness of the bronze but it should not exceed 5% in the high lead grades. An addition of 10% of tin to the 36% lead bronze has the effect of lowering the commencement of the miscibility gap to 20% lead. Nickel in quantities between 1 and 4%, is also used to augment hardness. It should be borne in mind that increased hardness reduces the ability of the bearing to adjust itself during the running-in period. Manganese can be substituted for tin and it enters into solid solution with both copper and nickel.

In the past, claims have been made that addition elements assist in preventing separation and segregation of the lead. This effect has been attributed to nickel and sulphur but any appreciable influence in this respect is considered doubtful. Nickel may result in the formation of a dendritic network at a somewhat earlier stage in the solidification process, so helping to confine the lead in the inter-crystalline spaces. Downie[10] claims that small residual traces of lithium also function to retard the separation of the lead. Other investigators[11, 12] advocate additions of 0·1 to 0·7% of mischmetal for retaining the lead in suspension. Mischmetal contains about 50% of cerium, the balance being a mixture of other rare-earth elements. It is possible that the alkali metals have a favourable influence on the primary crystallization.

Melting and Fluxing Technique

Copper is melted first either according to the procedure for high conductivity copper on p. 210 or as for tin–bronze on p. 214.

Remove all slag from the surface and thoroughly deoxidize by plunging phosphor copper cartridges. The lead (and tin if required) is preheated to about 200°C and introduced into the molten copper. Then cover the melt with a flux based on fluorspar and phosphate, which is capable of dissolving lead oxide, and bring the charge up to the appropriate super heating temperature in accordance with its lead content. The best structure for lead bronzes is achieved by preparing, in the first place, an 80–20 copper–lead melt which is stirred thoroughly, superheated to 1350°C and cast into pigs which are used subsequently as the main ingredient of the charges from which the castings will be poured.

Heavy oxidation of the lead must be avoided or irregular cavities containing a greyish powder may be present in the castings. Silicon and phosphorus residuals are also to be avoided as they combine with lead oxide to form heavy fluid silicates and phosphates which do not separate readily from the molten metal. About $\frac{1}{2}\%$ of the charge weight of a special lead-bronze flux is used during the melting down period and a further $\frac{1}{2}\%$ is plunged some 5 minutes before pouring. Suggested pouring temperatures are:

15% lead	1080°C to 1100°C
20% lead	1050°C to 1070°C
30% lead	1030°C to 1050°C

RECLAMATION AND REFINING OF LEAD AND TIN ALLOYS

General principles of recovery only can be discussed because individual requirements and conditions vary over a very wide field. There are two main classes of alloy, those in which tin predominates and those where lead is the main constituent. Both these classes may or may not contain also copper and antimony. Impurity elements too may be present such as aluminium, arsenic, bismuth, iron and sulphur. The procedure adopted for reclamation and refinement will depend, therefore, on the use intended for the recovered product, the elements to be conserved or removed and the scale on which the process is carried out.

Refining by oxidation is the oldest technique. It makes use of the different affinities between the elements for combining with oxygen, known as the heat of formation of oxides. For example, if air or steam is blown through a molten alloy of lead containing zinc, tin and antimony at a temperature of about 600°C, the last three elements will be oxidized out in the order given, leaving the lead which would then form oxide if the process was prolonged.

Sometimes, in order to speed atmospheric oxidation, sodium nitrate is used because it evolves nascent oxygen according to the reaction:

$$2\,NaNO_3 = Na_2O + N_2 + 5\,O$$

For lead alloys containing tin, antimony and arsenic, these elements oxidize in the sequence shown, at a temperature between 420° and 500°C. In the presence of a fused mixture of caustic soda and sodium chloride, the oxides formed are absorbed until the salts become saturated and thicken. They are then removed and replaced.

The other method of refining makes use of reactive metals which form inter-metallic compounds with the elements to be removed. These compounds have a higher melting point than the parent metal and freeze out forming a crust which allows of their separation. Parkes process is an example of this in which zinc is added to silver-bearing lead and a silver-zinc compound separates out. Typical removal agents for secondary lead, tin and their alloys are listed in the following table (Table 2) by G. E. Behr[13].

The residuals of zinc, sodium or aluminium which remain after their use as refining agents can be removed by blowing with air or steam or with sulphur and ammonium chloride.

TABLE 2.

Element to be removed	Removal agent
Copper	Sulphur or Aluminium
Zinc	Oxidation Sulphur or Ammonium chloride Chlorine Caustic soda–Lead chloride
Arsenic	Oxidation Zinc or Aluminium
Tin	Oxidation Caustic soda + Oxidizing agent Chlorine or Lead chloride
Antimony	Oxidation Aluminium or Sodium
Lead	Chlorine
Iron	Sulphur or Aluminium

After the scrap has been melted down and blown or not as the case may be, a layer of dirt will be present on its surface. Skimming will remove some of it but further dry drossing is needed to make it clean. A flux composed of resin and ammonium chloride, well stirred in, will bring most of the non-metallics to the surface. The caustic soda-sodium chloride flux, mentioned above, is also a good chemical cleansing medium, used at a rate of about 1 oz per 100 lb metal.

Removal of Copper

Type metals, solders, lead base bearing alloys and battery grid metal, all have a low limit specified for copper. Yellow sulphur is a much used agent for removing copper, but according to Behr[13] it is neither efficient nor economical in use. One pound of sulphur may remove from 2 to 4 lb of copper and in doing so create between 20 and 40 lb of dross, depending on the composition of the charge. Aluminium is a highly reactive metal which can be used effectively for the removal of iron, arsenic, copper and antimony from lead, tin and lead-tin alloys. Aluminium combines with the impurities mentioned in the order as set down. One pound of aluminium can remove 3 to 4 lb of these contaminating metals depending on the tin content of the charge, as the reaction is more efficient when the tin ratio is high.

Having calculated the weight of each impurity to be removed, the required amount of scrap aluminium is added to the melt. When the temperature has been raised to the point at which the aluminium melts, it is well stirred into the charge. The bath is allowed to cool down and the aluminium combined with the impurity elements thickens and forms a layer floating on the surface of the melt, from whence it is skimmed off for further recovery treatment if required. A good separation is possible because the aluminium layer containing the impurities has only a low solubility in lead-tin alloys near to their liquidus temperatures. The refined charge will contain traces of aluminium which can be removed, subsequently by blowing air or steam or by treatment with sulphur or ammonium chloride.

Removal of Zinc

Zinc is not a constituent of the lead and tin alloys under discussion but is often present as a contaminant when dealing with miscellaneous collections of white metal scrap. It is readily removed by oxidation treatments and can be separated, selectively, from tin-based alloys by ammonium chloride stirred in at a red heat. The table shows that chlorine can be used but it is a long process requiring special apparatus. Sodium hydroxide will remove zinc and tin at the same time and so is not suitable if tin must be conserved. Lead chloride will take out zinc, aluminium, sodium and any other metals which precede lead in the table of chloride heats of formation.

In *Metal Industry* of November 2nd, 1945, P. G. Forrester and J. W. Price give results of an investigation of the use of fluxes based on $NaCl : Na_2CO_3$ eutectic for removing zinc from tin and tin-base alloys. The melting point of the binary eutectic is about 660°C, so that a working temperature for the melt would be around 700°C. A working temperature of 600°C could be used by employing a ternary eutectic flux of $NaCl : KCl : Na_2CO_3$ composition which melts at 580°C. The percentage composition by weight of the two fluxes is as follows:

1. $61 \cdot 7 \%$ Na_2CO_3, $38 \cdot 3 \%$ NaCl.
2. $40 \cdot 2 \%$ Na_2CO_3, $23 \cdot 3 \%$ NaCl, $36 \cdot 5 \%$ KCl.

Tests were made on a Babbitt alloy containing 7% Sb, $3\frac{1}{2} \%$ Cu, remainder Sn, to which about 1% zinc had been added. Table 3 shows the percentage of zinc remaining in the alloy after various times of treatment.

TABLE 3.

Time (hours)	Initial	1	2	4	6	8	24
Binary flux	1·08	0·37	0·32	0·056	0·0025	—	0·0005
Ternary flux	1·08	—	0·38	0·014	—	0·002	—

Removal of Arsenic

Arsenic is not often present in troublesome amounts. Oxidation treatments will remove it but as tin will go first it is an impractical process for tin-based alloys. For these, zinc can be employed. At 450°C zinc, in a finely divided form is stirred in. One pound of zinc will combine with 1·7 lb arsenic and the compound separates out and thickens up at 250°C which enables it to be skimmed off. Further treatment of the skimmings in a sweating furnace yields metal and dry dross, the former being returned for retreatment. If an arsenic content as low as 0·01 % is required the process will need repeating.

As an alternative, aluminium may be used but as an arsenic remover it is not as selective as could be desired, in the simultaneous presence of tin and antimony. Behr[13] gives a word of warning here about the aluminium–arsenic compound. If this makes contact with moisture hydrogen is evolved and this in the presence of arsenic forms the toxic gas, arsine.

Removal of Tin

A molten flux of caustic soda and sodium nitrate has been mentioned already for this purpose. It transforms the tin in a lead-tin alloy to sodium stannate. When antimony is also present it will be removed as sodium antimoniate. If steam is used as an oxidant in place of the nitrate and, along with the caustic soda, brought into contact with the molten charge, only tin and arsenic are oxidized and not antimony or lead.

The use of chlorine gas as a tin remover has formed the subject of an American patent.[14] It reacts to form lead and stannous chlorides according to the reversible reaction:

$$PbCl_2 + Sn \rightleftharpoons SnCl_2 + Pb$$

The state of equilibrium is prevented by the use of an excess of chlorine to form $SnCl_4$ which is volatile and escapes from the system which causes the reaction to proceed from left to right, thereby removing tin from the melt. A development of the process has been the subject of further patents covering the use of metallic oxides in conjunction with lead chloride. At the stage shown by the above equation, lead oxide is introduced to the melt:

$$SnCl_2 + 2 PbO = SnO_2 + PbCl_2 + Pb.$$
$$Sn + 2 PbO = SnO_2 + 2 Pb.$$

Removal of Antimony

Oxidation at a low red heat and the use of aluminium have already been discussed. Aluminium is used in preference to oxidation only where it is necessary to conserve the tin. The use of sodium is recorded in a patent of 1930.[15] One pound of sodium removes about $2\frac{1}{4}$ lb of antimony and separates out as a viscous layer on the melt surface. If at this stage, caustic soda is added in quantity equal to about 10 times that of the sodium used, the separated sodium-antimony compound is dissolved. It is then possible to effect an excellent separation simply by ladling off the caustic soda melt. The latter when cool is treated with water which decomposes the antimony-soda compound and yields antimony of about 90% purity.

Where antimony is being removed from a melt containing tin, some tin is contained in the caustic skimmings. These are not then treated with water as the tin would be lost. In these circumstances the soda compound is smelted for full metallic recovery.

Removal of Iron

Iron is present in tin containing alloys as $FeSn_2$, a compound which reduces fluidity. Sulphur can be added to combine with the iron until the fluidity is restored. When using sulphur for copper removal the large amount of dross produced was mentioned. The same draw-back is present when sulphur is used to remove iron. In the table of removal agents aluminium is given as an alternative agent for elimination of iron from tin and its alloys. Aluminium removes not only the iron but any arsenic and copper, which may be present, as well.

Removal of Sulphur from Lead

Any sulphur present or remaining as a residue from refining treatment can be removed by oxidation with litharge or lead sulphate. The reactions are represented as follows:

$$2\,PbO + PbS = 3\,Pb + SO_2$$
$$PbS + PbSO_4 = 2\,Pb + 2\,SO_2$$

To remove 1 lb sulphur requires approximately 7 lb of PbO or $9\frac{1}{2}$ lb $PbSO_4$ A mixture of the two chemicals can be used.

REFERENCES

1. HUDSON, D. A., Inst. Brit. Foundrymen Proc., 56, 50–56 (1963).
2. SWINYARD, G., Inst. Brit. Foundrymen Proc., 56, 167 (1963).
3. RUTHERFORD, N. B., J. Inst. Metals, 80, 555 (1951–2).
4. RUDDLE, R. W., Inst. Brit. Foundrymen Proc., 46, B112 (1953).
5. TOWNSEND, D. W., et al., Inst. Brit. Foundrymen Proc., 57, 26 and 140 (1964).
6. RÖNTGEN, P., and MÖLLER, F., Metallwirtschaft, 13, 81 and 97 (1934).
7. RAUB, E., SCHOLL, A., Z. Metallk., 16, 9 (1927–29).
8. BAILEY, A. R., Metal Industry, 80, 519 and 526 (1952).
9. HUDSON, F., Foundry Trade Journal, 58, 155 (1938).
10. DOWNIE, C. C., Metal Treatment, 29, Jan. 2 (1962).
11. BUNGARDT, W., Fonderie, 231, 186 (1965).
12. AZZOLINI, A., and CORRADINI, D., La Fonderia Italiana, 12, Dec., 457 (1963).
13. BEHR, G. E., Metal Industry, 62, Apr. 23, 261 (1943).
14. BETTERTON, J. O., and PHILLIPS, A. J., U.S. Patent 2113643.
15. HANAK, R., U.S. Patent 1786908.

D. TREATMENT OF ALUMINIUM ALLOYS

CHAPTER 7

THE TREATMENT OF ALUMINIUM
AND ALUMINIUM ALLOYS

K. STRAUSS

The most important impurities in aluminium and aluminium alloys are hydrogen and aluminium oxide and although they are normally present only in very small quantities their influence on the properties of castings and wrought alloys is considerable. The literature dealing with these impurities which has been published over the last 30 years, is not always very precise and is sometimes contradictory. One reason for this is that the methods used for the determination of these impurities are not entirely reliable and do not always give reproducible results. This aspect will be dealt with later on in more detail but it should be emphasized here that efficient and reliable methods exist for the removal of these impurities from molten aluminium, so that even metal which is heavily contaminated with them can be used for making sound castings and wrought products.

Of all the gases which are normally in contact with aluminium it is only hydrogen which is soluble to any degree.[1]

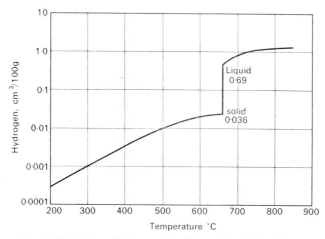

FIG. 1. Solubility of hydrogen in solid and liquid aluminium.

Nitrogen reacts with aluminium to form aluminium nitride AlN, oxygen reacts and forms aluminium oxide Al_2O_3, and both these compounds are practically insoluble in aluminium; the literature data on the solubility of

241

aluminium oxide in aluminium give values between 0·003 and 0·04%,[2] but this variation is no doubt due to the difficulty in taking homogeneous samples and carrying out exact aluminium oxide determinations.

The main source of hydrogen in aluminium is water vapour which reacts as follows:

$$2Al + 3H_2O \rightarrow Al_2O_3 + 6H$$

Water vapour is introduced through a variety of sources such as damp refractory material, damp furnace walls, water-containing fuels, damp tools and humid atmosphere. The hydrogen formed by decomposition is in its atomic state and as such it dissolves in aluminium; molecular hydrogen is soluble only to a very limited extent. The solubility of hydrogen in aluminium and its alloys is dependent on pressure, the temperature and composition of the alloy. A method to determine the equilibrium between hydrogen and the metal has been thoroughly investigated by Sievert and his co-workers.[3] They melted aluminium in a vacuum apparatus and degassed it under a high vacuum. They determined the "empty space" of the apparatus by filling it with an inert gas which is insoluble in aluminium, e.g. argon. The apparatus was again evacuated and hydrogen was introduced and maintained at the same pressure until solution equilibrium was attained. The difference in the volume of hydrogen and the volume of inert gas to fill the empty space indicates the amount of hydrogen dissolved by the aluminium, and from this follows Sievert's Law that the quantity m of hydrogen dissolved is proportional to the square root of the partial pressure p

$$m = K\sqrt{p}$$

The solubility of hydrogen in pure solid and liquid aluminium has been fairly well investigated; of all the published information the work done by Ransley and Neufeld[4] seems to be the most reliable and Table 1 gives the hydrogen content of solid and liquid metal between 300° and 850°C. Solubility is expressed in cm^3 per 100 g of metal (see Table 1).

TABLE 1.

	Temperature (°C)	Hydrogen (cm^3/100 g)
Solid metal	300	0·001
	400	0·005
	500	0·0125
	600	0·026
	660	0·036
Liquid metal	660	0·69
	700	0·92
	725	1·07
	750	1·23
	800	1·67
	850	2·15

The solubility of hydrogen in aluminium alloys as distinct from pure aluminium has been less well established and the published results show very large discrepancies. As a general rule the solubility increases or decreases with the ability of the alloying elements to dissolve hydrogen to a greater or lesser extent than the base metal. According to this rule the addition of magnesium should increase the solubility of hydrogen in an aluminium-magnesium alloy because of the high solubility of hydrogen in magnesium (20 cm^3/100 g). However, R. Eborall and C. E. Ransley[5] investigated the effects of magnesium on the solubility of hydrogen in solid aluminium and found that the magnesium content had no effect on the pick up of hydrogen. This, of course, is contradictory to practical experience that aluminium-magnesium alloys pick up more hydrogen than pure aluminium. In fact, in a recent publication K. O. Hornung[6] on the influence of chlorine on the hydrogen content of aluminium melts in presence of magnesium and sodium, shows that with an increase of the magnesium content up to 10% the hydrogen content increased from 0·2 cm^3/100 g to 0·65 cm^3/100 g. Sodium has a similar effect whilst the addition of copper and silicon, manganese and tin to aluminium lower the solubility of hydrogen; additions of chromium, iron and titanium on the other hand increase it.

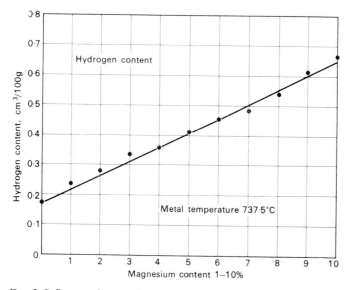

FIG. 2. Influence of magnesium content on the hydrogen content in Al.

METHOD FOR THE DETERMINATION OF GAS

The most common cause of scrap in aluminium castings is gas-porosity, that is the presence of gas in the form of cavities or pinholes. Because of the higher solubility of hydrogen in the molten metal the gas has a tendency to be released on solidification, provided of course that there is more gas present than can be kept in solution in the solid state. This gas is given off in the form of small bubbles which escape into the atmosphere. But if the surface of the casting has already solidified, or if the tension of the aluminium oxide layer on the surface prevents the escape of the gas, it is trapped in the form of holes of varying size and causes porosity. Similarly in the case of wrought alloys, gas is very often present either in solid solution, which can be expelled on heat treatment or in form of pinholes; in both cases the properties of the finished product will be affected and the exact determination of gas content is therefore of the utmost importance.

For the determination of gas one has to distinguish between qualitative and quantitative methods.

A. Qualitative Methods

There are a number of methods which will indicate the presence of gas in the metal without giving any information on the exact amount of gas which is present.

(1) X-ray radiography of castings

This is used on a large scale in industry for the non-destructive quality control of castings. Pin-holes caused by gas and porosity from other sources, such as shrinkage, show up in the X-ray film as dark spots because of the easier passage of the X-rays so that the exact location of the porosity can in most cases be easily established.

(2) Observation of the surface of test-specimen or castings on solidification in air

This is a well-practised method and known to most aluminium foundrymen and it gives a quick indication of the presence of too much gas. Some metal is poured into a dry, cylindrical mould approx. 3 or 4 in. in diameter and 3–4 in. deep. This can be made by hollowing out a refractory brick or it can be a well-baked sand core. If an excessive amount of gas is present, bubbles rise to the surface when the temperature of the metal is near its solidification point and from the number and size of these bubbles conclusions can be drawn as to the approximate amount of gas present.

(3) Solidification of a sample of aluminium in vacuum

In this method a small volume of molten metal, usually a few hundred grams, is allowed to solidify in a vacuum. The bubbles breaking through the surface and the number and size of holes in the solidified button give an indication of the amount of gas which was present in the molten metal. This method is known as the Straube-Pfeiffer method and the usefulness of the tests depends very much on the type of alloy, the amount of gas present and other factors, which make it rather unreliable. For this reason the British Non-ferrous Metals Research Association have developed a method based on the Straube-Pfeiffer but giving more reliable and reproducible results. Whilst the Straube-Pfeiffer method employs a high vacuum (1–5 mm) the BNFMRA method uses a vacuum of

60 mm which is established immediately after placing the sample into the autoclave so that it solidifies at a constant pressure. Unless there is an excessive amount of gas present the surface tension of the oxide film on top of the solidifying metal prevents the escape of the gas which is trapped in the button in form of holes of varying sizes. Determination of the density of the metal-button gives some indication as to the amount of gas in the metal. A ready-made instrument incorporating all these improvements and refinements and allowing quick and reliable tests to be made under foundry conditions is available on the market under the name of "Foseco Porotec".

FIG. 3. Foseco Porotec apparatus.

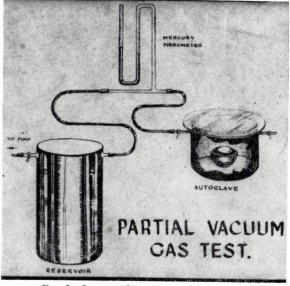

FIG. 3a. Layout of partial vacuum gas test.

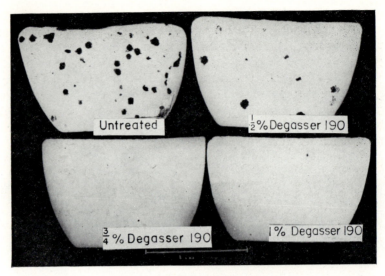

FIG. 3b. Some Porotec samples.

B. Quantitative Methods

(1) Vacuum extraction

This method has been developed by C. E. Ransley and his co-workers[5,7] to a fine art and with it hydrogen determinations of great accuracy can be made. It was originally developed for the determination of hydrogen in steel and Ransley and co-workers have adapted it for aluminium. The gas is extracted from a carefully prepared sample in a high vacuum and there is no need to melt the sample; heating to 500–550°C is sufficient to extract practically all the gas which is then collected and determined. During the last few years many refinements have been added to this method and there are a number of different, ready-made equipments available, with which highly accurate hydrogen determinations can be carried out even by only reasonably skilled laboratory personnel.

(2) Method of Dardel

This method is known as the "first bubble test" and it has been described in the literature by Y. Dardel in 1948.[8] It has been developed during the last 10 years into a quantitative method for determining hydrogen in aluminium. The sample of 300–400 g of metal is put into a vessel which is gradually evacuated. Through the glass in the lid it is possible to observe how the metal solidifies and when the first gas bubble breaks through the surface; at this moment the pressure and temperature are recorded and by means of equilibrium diagrams it is possible to determine directly the hydrogen content. This method is reasonably

Fig. 4. Picture of vacuum extraction apparatus.

accurate for high gas contents. If the gas content is low as in wrought alloys, results can be very misleading. This is particularly the case if the amount of suspended insoluble impurities, such as aluminium oxide particles, is low because these particles facilitate the formation of the gas bubble. An instrument developed on the principle of the initial bubble test is available on the market as "Hykon Tester" and very recently it has been greatly improved and is marketed as F.M.A. Alu–Schmelztester by a firm in Liechtenstein.

I

KEY

A. Extraction tube.
B. Greased conical joint.
C. Large-bore greased tap.
D. Liquid-air trap.
E. Two-stage mercury diffusion pump.
F. Dead space volume bulb.
G. Platinum filament.
H. Palladium tube.
J. Liquid-air-cooled side tube.
K. Pirani gauge.
L. McLeod gauge.
M, N. Greased taps to three-stage mercury diffusion pump for evacuating system.
O. Glass pip to relieve vacuum.

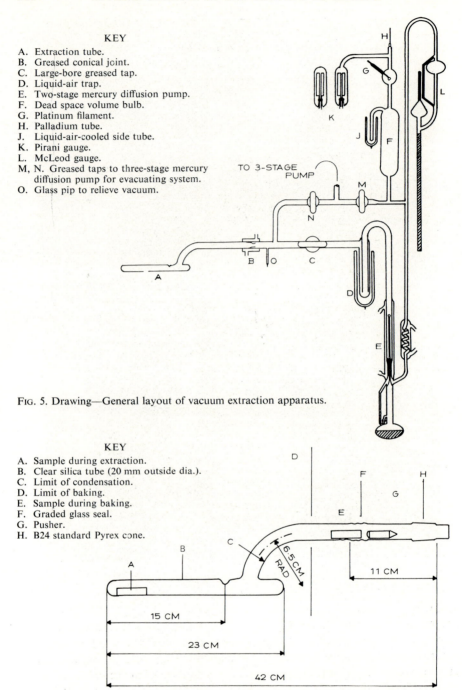

TO 3-STAGE PUMP

FIG. 5. Drawing—General layout of vacuum extraction apparatus.

KEY

A. Sample during extraction.
B. Clear silica tube (20 mm outside dia.).
C. Limit of condensation.
D. Limit of baking.
E. Sample during baking.
F. Graded glass seal.
G. Pusher.
H. B24 standard Pyrex cone.

6.5 CM RAD

11 CM

15 CM

23 CM

42 CM

FIG. 5a. Detailed drawing of silica extraction tube.

Figs. 5 and 5a by courtesy of C. E. Ransley and D. E. J. Talbot).

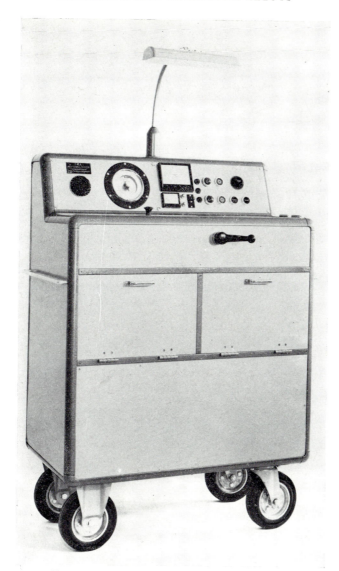

FIG. 6. F.M.A. hydrogen determination apparatus.
(By courtesy of Feinmechanik-Anstalt, Schaan).

(3) The Telegas Instrument

This was developed by C. E. Ransley[9] for the quick and accurate determination of hydrogen in commercial aluminium melts. A small quantity of nitrogen is repeatedly bubbled through a melt of aluminium, is collected under a small bell and recirculated until there is an equilibrium between the nitrogen and hydrogen contents. The gas mixture is then passed into a catharometer which measures the partial pressure and when this is constant, the bubbling of the nitrogen is stopped. Taking into consideration the temperature of the metal and

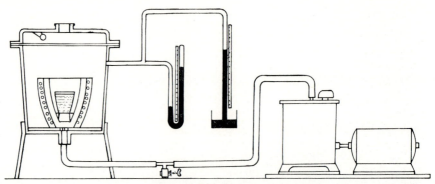

FIG. 6a. Layout of Fig. 6.

the solubility of hydrogen in the particular alloy, the hydrogen content can be determined with reasonable accuracy provided that the test has been carried out with great care by skilled personnel.[10] Because of this latter requirement and the difficulty of handling a rather delicate piece of apparatus under works conditions, the method has not found general acceptance in the industry.

(4) Other Methods

There are a few other methods by means of which hydrogen is determined quantitatively. In one[11] the sample of aluminium is dissolved in carbon disulphide to which bromine is added. Hydrogen, whether it is dissolved in the metal or adsorbed, is freed and hydrogen which is present as hydride forms

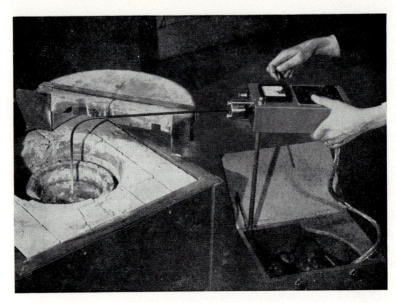

FIG. 7. Telegas instrument.
(By courtesy of Edwards High Vacuum Ltd.).

hydrogen bromide. By means of nitrogen the gases are first conducted through silver nitrate solutions to precipitate bromine and then over heated copper oxide to burn the hydrogen to water which is absorbed and weighed.

In the other method according to Sauerwald[12] chlorine is passed through the molten metal and the hydrogen–aluminium chloride mixture is passed over heated copper oxide as in the previous method. Hydrogen is burned to water, absorbed and weighed as such. These methods, however, are too long, cumbersome and not exact enough and have therefore not found general application.

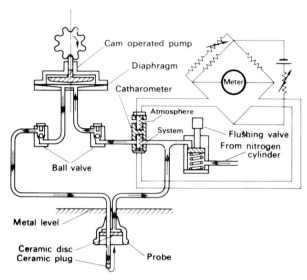

FIG. 8. Layout of Fig. 7.

ALUMINIUM OXIDE

Aluminium oxidizes readily at ambient temperature and produces a film of aluminium oxide which protects the metal from further oxidation. Oxidation increases with an increase in temperature and molten aluminium forms a dense film of oxide immediately.

The purer the aluminium the less the tendency to oxidize and super-pure aluminium will oxidize only very slowly and only gradually lose its bright surface. In molten aluminium the oxide is mainly formed by reaction with water vapour in the atmosphere according to the equation:

$$2Al + 3H_2O \rightarrow Al_2O_3 + 6H$$

which shows that the formation of the oxide goes hand in hand with the introduction of hydrogen into the metal. Other sources of aluminium oxide in aluminium melts are scrap, turbulent transfer of metal, careless stirring in furnaces and ladles.

After the metal is molten it has finely divided oxides and other non-metallic impurities suspended in the melt. The density of the aluminium oxide is higher than that of the molten aluminium, but it does not separate readily from the metal and remains suspended and floating in the melt. One reason for this is that the aluminium oxide particles are porous and contain some gas adhering to and trapped in the pores so that the specific gravity is lowered and is very near to that of aluminium.

Aluminium oxide is practically insoluble in molten aluminium; the solubility in the literature is given between 0·003 and 0·04%. The exact determination of oxide content is rather complicated because it is not evenly distributed in the metal, but very often forms aggregates and according to the way the sample is taken, it can vary within wide limits.

There are a number of methods available for the determination of aluminium oxide; the most popular one, and the one which gives reasonably reliable and reproducible results, uses the solubility of aluminium in bromomethanol. Aluminium oxide is insoluble and remains in the residue; it is separated from the other insoluble residue and the aluminium is determined according to well-known analytical methods.[13]

A modern method for the qualitative determination of aluminium oxide uses ultra sound.[14] If an ultrasonic wave is directed into liquid aluminium containing insoluble particles such as aluminium oxide in suspension, the sound wave is reflected from these particles and scatters with a loss of energy in the pulse; at the same time the amplitude of the bottom reflection diminishes. On letting the melt stand, the particles will settle out and the amplitude of the bottom echo increases. Therefore if a melt contains appreciable quantities of aluminium oxide the bottom echo is largely eliminated but it must be pointed out that crystals of $TiAl_3$ or $ZrAl_3$ which are insoluble in liquid aluminium below a certain temperature produce a similar effect.

THE REMOVAL OF ALUMINIUM OXIDE AND OTHER IMPURITIES

Aluminium oxide is a very stable compound which cannot be reduced to aluminium under ordinary melting conditions. For the cleansing of melts from aluminium oxide it is necessary to use salt mixtures or fluxes which will react with impurities either mechanically or chemically or both and bring them into the slag. There are "passive" fluxes which will protect only the surface of the molten aluminium from oxidation and more or less prevent the pick-up of hydrogen by the melt. There are more active fluxes which will react chemically with aluminium oxide and effectively clean the melt. Most fluxes have a lower density than the aluminium alloy so that they stay on top of the melt; but for maximum effect they should be brought into intimate contact with as much of the metal as possible. This is done either by hand with suitable tools by vigorous rabbling and stirring or sometimes by mechanical means whereby the metal is moving continuously as is the case in rotary furnaces. For the removal of aluminium oxide the presence of halides, particularly fluorides, has been thoroughly investigated by a number of authors because of its great importance for the electrolytic production of aluminium, in which the solubility of aluminium oxide in cryolite plays an important part. Pascal and Jouniaux[15] examined the reaction between cryolite and aluminium oxide and established a eutectic at 24% Al_2O_3 at 904°C.

H. Ginsberg and K. Resch[16] contributed a thorough examination of the systems $NaF–AlF_3–Al_2O_3$ and $NaF–AlF_3$ and concluded that the solubility of aluminium oxide increases with the sodium fluoride content. The reaction between cryolite and aluminium oxide in the presence of a surplus of NaF takes place as follows:

$$4NaF + 2Al_2O_3 \rightarrow 3NaAlO_2 + NaAlF_4$$
or
$$2NaF + Al_2O_3 \rightarrow NaAlO_2 + NaAlOF_2$$

Tronstad and Reimers[17] showed that at 600°C alumina is only slightly soluble in cryolite and therefore fluxes will chemically dissolve only little alumina; but as the quantities of alumina present in the melt are only of the order of 0·05% (or 50 g Al_2O_3 in a 100 kg melt) this solubility is sufficient to remove some of the aluminium oxide; most of it will be removed by the flux mechanically enveloping the particles and transporting them into the slag.

FLUXES FOR COVERING ALUMINIUM MELTS

These usually form a molten slag on the surface of the metal to protect it against oxidation and reaction with the humid atmosphere. The basis of these fluxes is a mixture of alkali-chlorides with or without the addition of fluorides. The eutectic mixture of NaCl–KCl melts at 658°C and the equilibrium diagram is shown in Fig. 9.

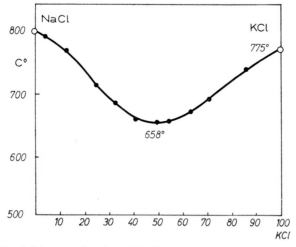

FIG. 9. Diagram of sodium chloride/potassium chloride mixtures.

The view that the removal of aluminium oxide is more a mechanical than a chemical action is supported by work carried out by Sully, Hardy and Heal.[18] In this publication they point out that the action of the flux is actually to strip the oxide films by surface-tension forces and they suggest that the flux penetrates the interface between alumina and aluminium. It must be pointed out, however, that the fluxes they used consisted only of chlorides or of chlorides with some calcium fluoride and as was shown more reactive fluorides are needed—such as sodium fluoride or cryolite—to react chemically with alumina.

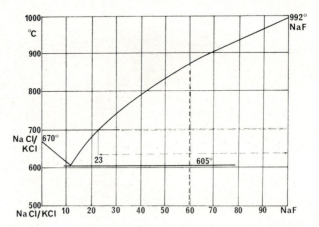

Fig. 10. Diagram of sodium chloride/potassium chloride/sodium fluoride mixtures.

The ternary diagram for NaCl, KCl and NaF is shown in Fig. 10.

These three compounds form the basis of most covering fluxes and they are suitable for almost all types of aluminium alloys, only some special alloys such as Al–Mg alloy and hypereutectic Al–Si alloys require the use of different fluxes. For effective protection from oxidation and for a thorough cleaning effect best results are obtained if the flux is molten, but very often a liquid

Fig. 11. Photograph of aluminium dross untreated.

slag is not desired because the skimming is troublesome and the molten salts hold considerable quantities of metal in the form of globules which are difficult to separate. The aim, therefore, is to produce a slag which will absorb alumina, which will contain as little metal as possible and which can be skimmed easily. Such a slag will be dry and should be powdery. The flux consists again mainly of chlorides to which, however, more fluorides and small quantities of oxidizing compounds such as sulphates or nitrates are added. The presence of these additives causes exothermic chemical reactions which will be explained in some detail.

Fig. 12. Photograph of aluminium dross treated with flux.

Sulphates or nitrates as oxygen-bearing compounds react with the very fine particles of aluminium which are intermixed with the slag and oxidize them to alumina, thereby increasing the temperature of the slag. This temperature increase facilitates the second exothermic reaction, namely the reaction between the aluminium particles and the fluorides. This has been very thoroughly investigated in the laboratories of "Foseco International" in connection with the development of exothermic feeding compounds, but the results have so far not been published and are available only as internal research reports.

At temperatures above 600°C fine aluminium particles react with fluorides, such as aluminium fluoride AlF_3 as follows:

$$AlF_3 + 2Al \rightarrow 3AlF$$
$$3AlF + 3O \rightarrow Al_2O_3 + AlF_3$$

The fluoride, either present as cryolite or as NaF (which will form AlF_3 in reaction with aluminium), reacts with aluminium and forms aluminium sub-fluoride. This is a gas and a most unstable and reactive compound; it reacts immediately with oxygen of the air, forming aluminium oxide and aluminium fluoride AlF_3 which then continues to take part in the reaction until all the fine

I*

aluminium particles are burned. The increase in temperature will allow the larger aluminium particles which are present in the slag to coalesce together after the fluorides have removed the oxide layer round these droplets and fall back into the melt, thus leaving dry, powdery skimmings free from metal.

The use of strongly oxidizing compounds such as sulphates and nitrates is sometimes discouraged for safety reasons. It has been reported in the literature that sodium sulphate, Na_2SO_4, in contact with molten aluminium resulted in violent explosions, but this happened only when the pure undiluted sulphate was used.[19] In the fluxes which are used for producing a dry dross the oxidizing additions are very much diluted and with these mixtures no explosion hazard exists.

A few fluorides, particularly the double fluorides, react exothermically with aluminium without the addition of oxidizing compounds and use is made of them for producing a dry dross. Sodium silicofluoride, Na_2SiF_6, for example, reacts with molten aluminium probably according to the equation:

$$Na_2SiF_6 + 2Al \rightarrow 2Na + Si + 2AlF_3$$

because both sodium and silicon are picked up by the melt. Another possible reaction is the decomposition of the silicofluoride under the heat of the metal into silicon tetrafluoride, SiF_4 (which is a gas), and sodium fluoride; the SiF_4 reacts exothermically with aluminium oxide and other oxides.

The exothermic reaction between fluorides and aluminium sometimes continues in the skimmings after they have been removed from the melt and it grows fiercer as the reaction continues until most of the aluminium present in the skimmings has been oxidized. To stop this reaction from getting out of hand— it is accompanied by intense heat and fumes—the skimmings should be spread out and cooled down as quickly as possible so that the temperature falls below the critical point at which fluoride and aluminium will react together. Sometimes intentional use is made of this reaction to free skimmings from metal by transferring the skimmings into a container and mixing them with a special flux rich in fluorides—sodium silicofluoride has proved to be particularly useful for this purpose. The ensuing exothermic reaction will increase the temperature of the skimmings to such an extent that the aluminium will melt and coalesce, run off and accumulate at the bottom of the vessel where it solidifies or from where it can be poured off.

FLUXES FOR ROTARY FURNACES

Metal melted in rotary furnaces needs special treatment. These furnaces are usually large, of several tons' capacity, and the type of scrap melted in them is often dirty and oily and contains a lot of non-metallic impurities. It is important that the metal is protected from oxidation and from the oil or gas flame as much as possible. This can be done by charging it into a molten flux bath so that each particle is at once surrounded by a layer of flux. The quantities of flux used are high, normally one-third to one-half of the weight of metal, and the melting point must be below the melting point of the aluminium. Such fluxes consist mainly of sodium chloride to which small quantities of potassium chloride can be added to lower the melting point and small quantities of fluorides are added to facilitate the removal of alumina.

This subject will be dealt with in greater detail in the section on remelting aluminium scraps, it is mentioned here only to underline the difference in the various fluxes for different uses.

FILTERING

A method for the removal of aluminium oxide and other non-metallic solid impurities which has gained wider acceptance in the industry during the last few years, is filtering the molten metal through a suitable filtering material. On a small scale such a process has been used for many years in aluminium die-casting foundries by pouring the metal through a glass cloth which was placed in the runner before it enters the die proper. Recent developments have considerably improved the process so that it can be applied today on an industrial scale with good results.

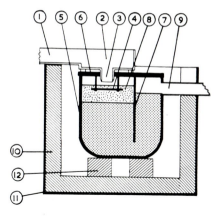

1	Entry Launder	7	Plumbago Baffle
2	Tundish	8	Fluid Flux Layer
3	Downspout	9	Exit Launder
4	Circular Spreader Plate	10	Refractory Furnace Lining
5	Plumbago Crucible	11	Sheet Steel Retainer
6	Lid	12	Refractory Stool

FIG. 13. Flux washing tank designed by Foseco.

The so-called flux-washing process[20] was developed from a method which was intended in the first place to degas the metal. Suspended oxides are often also the carriers of entrapped gas, therefore getting rid of the one, will also get rid of the other.

The treatment unit consists of an externally heated refractory washing chamber containing a deep layer of molten flux retained in the chamber by a refractory baffle which does not reach the bottom. Metal for treatment flows continuously over a spreader plate which is submerged in the flux bath so that the metal is broken up into a thin sheet or droplets and improves the metal–flux contact. The washing flux is a mixture of chlorides to which fluorides are added so that the melting point is well below the pouring temperature of the aluminium. This flux washing process, after a very promising start, ran into difficulties mainly of a mechanical nature which could not be solved entirely satisfactorily, so it is in use only on a very limited scale.

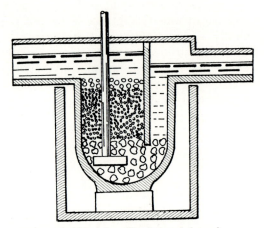

FIG. 14. Aluminium filtering and degassing.

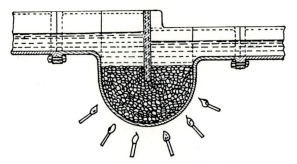

FIG. 15. Aluminium filtering and degassing.

In another method[21,22] molten aluminium is poured through a bed of granular refractory material while chlorine or an inert gas is bubbled at the same time.

From investigations which are not yet fully completed, it can be shown that replacement of the granulated refractory filter material by more chemically active materials, such as fluorides which are wetted by aluminium, produces very much better results since they are more effective in absorbing aluminium oxide than the inert materials. Impregnation of the refractory materials with calcium and magnesium fluoride produces a particularly effective filter material.[23]

Good results have been reported by filtering aluminium through a porous refractory plate with pores of about $\frac{1}{8}$ in. in diameter which is built into a special crucible so as to divide it into two chambers, one a charging compartment and the other the ladling compartment.[24] If the filter plate is preheated, the metal filters well and only a small metallostatic head is required to start the operation.

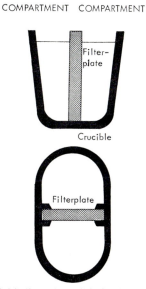

FIG. 16. Special ladle and crucible for filtering aluminium.
(By courtesy of Kissling and Wallace[24]).

FLUXES FOR THE MODIFICATION OF ALUMINIUM–SILICON ALLOYS

Modification is a process which was invented by A. Pacz in 1920[25] for the treatment of eutectic or near eutectic Al–Si alloys. The eutectic is at 12·6% Si with a melting point of 577°C and this alloy is treated either with metallic sodium or with salt mixtures containing sodium fluoride, NaF, which will react with the metal and form sodium and aluminium fluoride. This treatment with either sodium or sodium-containing flux has the effect that the structure of the alloy containing rather coarse grains is changed into a very fine, silky-looking structure which is accompanied by a substantial increase in tensile strength and elongation. The responsible constituent in salt modification is the sodium fluoride which reacts with the metal as follows:

$$6NaF + Al = Na_3AlF_6 + 3Na$$

The sodium is taken up in the melt and aluminium as sodium aluminium fluoride goes into the slag.

The melting point of sodium fluoride is high (992°C) and it will react only when it is in the fused or molten condition and therefore additions of other salts are made in order to lower the melting point of the salt mixture. Numerous combinations have been investigated, particularly the systems NaF–NaCl and

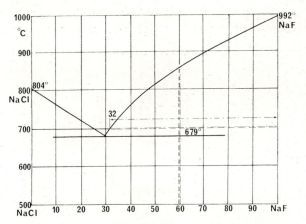

Fig. 17. Sodium chloride/sodium fluoride diagram.

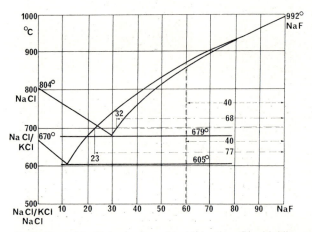

Fig. 18. Sodium chloride/potassium chloride/sodium fluoride diagram.

NaF–NaCl–KCl. The eutectic mixture of 30% NaF and 70% NaCl fuses at 679°C and a mixture of 12% NaF and 88% NaCl–KCl (1:1) fuses at 605°C and the reaction between the metal and these salt mixtures takes place at 720–750°C quickly and effectively.

By using low melting modification salts there is sometimes severe attack on the graphite crucibles or the refractory of the furnace walls; therefore one tries to keep this attack down to a minimum by stiffening the slag with additions which will not melt and will not interfere with the modification process. This will prevent the penetration of flux into the crucible- or furnace-walls and it will facilitate the skimming process.

FIG. 19. Picture of Foseco Navac containers.

FIG. 20. Aluminium silicon alloy 13% not modified.

Modification can also be carried out by means of metallic sodium; other metals, such as lithium, potasium, beryllium and calcium, also have some modifying effect but they are not as efficient. Sodium has certain disadvantages.

It is hazardous to handle for the inexperienced, it oxidizes quickly on exposure to air and has therefore to be kept under oil or kerosene, it reacts violently with water and is very hygroscopic and it therefore introduces gas into the metal and it can itself dissolve considerable quantities of hydrogen. Most of these disadvantages have been eliminated by the introduction of vacuum-melted sodium which is sealed in airtight aluminium containers which are simply thrown on to the metal and plunged. Not only is the application very much simplified and less hazardous, the cutting up of the sodium is eliminated and the introduction of gas by the sodium is decreased to a large extent.

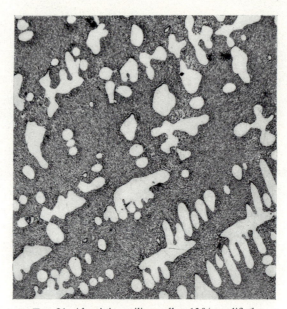

Fig. 21. Aluminium silicon alloy 13% modified.

TREATMENT OF ALUMINIUM–MAGNESIUM ALLOYS

Aluminium alloys which contain between 3% and 10% magnesium are particularly liable to oxidation and gas absorption and for this reason special fluxes similar to those which are used for the treatment of magnesium alloys are used. These fluxes are normally mixtures of various chlorides and fluorides with a substantial part of the chlorides as anhydrous magnesium chloride, $MgCl_2$. This forms low melting point mixtures with potassium chloride which are then thickened by additions of fluorides such as magnesium fluoride, MgF_2, or calcium fluoride, CaF_2. If calcium fluoride is used, it reacts with the magnesium chloride when the flux is molten and forms almost quantitatively (because there is usually a large surplus of $MgCl_2$) magnesium fluoride and calcium chloride:

$$MgCl_2 + CaF_2 \rightarrow MgF_2 + CaCl_2$$

Non-metallic impurities, such as magnesium oxide, magnesium nitride and aluminium oxide, are wetted and enveloped by the liquid flux and brought into the slag.

It is very important that these fluxes do not contain any sodium compounds of any description. Sodium salts such as NaCl or NaF react with magnesium and introduce sodium into the melt. The influence of sodium on the embrittlement of Al–Mg alloys has been thoroughly studied by C. R. Ransley and D. E. J. Talbot.[26] They found that the hot working properties of Al–Mg alloys are severely impaired by the presence of traces of sodium of the order of 0·001%. This is so because the sodium is present in aluminium in presence of magnesium in the elementary state, according to the equation:

$$[\text{NaAlSi}] + 2\text{Mg} \rightleftharpoons \text{Mg}_2\text{Si} + \text{Na "free"}$$

In casting alloys of up to 10% Mg too, sodium has an embrittling effect and makes the alloy hot—short and therefore fluxes must be free from sodium.

DEGASSING OF ALUMINIUM ALLOYS

There are a number of methods for degassing aluminium alloys which will be considered in more detail.

1. Pre-solidification.
2. Waiting period in furnace in water-vapour free atmosphere.
3. Treatment with gases such as N and Cl.
4. Treatment with chlorides.
5. Melting in vacuum.
6. Treatment with ultrasonics.

1. Pre-solidification

This method is based on the difference in solubility of hydrogen in the liquid and solid metal. As pointed out before, the solubility decreases as the temperature of the metal decreases to become very low at the point of solidification. The liquid metal, therefore, is slowly cooled and near the freezing point it will lose most of its gas so that the solidified metal will contain very little gas. The metal is then heated as rapidly as possible up to the pouring temperature and cast; during this remelting period only small amounts of gas will be re-absorbed.

2. Waiting Period in Furnace in Water-vapour Free Atmosphere

C. L. Brooks[27] examined the gas content of pure aluminium and of an Al–Zn–Mg–Cu alloy after standing for lengthy periods in an atmosphere of varying water-vapour contents and the results are summarized in Fig. 22(a) and (b).

If molten aluminium is exposed to an atmosphere low in water-vapour the hydrogen diffuses out of the metal until equilibrium is reached between the partial pressure of hydrogen in the molten metal and the surrounding dry air. The time of degassing is directly proportional to the surface area, which means that in furnaces with large area and low depth of metal the degassing process proceeds quicker than in deep crucible furnaces with small diameter. If the

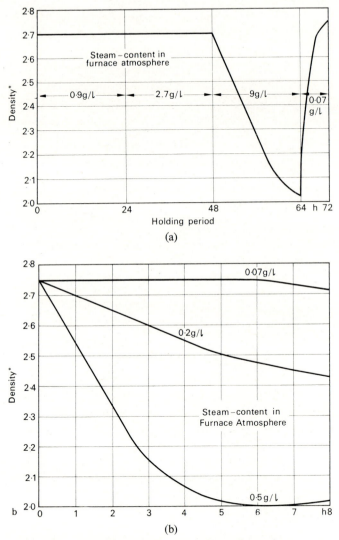

FIG. 22. Gas pick-up of melt of aluminium from steam in the furnace atmosphere. Weight 30 kg, metal temperature 730°C. (a) Metal surface remained undisturbed except whilst taking sample. Commercial aluminium 99%. (b) Effect of disturbing metal surfaces. The metal surface was broken several times a minute with a mechanical device. Al-Zn-Mg-Cu alloy.

(By courtesy of C. L. Brooks[27]).

metal is covered with a dense aluminium oxide film, the diffusion of gas is slowed down; it will help to disturb the oxide film from time to time and expose a blank surface of the metal through which the hydrogen will escape quicker.

These two methods for degassing are only very rarely used because of the length of the waiting period and the subsequent loss of production.

3. Treatment with Gases

This can be subdivided into treatment with inert gases such as nitrogen, helium and argon and treatment with active gases such as chlorine, fluorine and compounds containing the two.

During the degassing process the gas is blown into the metal in a steady stream; the gas forms into bubbles which will rise through the metal to the top and escape into the atmosphere. The partial pressure of hydrogen in the bubble as it forms in the metal is nil, but as it passes through to the top, the hydrogen dissolved in the metal diffuses into the bubble until the partial pressure in the bubble and in the metal are at an equilibrium, or until the bubble floats to the top and carrier-gas and hydrogen escape into the atmosphere. This process is repeated with every bubble until the partial pressure of dissolved hydrogen in the metal is very low and there is only little hydrogen left in solution. At this point the degassing process is broken off, because it is practically impossible to remove the last traces of gas by this method. Even so, the treatment with inert gases in practice is a lengthy process and compared with other more active gases (see next paragraph) is reasonably successful only after prolonged gassing. Furthermore, if the metal temperature is rather high (over 700°C) nitrogen is not so completely inert a gas and it can form aluminium nitride, AlN, which is a solid, insoluble and suspended in the melt. If the alloy contains magnesium there is an added possibility that magnesium nitride Mg_3N_2 is formed at a temperature of 670°C and above; this magnesium nitride too is insoluble in aluminium and remains in suspension.

Gases blown into the metal should form small bubbles because their surface area and therefore their contact with the metal is greater than with larger bubbles. Smaller bubbles also rise more slowly to the surface than larger ones and therefore have a better degassing effect.

Chlorine is one of the active gases and the following reaction takes place when it comes into contact with molten aluminium:

$$2Al + 3Cl_2 = 2AlCl_3$$

Aluminium chloride at ordinary temperature is a solid but it is volatile above 180°C and it is present in gas form in the molten metal. As in the case of an inert gas the partial pressure of hydrogen in the aluminium chloride bubble is zero at the moment when the bubble is formed, but hydrogen immediately starts diffusing into it until the partial pressure in metal and bubble is the same or until the bubble has escaped from the metal. The reaction between chlorine and the metal is quantitative so that scarcely any free chlorine exists in presence of molten aluminium. It has been erroneously suggested that chlorine reacts with dissolved hydrogen to form hydrogen chloride. Such a reaction is thermodynamically impossible because hydrogen chloride would react immediately with aluminium and form hydrogen and aluminium chloride, which is where a start was made. It is essential for an efficient degassing effect that the chlorine is introduced into the melt in form of small bubbles, which will form small bubbles of aluminium chloride, so that the contact with the metal is over as large a surface as possible. Another factor for the efficiency of the degassing process is the depth of the metal bath; the chlorine should enter it at the bottom of the furnace or crucible so that the distance the bubble has to travel is as long as possible. For thorough degassing it is necessary to use up to 0·6% Cl in relation to the quantity of metal to be degassed. But this depends on a number of factors, such as the type of alloy, the type of furnace, the method of introduction, the metal temperature, the depth of the metal bath and the humidity of the atmosphere. As an example, on a day on which humidity is high, more chlorine has to be used over a longer period than if the atmosphere is relatively dry; furthermore, a shallow bath needs more chlorine than a deep bath in which the aluminium chloride bubbles have to travel a longer distance.

A number of investigators have tried to answer the question why degassing with chlorine is so much more efficient than degassing with nitrogen. The answers are not unanimous but one explanation is that chlorine forms a very large number of small aluminium chloride bubbles which have a much larger combined surface area than bubbles of nitrogen. Degassing with chlorine is efficient but this method has certain disadvantages, the main one is the toxicity of the gas and the elaborate safety- and ventilation-system necessary for handling it without health hazard. Then there is the loss of aluminium which is about 0·15% after 0·6% of chlorine has passed through the melt, and if the alloy contains magnesium, this element is lost in the form of magnesium chloride in preference to aluminium, and this loss can be so severe that magnesium metal has to be added in order to compensate for it. Attempts have therefore been made to dilute the chlorine with nitrogen and mixtures of 10% chlorine with 90% nitrogen are more effective than 100% nitrogen, but not quite as good as pure chlorine.

The following diagrams (Fig. 23a, b, c), according to Brooks,[28] give comparative results of degassing with pure chlorine, pure nitrogen and a mixture of the two gases. The density determinations were done according to the improved Straube-Pfeiffer method.

4. Treatment with Chlorides

The treatment of aluminium with metal chlorides is as old as the aluminium industry itself. It is based on the reaction between the aluminium and the chloride during which the chloride is decomposed and aluminium chloride and the metal are formed. With the exception of the alkali—and alkaline earth metal chlorides—practically all chlorides react with aluminium as follows:

$$3ZnCl_2 + 2Al \rightarrow 2AlCl_3 + 3Zn$$
$$BCl_3 + Al \rightarrow AlCl_3 + B$$
$$3CuCl_2 + 2Al \rightarrow 2AlCl_3 + 3Cu$$

In each case gaseous aluminium chloride is formed which bubbles through the melt and hydrogen which is dissolved in the melt, will diffuse into these bubbles until partial pressure equilibrium is reached—exactly as described in chlorine degassing. At the same time the metal of the metal chloride is taken up by the

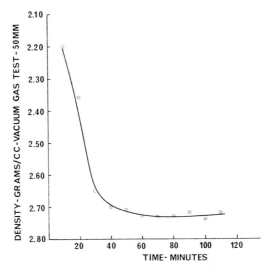

FIG. 23a. Degassing with chlorine gas. 200 kg 7075 alloy, open-hearth furnace. Temperature 705°C, gas flow 140 l./hr.

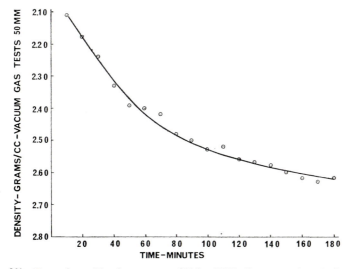

FIG. 23b. Degassing with nitrogen gas. 200 kg 7075 alloy, open-hearth furnace. Temperature 705°C, gas flow 140 l./hr.

aluminium. With zinc chloride, therefore, zinc will go into the aluminium, similarly with copper chloride, manganese chloride, titanium chloride, copper, manganese or titanium, respectively, are alloyed to the aluminium. Some of these metals are often unwanted impurities and careful selection of the degassing chloride is therefore necessary. Furthermore, most of the metal chlorides are more or less hygroscopic; they must be used in the anhydrous state and they absorb water very quickly, so that there is a danger of introducing more hydrogen than is taken out. Other chlorides, such as aluminium chloride and phosphorus pentachloride, decompose in contact with humid air and give off clouds

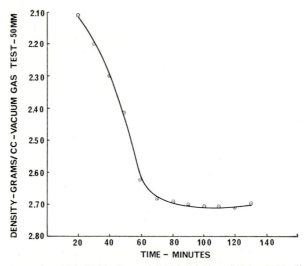

FIG. 23c. Degassing with 90% nitrogen, 10% chlorine. 200 kg 7075 alloy, open-hearth furnace. Temperature 705°C, gas flow 140 l./hr.
(Figs. 23a, b and c by courtesy of C. L. Brooks[28]).

of hydrochloric acid and are therefore very unpleasant to handle. Because of these disadvantages the use of anhydrous chlorides has been practically abandoned. In the search for chlorine compounds which would have a good degassing effect, which are reasonably economical to use, which are not hygroscopic, and do not introduce unwanted constituents into the metal, the choice fell on two organic, fully chlorinated hydrocarbons, carbon tetrachloride, CCl_4, and hexachloroethane, C_2Cl_6, of which hexachloroethane particularly comes very near to fulfilling most of the stipulated requirements.

Carbon tetrachloride is a liquid at ambient temperature, it looks like water and has to be soaked into asbestos wool so that it can be plunged into the metal; a number of patents have been taken out on the use of carbon tetrachloride as a degassing agent, but none of these are used on a commercial scale. Another method of application is by passing a stream of nitrogen gas through warm carbon tetrachloride; because of the low boiling point (67°C) the nitrogen carries some carbon tetrachloride vapour into the metal thus exerting a combined degassing effect.

Hexachloroethane, on the other hand, is a solid; the well-defined crystals sublimate at 183°C; they are not hygroscopic and are insoluble in water but easily soluble in most organic solvents. When the first tests with this material as a degasser were carried out in 1934, it was not available as a commercial product, but as the demand increased, plants for its large-scale manufacture were installed in a few countries.

If hexachloroethane is plunged into molten aluminium, all or some of the following reactions take place:

$$Al + C_2Cl_6 \rightarrow Al + C_2Cl_4 + Cl_2$$
$$2Al + 3Cl_2 \rightarrow 2AlCl_3$$
$$Al + C_2Cl_4 \rightarrow Al + CCl_4 + C$$
$$Al + CCl_4 \rightarrow Al + 2Cl_2 + C$$
$$Al + C_2Cl_6 \text{ (solid)} \rightarrow Al + C_2Cl_6 \text{ (gaseous)}$$

At a metal temperature of about 720°C, 30% of the hexachloroethane is decomposed into chlorine which reacts with aluminium, forming gaseous aluminium chloride which acts as described under chlorine degassing.

The other decomposition products, with the exception of carbon, are gases at the temperature of the molten metal. They bubble through the melt and the bubbles take up hydrogen exactly like aluminium chloride or nitrogen; approximately 94% of the hexachloroethane passes as gas through the metal. The higher the temperature the more chlorine is formed on decomposition.

Hexachloroethane can be easily handled; its low toxicity and the ease of application, and a degassing efficiency comparable to that of chlorine, make it a degassing agent which is generally accepted in aluminium foundries and wrought-aluminium works alike. It can be tabletted easily and mixed with other salts which will delay decomposition and produce smaller bubbles; it can also be applied with grain refining salts to produce a fine grain at the same time as it degasses (see grain refinement, p. 272). Table 2 is a list of the most common degassing agents and their relative efficiency, advantages and disadvantages.[29]

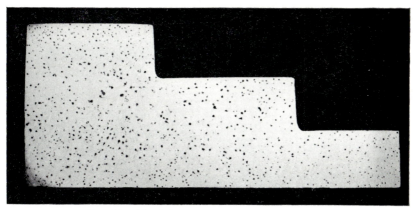

FIG. 24a. Aluminium step casting before degassing.

For the degassing of large quantities of aluminium, particularly in wrought-aluminium works, continuous chlorination processes have been developed. Special furnaces have been installed in the course of the last 10–15 years in a number of aluminium rolling mills. Chlorine gas is pumped through the metal continuously at the rate of 5–9 kg chlorine per ton of metal, depending on the

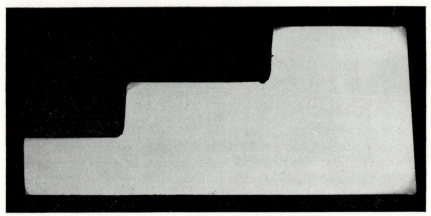

FIG. 24b. Same casting after degassing with hexachloroethane degasser.

TABLE 2.

	Chlorine, Cl_2	Hexachloro-ethane, C_2Cl_6	Aluminium chloride, $AlCl_3$	Nitrogen, N
Removal of hydrogen	very good	good to very good	good to very good	good
Removal of aluminium oxide	very good	good	good	little
Protective gas—atmosphere over melts	good	good	good	little
Type and metal content of skimmings				
Type	dry	dry	dry	wet
Metal content	very low	low	low	high
Can it be used?				
(a) for ladle degassing	(a) yes	(a) yes	(a) yes	(a) yes
(b) for continuous degassing	(b) yes	(b) no	(b) no	(b) yes
Control over quantities to be applied	very good	very good	good	very good
Relative costs of degassing agents	low	reasonably low	reasonably low	low
Main disadvantages	toxic and corrosive, elaborate ventilation necessary	somewhat corrosive, suitable for ladle degassing	corrosive, very hygro-scopic and unpleasant to handle	high metal content in skimmings

type of alloy to be treated and gas contents as low as 0·005–0·08 cm³ per 100 g of metal can be obtained. This treatment eliminates to a large extent the re-absorption of hydrogen between degassing and pouring of the metal.[30]

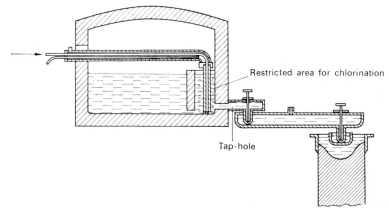

Restricted area for chlorination

Tap-hole

FIG. 25. American patent for the continuous degassing of aluminium with chlorine.

5. Treatment in a Vacuum

For the solubility of gases in aluminium Sievert's Law applies; this means that it is dependent on temperature and pressure. If either are decreased, solubility is decreased too. If, therefore, the molten metal is subjected to a vacuum, the hydrogen is released in molecular form; this is the basis for degassing in vacuum. The gas escapes in the form of bubbles which can be seen as they break through the oxide film on the surface. The speed with which the hydrogen is ejected depends on the degree of vacuum and on the denseness of the oxide film. Generally, degassing by vacuum takes much longer than the methods discussed before, particularly the degassing with chlorine or chlorides and the special plants required for it are expensive and complicated. It is therefore not used extensively in practice but it is a useful tool for research and for qualitative and quantitative hydrogen determinations. By a combination of treatments in a vacuum and bubbling through the melt at the same time an inert or reactive gas, the degassing process can be speeded up considerably.

6. Treatment with Ultrasonic Waves

This method for degassing aluminium is of little practical interest but small quantities of metal can be effectively treated. It consists of applying ultrasonic waves to a melt whereby the wave frequency is not critical and can vary between 500 and 2000 cycles per second. The gas diffuses through the metal due to the disturbance caused by the waves and in the presence of much gas bubbles can also form.

7. Treatment with Compounds and Elements Forming Hydrides

Some elements combine with hydrogen to form stable hydrides; the hydrogen cannot therefore diffuse freely in the metal and cannot form bubbles any more and thus porosity is prevented. There are a number of elements which can form hydrides, the most important ones are zirconium, calcium, sodium, titanium and lithium. Some of these hydrides are insoluble in molten aluminium and can be removed by skimming or they remain suspended in the melt. They hold the hydrogen in form of stable compounds so that it cannot appear as a gas bubble. The best known and most thoroughly studied example of an insoluble hydride is that of zirconium.

Most of the investigations have been carried out in connection with aluminium-magnesium alloys.[31] Zirconium is added as potassium-zirconium fluoride, K_2ZrF_6, which on contact with the molten metal decomposes and forms zirconium metal and magnesium fluoride. The zirconium metal enters the alloy and part of it combines with hydrogen to zirconium hydride, the other part forms the aluminium-zirconium compound, Al_3Zr, which is practically insoluble in aluminium and precipitates as fine crystals which act as nuclei for crystallization and produce grain refinement of the alloy (see grain refinement).

According to Odina and Sharov,[31] 0·1g of zirconium can combine with 18·4 cm^3 of hydrogen at 700°C. From a melt of 100 kg of aluminium containing 1·0 cm^3/100 g hydrogen (which is well above the average hydrogen content) 1000 cm^3 of hydrogen will have to be removed. Theoretically, according to the above literature, this can be done with 5·5 g of zirconium or 17·5 g K_2ZrF_6. Allowing for an efficiency of only 10% only 175 g of K_2ZrF_6 would be needed for degassing, a figure which can only be accepted with the greatest reserve.

Sodium, apart from its use for the modification of aluminium-silicon alloys, can also serve as a degassing agent for aluminium alloys by dissolving large quantities of hydrogen and forming sodium hydride; however, the use of sodium has disadvantages—a small surplus of it in the metal will react with moist air and pick up gas.

Titanium, too, has a high solubility for hydrogen and forms hydride, thus it has been used for degassing aluminium melts. K. Anderko[33] reports on a method, where melts of Al–Cu–Si–Ni–Mg alloys and Al–Si–10Mg alloy, both of which are very prone to gas pick-up, are treated with 0·5–2·0% titanium sponge which will dissolve hydrogen and keep it in solution. The hydrogen content of a melt of Al–Si–Mg alloy which before treatment was 0·5 cm^3/100 g was reduced to 0·2 cm^3/100 g after treatment with 4% titanium sponge.

GRAIN REFINING OF ALUMINIUM AND ALUMINIUM ALLOYS

The beneficial effects of fine grain size on the mechanical properties, shrinkage and hot-tearing characteristics of aluminium alloys has become more widely appreciated within recent years. That these fine grain sizes could be obtained by additions of certain elements to the molten metal has been known for some time and utilized in practice. But the exact mechanism of grain refinement and the optimum additions and combinations of the grain refining elements have been studied only within the last 20–30 years. Since then, however, a great deal of literature has been published on this subject and the theory and practice of this process have been thoroughly investigated.

This does not mean that the theory of grain refinement has been completely established and that research workers are unanimous on their results, but in practice the best ways of obtaining optimum grain refinement are fairly well established.

There are two ways in which grain refinement can be obtained:

(a) grain refinement through "own nuclei",
(b) grain refinement through "foreign nuclei".

As far as (a) is concerned this applies only to pure and super-pure aluminium. By heating pure aluminium just above the liquidus temperature and immediately solidifying it, a very fine grain is developed. A similar effect can be produced by inserting small pieces of aluminium sheet into a melt which is just above the melting temperature.[34] In both cases some dendrites are formed which act to nucleate the metal bath. Quick solidification of the metal as is the case in die-casting, continuous casting and pressure-diecasting will also result in a fine grain. Solidification under pressure has a pronounced influence on the grain size of the metal, density is increased whilst the grain becomes smaller and physical properties improve. Treatment with ultrasonic waves has a similar effect.

(b) Whilst grain refinement through "own nuclei" applies only to pure aluminium, grain refinement through "foreign nuclei" can apply to pure aluminium as well as to practically all types of aluminium alloys and this is therefore the method which is used in practice on a large scale for casting and wrought metals.

In the aluminium field one of the earliest suggestions was that elements resulting in grain refinement usually form a peritectic system with the base metal and that grain refinement occurs when the peritectic composition is exceeded. It was suggested that the intermetallic particles produced by the peritectic reaction are precipitated as a fine shower and act as nuclei for the precipitation of the solvent metal—aluminium. Work by Crossley and Mondolfo[35] and also independent work by Kondic[36] and Atterton and his colleagues[37] tends to support this idea.

However, Eborall[38] studied the influence of a large range of additions on the grain refinement of aluminium alloys and observed that certain elements which form peritectic systems resulted in no grain refinement, e.g. the aluminium-chromium system. Grain refinement was also observed with contents of certain elements, notably titanium and zirconium, below the percentage necessary to produce primary particles by the peritectic reaction. Cibula[39] in a very extensive study of grain refinement was able to eliminate the undercooling of high-purity aluminium by small additions of certain elements such as titanium, boron, zirconium, vanadium, columbium and others, and this strongly suggests that nucleation of aluminium is caused by solid particles. Centrifuging of molten metals and X-ray examination supplied considerable evidence that the active nuclei resulting from boron additions are aluminium boride particles (AlB_2). Cibula suggested that such nuclei were suited for aluminium nucleation since the metal atoms in the boride lattice are arranged in simple hexagonal lattices in the close-packed planes of which the interatomic distances are similar to the interatomic distance in solid aluminium. Refinement of aluminium alloys by additions of titanium were shown to be due almost certainly to the formation of titanium carbide, TiC., carbon being a minor constituent in most aluminium alloys. As for the borides the lattice dimensions of titanium boride are similar to those of solid aluminium. Combined additions of titanium and boron lead to considerably greater refinement than the single additions of either and Cibula suggested that this is due to the formation of a mixed boride, probably titanium boride (TiB_2) with other intermetallic compounds in solid solution.

Cibula also investigated the stability of grain refinement on remelting and he concluded that the loss of refinement on remelting and superheating should be less in alloys refined by boron than in those refined by titanium. His theory is that when a metal refined by titanium is remelted, the nuclei of titanium carbide dissolve but do not readily re-precipitate on cooling due to the great dispersion of titanium and carbon atoms which make nucleation with titanium carbide more difficult. Similarly, in melts treated with boron the aluminium boride also dissolves if heated above the aluminium-boron liquidus (660°C) but the re-formation of aluminium boride on cooling should be less liable to suppression since aluminium is the main constituent of the melt.

In contrast to Cibula, Atterton and collaborators[37] have found that titanium leads to greater grain stability than boron on remelting and their results suggest that increase of grain size on remelting is due to coalescence of nuclei. It is generally accepted that the particles responsible for grain refinement are very fine, probably sub-microscopic, and that an increase in grain size on remelting may well be due to coalescence of nuclei so that only a reduced number of them are available for grain refinement and therefore grain size increases. Kondic[40] has shown that in spite of the presence of sufficient titanium, repeated remelting results in grain growth even if the titanium content of the melt remains constant. The grain refining effect can be reactivated by adding further small quantities of titanium to the melt after every remelting process.

In practice it has been found that the best results for grain refinement can be obtained by a combined treatment with titanium and boron and the refinement is appreciably greater than with much larger quantities of the single elements. The nuclei mainly responsible for the refinement are most probably titanium boride, TiB_2, crystals but other particles such as aluminium boride, AlB_2, and titanium carbide, TiC, will contribute towards a fine grain. The formation of these intermetallic compounds depends on the type of the alloy to be treated, on the temperature of the metal and on the ratio of titanium to boron. One of the earliest methods of adding titanium and boron, which is still much used today, is the application of these two elements in form of their respective salts, potassium titanium fluoride and potassium borofluoride, K_2TiF_6 and KBF_4 respectively. The salts are very often tabletted and plunged into the molten metal. They decompose in contact with the metal and introduce the elements in a state of high purity.

Potassium titanium fluoride is reduced by aluminium and forms titanium and aluminium fluoride, or if magnesium is present in the alloy, some magnesium fluoride is formed.

Potassium borofluoride decomposes under heat forming gaseous borofluoride BF_3, which on bubbling through the melt will react with aluminium and magnesium forming boron and the respective metal fluorides.

$$K_2TiF_6 + 2Al \rightarrow Ti + 2AlF_3 + 2K$$
$$KBF_4 \xrightarrow{\text{heat}} KF + BF_3$$
$$BF_3 + Al \rightarrow B + AlF_3$$

Titanium and boron together will form titanium boride, TiB_2, besides some aluminium boride, AlB_2, and titanium carbide, TiC, is formed; the reaction which takes place can also be expressed by the equation:

$$3K_2TiF_6 + 6KBF_4 + 14Al \rightarrow 3TiB_2 + 14AlF_3 + 12K$$

Borofluoride gas can also react with aluminium oxide and form aluminium fluoride:

$$2BF_3 + Al_2O_3 \rightarrow 2AlF_3 + B_2O_3$$

The ratio of titanium compound to boron compound in the mixture is usually so chosen that the maximum amount of titanium boride crystals are obtained; it is a fact that the yield of titanium from potassium titanium fluoride is generally much higher than the yield of boron from potassium borofluoride; the first can be as high as 90% and is of course due to the fact that the reaction with aluminium is a straight reducing reaction without the formation of gaseous products; in the case of potassium borofluoride, the yield is much lower (about 50%) partly due to the formation of gaseous BF_3, some of which will escape into the atmosphere without reacting. Efforts have been made to increase the yield, especially that of boron and to produce nuclei of very small particle size. W. Thury[41] adds aluminium—or magnesium-powder—to the mixture of double-fluorides and claims that these exothermic, more reactive combinations will react at lower temperatures, give better yields and produce nuclei of smaller and more consistent size.

Sometimes hexachloroethane is added to the tablets of mixed fluorides; this helps the breaking up of the tablets and the decomposition of the fluorides; at the same time the carbon resulting from the decomposition of the hexachloroethane helps to form some titanium carbide which adds to the efficiency of grain refinement. With alloys already containing titanium as part of the alloy it is generally only necessary to add boron to obtain satisfactory grain refinement.

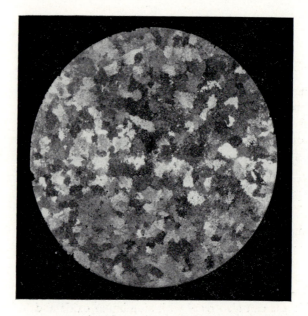

Fɪɢ. 26a. Aluminium alloy before grain refining.

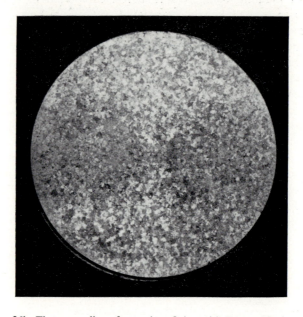

Fɪɢ. 26b. The same alloy after grain refining with Foseco Nucleant 2.

Where metal is treated in furnaces with electric resistance heating some caution is necessary because the fumes arising from the decomposing fluorides may attack the metal of the resistance coils and shorten their life; an extraction fan at the side of the furnace will prevent the fumes reaching the coils.

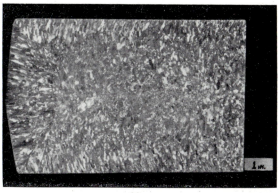

26c.

26d.

FIG. 26c and d. Chilcast aluminium alloy before and after treatment with 0·25% Nucleant No. 2. (*Note:* large columnar crystals at edges of untreated ingot.)

In recent years the addition of titanium and boron in form of "hardeners". i.e. alloys with aluminium, has gained acceptance and is used today on a considerable scale especially in the wrought-aluminium industry. A range of hardeners is available, of which the one containing 5% titanium and 1% boron is the most widely used. According to D. Rowe[42] this is an effective method of grain refining but there are some disadvantages:

1. Because TiB_2 is much denser than pure aluminium it tends to precipitate at the bottom of the melt.

2. The formation of hard boride particles sometimes leads to reduced die life in extrusion dies.

GRAIN REFINING OF HYPEREUTECTIC ALUMINIUM–SILICON ALLOYS WITH 17–25% SILICON

Hypereutectic aluminium-silicon alloys have been developed in Europe in recent years mainly for use as piston alloys for internal combustion engines. They contain between 17% and 25% silicon besides additions of copper, nickel, magnesium and other metals, and the reason for their suitability as piston alloys is their low thermal expansion and their good antifriction properties.

In these alloys the silicon precipitates on solidification in large irregularly shaped and distributed crystals which have an adverse effect on the mechanical properties, workability and machinability. These large primary silicon crystals have, therefore, to be refined in order to obtain a fine and more uniformly distributed structure to confer maximum physical properties. This cannot be done with titanium and boron; as already discussed, these alloys need special treatment with a grain refining agent which has a similar crystal structure and lattice spacing as silicon. It has been found that aluminium phosphide fulfils these requirements[43] very well and that excellent grain refining can be obtained.

Aluminium phosphide can be introduced in a number of ways; many phosphorus compounds react with aluminium to form aluminium phosphide, AlP. This is practically insoluble in aluminium at the casting temperature of the alloy and forms the nuclei for crystallization.

Phosphorus pentachloride PCl_5 is used in some foundries in Germany. It decomposes in the melt and forms aluminium phosphide which causes grain refinement. This compound, however, is very unpleasant to handle; it starts decomposing at 100°C, it is very hygroscopic and it gives off clouds of poisonous fumes if exposed to the atmosphere and if applied to the metal. For these reasons strict safety precautions are necessary for safe handling and it needs an efficient ventilation system. The water-free and suitably packed product is expensive, but if all safety measures are taken, phosphorus pentachloride is an efficient grain refiner.

Another way of introducing phosphorus is by way of 10 or 15% phosphor copper; this works without any side reaction—it dissolves in the melt and forms aluminium phosphide but it introduces copper into the alloy which in the case of copper free alloys is undesirable.

The easiest and most convenient way of introducing phosphorus into the melts is by way of red phosphorus. This is obtainable as a fine, dry powder; it is neither hygroscopic nor toxic. It burns easily but once it is mixed with other suitable chemicals the fire hazard is much reduced. For optimum grain refining results the size of the phosphorus particles should not be more than 60 microns and the phosphorus is usually carried in a mixture of potassium chloride and potassium titanium fluoride.[44] It is of great importance that no sodium or sodium salts should come into contact with the metal; this would lead to the formation of sodium phosphide, Na_3P, which forms in preference to aluminium phosphide but does not contribute to grain refinement. The use of either metallic sodium or of sodium salts must therefore be carefully avoided.

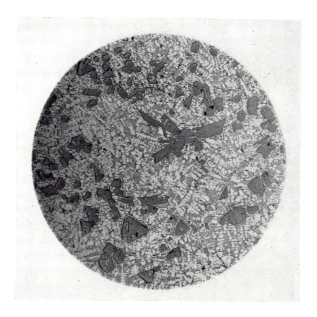

FIG. 27a. Hyper-eutectic aluminium silicon alloy not refined.

FIG. 27b. Same alloy refined with phosphorus.

K

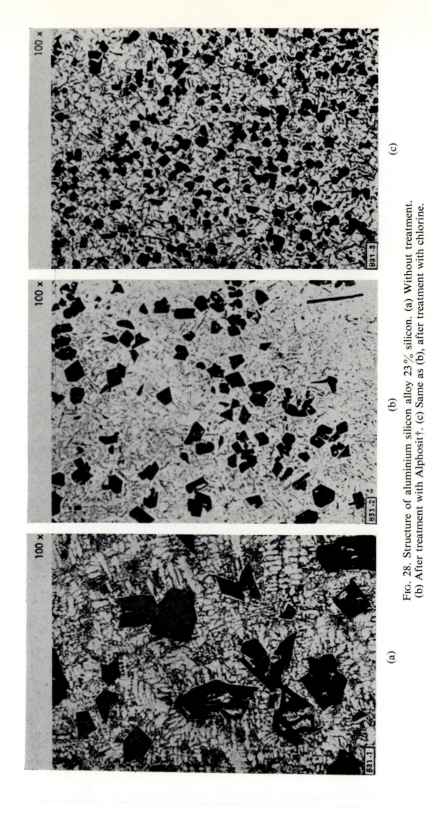

FIG. 28. Structure of aluminium silicon alloy 23% silicon. (a) Without treatment. (b) After treatment with Alphosit†. (c) Same as (b), after treatment with chlorine.

(By courtesy of W. Thury and H. Kessler).

† Trademark for flux according to DBP 1035911.

Improved grain refining results can be obtained if the metal, after grain refining treatment, is treated with chlorine either from a cylinder or in form of hexachloroethane. The chlorine seems to activate the aluminium phosphide particles and produce many small crystals which act as nuclei for crystallization. According to DBP 1035911[44], if a melt which has been grain refined is left standing at casting temperature for prolonged periods, some of the grain refinement effect is lost, but it can be reactivated by subsequent treatment with chlorine without further additions of phosphorus.

GRAIN REFINEMENT OF EUTECTIC ALUMINIUM–SILICON ALLOYS

The use of metallic sodium and sodium-salts for the grain refinement or "modification" of 11–13% aluminium-silicon alloys has already been discussed in the section dealing with fluxing of aluminium alloys. The reason for the change from a coarse to a fine structure after the addition of sodium has been investigated by numerous research workers, and the results have been published in numerous research reports, but even today there are still a number of problems which await complete clarification.

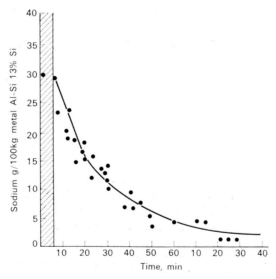

FIG. 29. Loss of sodium in 13% aluminium silicon alloy.
(By courtesy of C. Mascré and A. Lefèbvre).

The degree of modification depends on the amount of sodium and on the speed of solidification; for sand castings which cool much slower more sodium is necessary to produce a fine grain structure than for die castings or pressure die castings. A well-modified sand casting should have a residual sodium content of 0·015–0·02%; for die casting it can be much lower, depending on the thickness of sections. Because sodium oxidizes easily, the sodium content of the metal will gradually diminish; the rate of loss of sodium is dependent upon both time and temperature. The pouring of the metal must therefore take place within a limited period of time after the sodium has been added in order to obtain the finest possible grain. Figure 29 shows the loss of sodium over a period of 90 minutes at constant temperature.[45]

It is therefore the general practice to add a small surplus of sodium in the case of sand castings to allow for the inevitable loss during the period of treatment and casting.

The unmodified structure of eutectic aluminium-silicon alloys occurs either in lamellar or in granular form. The presence of traces of phosphorus favours the granular structure; the lamellar structure can be changed into the granular structure by the addition of traces of phosphorus. When sodium is added, the phosphorus is first of all eliminated by forming sodium phosphide, the structure becomes first lamellar and then modified (Fig. 30(a) and (b)).

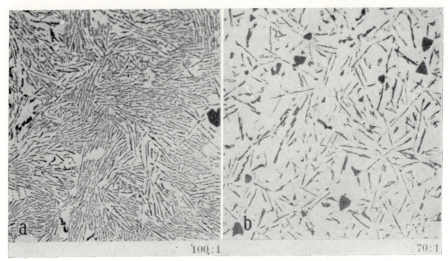

FIG. 30a. Microstructure of lamellar Al-Si alloy.
FIG. 30b. Microstructure of granular Al-Si alloy.
(By courtesy of D. Altenpohl[48]).

When the metal is kept molten over long periods in holding furnaces modification tends to go off unless fresh metallic sodium or modification salts are added from time to time. In order to overcome this difficulty, fused or compressed blocks of modifying mixtures have been developed which are placed on top of the metal and which will keep shape over long periods and will disintegrate only slowly. Small quantities of sodium are continuously released into the metal which are sufficient to maintain modification; such a block may last from 3 to 5 hours.[46]

In the search for additions which would produce a more permanent modification many metals and non-metallic compounds have been tried. Most elements in the first and second group of the periodic system have some effect but none as good as sodium. Only during the last few years has the influence of strontium been more thoroughly investigated. It was found that strontium produces a modified structure very similar to that of sodium; the mechanical properties of a strontium-modified metal are similar to those with sodium and so is the fluidity. The main difference is that the modification with strontium will persist for a few hours and that metal modified with strontium can be remelted a number of times without losing the fine structure.[47]

Under ideal conditions in a moisture-free atmosphere, sodium will act as a degassing agent, but in a humid atmosphere sodium can greatly increase the danger of hydrogen pick-up. In a study on the influence of sodium on the gas content of aluminium alloys, I. V. Volkhontsev and D. P. Lovtsov[49] found that in a moist furnace atmosphere, metal containing sodium will pick up hydrogen faster and in larger quantities than metal without sodium.

TREATMENT OF ALUMINIUM ALLOYS FOR DIE CASTING

The fluxing and degassing of die casting alloys is carried out in very much the same way as for sand castings. Sometimes either through faulty die design or through inadequate feeding of a thick section through a thin one the casting shows cracks, draws and shrinkage cavities which cause it to be scrapped. This happens mainly when the metal is almost gas-free and the trouble can often be rectified by the introduction of carefully controlled quantities of hydrogen in the form of water vapour. The dissociated hydrogen dissolves in the molten metal and causes a very slight increase in its volume, but because of the very rapid solidification of die castings the gas will remain in solution (providing it is not greatly in excess). Even if it is somewhat above the solid solubility point the casting can be made without the previously mentioned difficulties. The introduction of hydrogen can be carried out in different ways; it has been tried by bubbling hydrogen through the metal from a cylinder of hydrogen, but as the gas is then in its molecular state only very small quantities will dissolve. In order to dissolve, the gas must be in its atomic form and this is obtained by the introduction of water-vapour. This decomposes in the metal and forms hydrogen in *status nascendi* which readily dissolves.

In former days some die casters used green wood as the source of water-vapour; this certainly introduced hydrogen into the metal but the amount of gas was difficult to control and the burning wood produced a lot of dross with corresponding metal losses.

For many years now carefully controlled quantities of water are introduced into the metal by means of chemicals which contain crystal water as part of the molecule. Some chemicals contain more than 50% water (such as borax $Na_4B_2O_7 \cdot 10H_2O$) but they are dry, free running crystals which can be plunged into the liquid metal without danger; they give off their water as vapour slowly over prolonged periods and in this way the gas pick-up can be easily regulated and controlled. There are a large number of chemicals which contain appreciable amounts of crystal water but not all of them are suitable for this particular use; they may react too violently with the molten metal, they may oxidize it too heavily or they may introduce unwanted impurities and for these reasons the choice is limited and these compounds must be carefully selected.

THE REMELTING AND TREATMENT OF ALUMINIUM SCRAP

1. The Melting of Scrap

The recovery of aluminium from aluminium scrap has assumed very great importance and an industry of respectable size has been built up in the course of the last 60 years. This so-called "secondary aluminium industry" uses for its raw materials the components made of aluminium which return to the scrap market after they have become useless in the course of time.

The recovery from scrap of the maximum amount of usable metal needs special knowledge and experience and the men in charge of aluminium refineries are specialists in their field. Their task is, by the removal of impurities, either metallic or non-metallic, and by using the most suitable melting method for a particular type of scrap, to transform it into clean metal which can be used in the same way and for the same purpose as virgin metal.

There are two classes of aluminium products, into which the entire production of aluminium is channelled:

1. Those which disappear or are destroyed in the course of use; into this class belong aluminium powder and granules used for deoxidation of steel, or for the thermit-process, or for heat and corrosion protection, or for paint-powder in aluminium paints. Although the total tonnage used up in these processes is quite considerable, it is only a small fraction of the total production of aluminium metal.
2. Those products for which the aluminium metal is kept intact and which have a comparatively short life, such as aluminium foil for packing and wrapping, tubes, containers and goods with a much longer life like pots and pans, automobile and aeroplane parts, cables for electric installations, ship fittings and interior and exterior architecture. The greater part of these goods after they have served their purpose return to the market in form of scrap and are remelted. The quantity of scrap increases from year to year and is in proportion to the increase in total aluminium production and runs into several hundred thousands of tons throughout the world. It can be seen therefore that the selection of the right type of furnace and the correct methods for cleaning and purifying the molten metal are of considerable importance to the secondary metals industry.

Aluminium scrap can be in the form of large sheets and big castings down to very fine turnings and sawings, it can be pure aluminium or aluminium alloys; it can be contaminated with iron and other heavy metal and it may contain oil and grease, paint and paper. There are methods for sorting scrap and removing iron by magnets, of removing oil by treatment with degreasing agents and by pressing scrap into briquettes for easier handling before remelting, but a detailed description of these methods can be found in the relevant literature. Furthermore, the scrap may consist of several different aluminium alloys and there are several quick methods available, using spot analyses, for the identification of these alloys.

The selection of the furnace for remelting the scrap depends on the type of scrap, whether it is large or in briquettes or in form of turnings and other fine material. Large scrap is usually melted in reverberatory-type furnaces of several tons' capacity which are gas or oil fired; as the amount of aluminium oxide is usually rather high, low melting-point fluxes are normally used. They contain sodium and potassium chloride with additions of fluoride; sometimes calcium chloride is added for lowering the melting point and viscosity of the flux; a typical composition of a flux of this type is: 50% calcium chloride, $CaCl_2$, 30% sodium chloride, NaCl, and 20% potassium chloride, KCl, which has a melting point of $ca.$ 500°C. Apart from, or instead of, $CaCl_2$, calcium chloride, the flux may also contain some magnesium chloride, particularly if the aluminium alloy contains magnesium as part of the alloy. These fluxes are satisfactory, take up impurities and perform well as long as the furnace is kept hot; when it cools down, however, the calcium chloride or magnesium chloride being very hygroscopic, will attract water and get wet and the dampness may also affect the brickwork of the furnace.

The chief function of the flux is to prevent oxidation, to remove aluminium oxide and to form skimmings as free from metal as possible. This is all one can expect of a good remelting flux—other operations such as degassing and the removal of unwanted impurities, are the function of special fluxes or special treatment.

The remelting of fine scrap, such as turnings, chippings, sawings and skimmings, is nearly exclusively carried out in large rotating furnaces, a practice

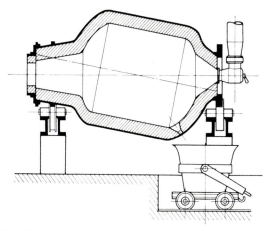

FIG. 31. Rotary furnace for melting fine aluminium scrap.

which has been developed over the last 30–40 years. To prevent extensive oxidation of the large surface area of the scrap it is introduced into a molten bath of flux, whereby it is immediately covered with the flux and protected from further oxidation. The flux usually consists of sodium chloride with small additions of cryolite or some other fluoride. The melting point of NaCl is at 800°C and additions of KCl are made to lower the melting point of the salt bath; a mixture of 45% sodium chloride, 45% potassium chloride and 10% cryolite has a melting point of 607°C. Depending on the fineness and the degree of oxidation of the scrap material, the amount of salt mixture used is between 10 and 40% but the salt can stay in the furnace until it has taken up so many impurities that it is too thick and contaminated for further use.

Paper-backed aluminium foil has to be treated in a different way before the metal can be charged into the salt bath of the rotary furnace. The paper-backed foil which can contain up to 50% of paper is heated in special muffle furnaces under exclusion or part exclusion of air to a temperature at which the paper will char but without oxidizing the foil to any extent. The carbon from the paper which is now in flake or powder form can be separated from the metal by sieving or by air separation in a cyclone.

Virgin aluminium and aluminium alloys are usually free from metallic impurities and only contain small quantities of non-metallic impurities of hydrogen and aluminium oxide. In contrast scrap aluminium very often contains appreciable quantities of both non-metallic and metallic impurities which will have to be removed if metal has to be produced which conforms to specified requirements. For their removal a large number of processes have been proposed of which only a limited number are really satisfactory and used on an industrial scale.

As with virgin metal the most common non-metallic impurities are hydrogen and aluminium oxide; but according to the origin and type of scrap and to conditions of storage these impurities are generally present in larger amounts and need more thorough treatment for their removal. In heavily oxidized scrap, such as wet turnings, sheet and foil or badly corroded machine parts, the aluminium oxide content can be as high as 0·10–0·15%. The remelting and removal of aluminium oxide from such material has been discussed in the previous section.

The hydrogen content of scrap melts is usually much higher than that of virgin melts but the methods for removing the gas are very similar to those described in the section on degassing. Chlorine, organic chlorides and mixtures of chlorine and nitrogen are used for the degassing treatment but because of the higher gas content the treatment has to be over longer periods and larger quantities of the purging gases have to be used. For the chlorination of the molten metal special converters have been developed into which the metal is transferred from the melting furnace and where it is subjected to chlorine degassing for long periods. These converters usually can hold 1–3 tons of metal and either Cl or a mixture of 90% N and 10% Cl are passed through it until the metal is reasonably free from gas; up to 3–6 kg of Cl per ton of metal is sometimes necessary, but as the reaction and formation of aluminium chloride is exothermic no additional heat is required to keep the metal molten during the degassing period.

Badly contaminated scrap very often contains other non-metallic impurities, such as carbides, nitrides and sulphides. The treatment with chlorine or organic chlorides will also clean the metal also from these impurities either by chemical reaction or by purely mechanical agitation whereby they are removed into the slag and skimmed off.

Fig. 32. Rotary converter for chlorine and nitrogen refining.
(By courtesy of W. Dautzenberg[51]).

Removal of Metallic Impurities

If aluminium scrap of unknown origin is to be remelted and if it is a mixture of a number of different alloys, the remelted metal can contain metallic constituents whose presence in the alloy is undesirable. Sometimes an element which is desirable in one alloy can be detrimental if present in another and therefore it will have to be removed to make good secondary metal. The elements most commonly found are magnesium, iron, silicon, copper, zinc, manganese, chromium, lead, titanium, zirconium, tin, sodium and calcium. For the removal of these elements a number of methods are available which are based partly on chemical, partly on physical processes.

Table 3 gives the heat of formation of the oxides, chlorides, fluorides, nitrides and sulphides of those metals which are usually alloyed with aluminium or are present as impurities.[50]

κ*

TABLE 3.

Element	Oxides Formula	Oxides Heat of formation	Chlorides Formula	Chlorides Heat of formation	Fluorides Formula	Fluorides Heat of formation	Nitrides Formula	Nitrides Heat of formation	Sulphides Formula	Sulphides Heat of formation
Al	$\frac{1}{3}Al_2O_3$	133·3	$\frac{1}{3}AlCl_3$	56·2	$\frac{1}{3}AlF_3$	118·6	AlN	76·5	$\frac{1}{3}Al_2S_3$	57·6
B	$\frac{1}{3}B_2O_3$	102·0	$\frac{1}{3}BCl_3$	32·4	$\frac{1}{3}BF_3$	—	BN	60·7	$\frac{1}{3}B_2S_3$	19·0
Be	BeO	143·1	$\frac{1}{2}BeCl_2$	59·0	$\frac{1}{2}BeF_2$	120·6	$\frac{1}{5}Be_3N_2$	67·3	BeS	55·9
Bi	$\frac{1}{3}Bi_2O_3$	46·0	$\frac{1}{3}BiCl_3$	30·2	$\frac{1}{3}BiF_3$	70·6			$\frac{1}{3}Bi_2S_3$	14·1
Ca	CaO	151·5	$\frac{1}{2}CaCl_2$	95·4	$\frac{1}{2}CaF_2$	145·2	$\frac{1}{5}Ca_3N_2$	52·5	CaS	110·0
Cd	CdO	61·1	$\frac{1}{2}CdCl_2$	46·5	$\frac{1}{2}CdF_2$	83·2	$\frac{1}{5}Cd_3N_2$	-19·3	CdS	34·5
Ce	$\frac{1}{3}Ce_2O_3$	144·9	$\frac{1}{3}CeCl_3$	84·3	$\frac{1}{3}CeF_3$	88·7	CeN	78·0	CeS	118·0
Cr	$\frac{1}{3}CrO_3$	46·2	$\frac{1}{3}CrCl_3$	43·5	$\frac{1}{3}CrF_3$	64·0	CrN	29·4	CrS	
Cu	CuO	37·1	$\frac{1}{2}CuCl_2$	24·6	$\frac{1}{2}CuF_2$	84·0	Cu_3N	-17·8	Cu_2S	19·6
Fe	FeO	63·2	$\frac{1}{3}FeCl_3$	31·9	$\frac{1}{2}FeF_2$	64·8	Fe_2N	0·9	FeS	22·8
H	H_2O	68·3	HCl	22·0	HF	134·5	H_3N	11·0	H_2S	4·8
K	K_2O	86·4	KCl	104·2	KF	146·3			K_2S	102·4
Li	Li_2O	142·5	$LiCl$	96·9	LiF	133·0	Li_3N	47·0	Li_2S	107·4
Mg	MgO	143·7	$\frac{1}{2}MgCl_2$	76·7	$\frac{1}{2}MgF_2$	95·0	$\frac{1}{5}Mg_3N_2$	55·1	MgS	83·0
Mn	MnO	92·0	$\frac{1}{2}MnCl_2$	57·6	$\frac{1}{2}MnF_2$	62·0	Mn_4N	30·3	MnS	49·0
Mo	$\frac{1}{3}MoO_3$	59·4	$\frac{1}{5}MoCl_5$	18·1	$\frac{1}{6}MoF_6$	136·5	Mo_2N	16·6	$\frac{1}{3}Mo_2S_3$	31·0
Na	Na_2O	100·7	$NaCl$	98·6	NaF	79·0			Na_2S	92·4
Ni	NiO	57·5	$\frac{1}{2}NiCl_2$	36·5	$\frac{1}{2}NiF_2$	79·2	Ni_3N	-0·2	NiS	22·2
Pb	PbO	52·4	$\frac{1}{2}PbCl_2$	42·9	$\frac{1}{2}PbF_2$	72·4			PbS	22·5
Sb	$\frac{1}{3}Sb_2O_3$	55·7	$\frac{1}{3}SbCl_3$	30·4	$\frac{1}{3}SbF_3$	96·0			$\frac{1}{3}Sb_2S_3$	13·5
Si	$\frac{1}{2}SiO_2$	108·5	$\frac{1}{4}SiCl_4$	41·6	$\frac{1}{4}SiF_4$	—	$\frac{1}{7}Si_3N_4$	44·1	$\frac{1}{2}SiS_2$	24·5
Sn	$\frac{1}{2}SnO_2$	68·4	$\frac{1}{2}SnCl_2$	41·8					SnS	25·1
Ti	$\frac{1}{2}TiO_2$	112·8	$\frac{1}{4}TiCl_4$	47·9	$\frac{1}{4}TiF_4$	98·1	TiN	80·4	TiS_2	—
V	$\frac{1}{5}V_2O_3$	98·0	$\frac{1}{3}VCl_3$	44·6	$\frac{1}{5}VF_5$	70·4	VN	51·9		
Zn	ZnO	83·2	$\frac{1}{2}ZnCl_2$	49·7	$\frac{1}{2}ZnF_2$	91·3	$\frac{1}{5}Zn_3N_2$	2·6	ZnS	48·2
Zr	$\frac{1}{2}ZrO_2$	129·8	$\frac{1}{4}ZrCl_4$	58·7	$\frac{1}{4}ZrF_4$	114·2	ZrN	87·3		—

As can be seen only the oxides of calcium, magnesium, beryllium and zirconium have higher heats of formation than aluminium; of the chlorides only those of sodium, calcium and magnesium and of the fluorides only calcium, sodium and magnesium. As to the nitrides it can be seen that zirconium, titanium, beryllium and magnesium nitrides form before aluminium nitride.

The difference in the heat of formation is utilized for the selective removal of these elements from molten aluminium, but as for most of these metals quantities are comparatively small, it follows that some aluminium is always lost in the attempts to remove the other metals from the melts.

Magnesium is nearly always present in scrap aluminium in varying amounts, because most aluminium alloys contain magnesium as part of the alloy (some up to 10%), and it is also possible that some magnesium scrap is mixed with the aluminium. The complete removal of magnesium is only rarely necessary, in most cases it is sufficient to bring the magnesium content down to a figure which falls within the specification of the alloy to be produced.

Some magnesium is lost through oxidation in melting the metal; this loss is increased with an increase in temperature. The higher the magnesium content— say 1% or more—the more magnesium is oxidized until the reaction slows down at 0·6% or less when an increasing amount of aluminium is oxidized at the same time and the reaction becomes less selective.

This oxidation or "burning up" of magnesium was carried out on a commercial scale years ago when large scrap, such as sheets with a high magnesium content, was heated in special furnaces to 800–900°C; it could not be prevented, however, that considerable amounts of aluminium were also lost through oxidation; the process therefore became too costly and is not used any more for commercial production.

Looking at Table 3 again it can be seen that there is an appreciable difference in the affinity of sodium, calcium and magnesium to chlorine compared with aluminium. Chlorine reacts preferentially with magnesium and this is utilized for the removal of magnesium from melts of aluminium on a commercial scale.[51] For the treatment of large melts special furnaces have been constructed; they are rotating converters into which chlorine gas is blown through fine pores so that the gas comes in intimate contact with the metal. The following reaction takes place:

$$Mg + Cl_2 \rightarrow MgCl_2$$

The temperature of the metal should be approximately 800°C and the higher the temperature the more vigorous and quicker the reaction, but upper limits are set for practical and economic reasons. At 800°C chlorine reacts selectively with magnesium and only little aluminium is lost through forming aluminium chloride if the magnesium content is reasonably high, e.g. over 0·5%. Under these conditions most of the chlorine is used up to form magnesium chloride which above 710°C is liquid and rises through the melt and collects solid impurities such as aluminium oxide, aluminium nitride and siliceous compounds on the way to the top, by occlusion. For the removal of 1% magnesium (10 kg) from

1 ton of aluminium at 800°C approximately 30 kg of chlorine are necessary. As the magnesium content drops and the temperature gets lower, more than the theoretical amount of chlorine is necessary and more aluminium is lost through formation of aluminium chloride which at that temperature is gaseous and escapes into the atmosphere. Hand in hand with the removal of magnesium goes a thorough degassing and general cleansing of the metal from other impurities such as carbides, nitrides, calcium and sodium. Similar results to chlorine can be obtained with hexachloroethane but the application of large quantities of this compound is not quite so simple.

The reaction between heavy metal chlorides and magnesium is similar to that of chlorine and takes place according to the equation:

$$MnCl_2 + Mg \rightarrow MgCl_2 + Mn$$

It is reasonably selective at high temperature but the quantities of chlorides are high and therefore the process is costly. Furthermore, rather large quantities of the metal forming the chloride are introduced into the melt. To remove 1% (10 kg) magnesium from 1 ton of aluminium theoretically 52 kg of manganese chloride, $MnCl_2$, are necessary and this will introduce about 23 kg of manganese or 2·3% which would make the alloy unsuitable for most purposes.

Sodium chloride also removes magnesium at temperatures of 800°C and above, and this process is carried out on a large scale when aluminium swarf is remelted under a salt cover in rotating furnaces; magnesium is removed and goes into the slag as magnesium chloride.

$$2NaCl + Mg \rightarrow MgCl_2 + 2Na$$

But the process will only operate in the right direction as long as the magnesium chloride content in the salt bath does not exceed approximately 5%. As it reaches this concentration the reaction stops and no more magnesium will be removed. To keep the concentration of magnesium chloride below the critical 5%, large quantities of salt are necessary; to remove 1% magnesium from 1 ton of alloy approximately 1 ton of sodium chloride would be required and such a process would be cumbersome and costly.

The fluorides of calcium, sodium and magnesium have a higher heat of formation than aluminium it follows; therefore, that it should be possible to remove magnesium from aluminium melts by means of fluorine or gaseous organic fluorine compounds;[52] in practice, however, no use is made of this possibility. However, fluorides also react with magnesium as follows:

$$3Mg + 2AlF_3 \rightarrow 3MgF_2 + 2Al$$

This is a straight metal-replacement reaction and it proceeds practically quantitatively without the need for over-heating the metal and unlike the sodium chloride treatment it does not come to a stop. To remove 1% (10 kg) magnesium from 1 ton of aluminium, 23 kg aluminium fluoride are required and as fluorides are generally rather expensive the magnesium removal is costly. In place of aluminium fluoride other fluorides can be used such as sodium fluoride and cryolite, and to lower the melting point for quicker and more efficient reaction they are often mixed with sodium—and potassium—chloride and calcium chloride.

Chlorine, chlorides and fluorides are therefore the usual reagents for the removal of magnesium and other metals with similar heat of formation which are sometimes present; the most frequently met are sodium, calcium and beryllium. Only very rarely is it necessary to remove magnesium completely down to $0 \cdot 0\%$; in most cases is it sufficient to adjust the magnesium content to approximately $0 \cdot 6\%$, which is the average percentage for many aluminium alloys.

Titanium, which is present in small quantities in many aluminium alloys, needs to be removed only very rarely. It can be eliminated by treating it with a mixture of chlorine and nitrogen[53] whereby it goes into the slag as titanium nitride. Another way to remove titanium, especially from high-purity aluminium with high electrical conductivity, is to add boron either as hardener or as potassium borofluoride in sufficient quantities so that all the titanium is precipitated as titanium boride, which is insoluble in aluminium and, because of its higher density, will settle out.

Attempts to remove silicon from aluminium melts or to reduce the content have not met with success; only by electrolysis is it possible to produce aluminium free from silicon (see three-layer electrolysis). Similarly it is not possible to remove zinc by chemical reaction from molten aluminium, but because of the high vapour pressure and the ease with which zinc will evaporate in vacuum it is possible to remove it by distillation in vacuum; the boiling point of zinc at 100 mm pressure is 553°C whilst that of aluminium at 100 mm is 2080°C; zinc therefore will readily separate from aluminium under these conditions but the plant for vacuum distillation is expensive and this process is therefore not carried out on any commercial scale.

When cadmium, lead, bismuth and antimony are present as impurities an addition of metallic calcium will form compounds with these elements which are insoluble in aluminium and which will rise to the surface where they can be skimmed off.[54] Metallic sodium forms a foam with antimony, lead, bismuth and tin which can be skimmed off.[55]

One of the most common impurities in aluminium is iron. When it is present in the free state as iron turnings, filings or grindings, it can be separated by passing the aluminium scrap over magnets; these magnet machines are used in practically every aluminium refinery. But if the iron is alloyed to the aluminium no simple chemical methods exist for its separation and in spite of numerous patents and processes which have been proposed, none is really satisfactory.

The only successful method to remove iron from aluminium melts involves an electrolytic process whereby heavily contaminated metal, which may contain, besides iron, also impurities of silicon, copper, nickel, manganese and zinc up to a total of 15%, can be refined into aluminium of the highest purity of 99·990–99·999%.

This process, called three-layer electrolysis, employs a molten bath in three separate layers. The lower layer which forms the anode is the impure aluminium to be purified to which up to 30% copper has been added to make the alloy heavy so that it will stay at the bottom. Then follows a molten flux, the density of which must be higher than that of pure aluminium and the melting point of which must be above that of aluminium. A number of patents have been granted for the composition of this flux[56] which generally includes a proportion of barium chloride or barium fluoride to provide the required density in addition to sodium and aluminium fluorides. The third layer on top of the molten flux is pure aluminium, which forms the cathode of the cell. For conducting current to the anode pure carbon is used and the aluminium dissolves from the impure bottom layer and passes through the layer of flux where it is further purified and accumulates on top with a carbon electrode forming the cathode. The purity of this metal is 99·99+. As the percentage of impurities in the bottom layer increases it will have to be withdrawn and replaced by new metal to be refined. As the capital costs of the refining cells are high and the consumption of energy is *ca.* 20 kW per kg of refined aluminium plus the costs of the carbon electrodes, it is clear that this process can only be carried out economically on a large scale in electrolytic reduction works where electric power is cheap; but for the production of super-pure aluminium from badly contaminated scrap, this process is of great interest and importance.

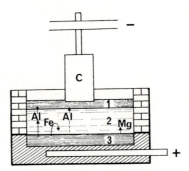

FIG. 33. Three-layer electrolysis.[57]

REFERENCES

1. RANSLEY, C. E., and NEUFELD, H., *J. Inst. Metals* **74**, 599 (1948).
2. KOSTRON, H., *Z. Metallkunde* **43**, 269–84 (1952).
3. SIEVERT, A., *Z. Metallkunde* **21**, 37 (1929).
4. See No. 1.
5. EBORALL, R., and RANSLEY, C. E., *J. Inst. Metals* **71**, 525 (1945).
6. HORNUNG, K. O., *Giesserei Techn. Wiss. Beihefte* **18**, Heft 4, 237 (1966).
7. RANSLEY, C. E., and TALBOT, D. E. J., *Z. Metallkunde* **46**, 328 (1955).
8. DARDEL, Y., *Metal Technology*, T.P. 2484 (1948); *Metal Industry* **76**, 203 (1950).
9. RANSLEY, C. E., *Inst. Metals Monograph* **24**, 74–81 (1958).
10. GINSBERG, H., and KEESE, W., *Aluminium* **39**, 356 (1963).
11. CHRETIEN, P. E., NIPPER, H. A., and PIWOWARSKY, E., *Aluminium-Archiv.* **23** (1939).
12. SAUERWALD, F., *Z. Anorg. Chem.* **256**, 217 (1948).
13. NOWOTNY, H., and PONAHLO, H., *Aluminium Ranshofen, Mitt. Sonderheft* **3**, 13–27 (1955).
14. MOUNTFORD, N. D., and CALVERT, R., *J. Inst. Metals* **88**, 121 (1959).

15. PASCAL, P., and JOUNIAUX, A., *Z. f. Elektrochemie*, **19**, 610 (1913).
16. GINSBERG, H., and RESCH, K., *Erzmetall,* Bd. 13, H 11, 523 (1960).
17. TRONSTAD, L., and REIMERS, J. H., *Aluminium* **21**, 834 (1939).
18. SULLY, A. H., HARDY, H. K., and HEAL, T. J., *J. Inst. Met.* **82**, 49 (1953-4).
19. KOHLMEYER, E. J., *Aluminium* **24**, 361, (1942).
20. REEVE, M. R., *Foseco Developments,* No. 7, 18-26 (1961).
21. BRONDYKE, K. J., and STROUP, P. T., U.S. Patent 2863558 (Dec. 1958).
22. HESS, P. D., BRONDYKE, K. J., and JARRETT, N., U.S. Patent 3039864 (June 1962).
23. SPASSKII, A. G., PIKUNOV, M. V., KURDYUMOV, A. V., and LEBEDEV, E. A., *Russian Castings Production* 546 (1961)
24. KISSLING, R. J., and WALLACE, J. F., *Foundry*, 91, **76** (1963).
25. PACZ, A., U.S. Patent No. 1387900.
26. RANSLEY, C. R., and TALBOT, D. E. J., *J. Inst. Metals* **88**, 150 (1959).
27. BROOKS, C. L., Lecture at an A.I.M.E. meeting in San Francisco (1959).
28. BROOKS, C. L., *Determination of Gas in Molten Aluminium* Metall. Research Lab., Reynolds Metals Co (Feb. 1959).
29. DORE, J. E., *A.S.M.* Feb. 1959, *Foundry Trade Journal* **106**, 301 (1959).
30. Kaiser Aluminium and Chem. Corp., U.S. Patent 2821472.
31. ODINA, M. F., and SHAROV, M. V., *Liteinoye proiz* No. 9, 26-7 (1951).
32. SOKOL'SKAYA, L. I., *Gases in Light Metals,* Pergamon Press, 135 (1961).
33. ANDERKO, K., *Giesserei* **54**, No. 3, 73 (1967).
34. SCHEIL, E., *Giesserei* **43**, 233-40 (1956).
35. CROSSLEY, F. A., and MONDOLFO, L. F., *J. Metals* **3**, 1143 (1951).
36. KONDIC, V., unpublished work.
37. ATTERTON, D. V., *The Australian Inst. Met. Symposium on 1st Nov.,* Preprints 30-38 (1955).
38. EBORALL, M. D., *J. Inst. Metals* **76**, 295 (1949-50).
39. CIBULA, A., *J. Inst. Metals* **76**, 321 (1949); *ibid.* **80**, 1 (1951-2).
40. KONDIC, V., and SHUTT, D., *J. Inst. Metals* **78**, 105 (1950).
41. THURY, W., *Zeitschr. f. Metallkunde* **46**, No. 7 (1955), Austrian Pat. 204291.
42. ROWE, D., *Light Metal Age* **15** (June 1963).
43. GÜRTLER, G., *Zeitschr. f. Metallkunde* **44**, 503 (1953).
44. Aluminiumwerke Nürnberg, Auslegeschrift 1035911, sole licensee Foseco International Limited. THURY, W., and KESSLER, H., *Zeitschr. f. Metallkunde* **46**, 846 (1955).
45. MASCRÉ, C., and LEFÈBVRE, A., *Fonderie* 166, 484 (1959).
46. RIEDELBAUCH and STOFFREGEN, Brit. Pat. No. 990125.
47. THIELE, W., and DUNKEL, E., *Giesserei* 768 (1966), Metallgesellschaft Auslegeschrift 1255 928.
48. ALTENPOHL, D., *Aluminium and Aluminiumlegierungen,* Springer Verlag, Berlin, 187 (1965).
49. VOLKHONTSEV, I. V., and LOVSTOV, D. P., *Russian Casting Production* 443 (Oct. 1966).
50. SMITHELLS, C. J., *Metals Reference Book,* 4th ed., Vol. 1, 241 (1967).
51. DAUTZENBERG, W., *Metall,* Heft 7/8, 125 (Apr. 1950).
52. STROUP, P. T., and Aluminium Co. of America, U.S. Pat. 1998467 (1935).
53. Vereinigte Aluminium-Werke A. G., D.R.P., 547582.
54. HARDY, C., *Metal Progress* **40**, 70/71 (1941).
55. STROUP, P. T., and Aluminium Co. of America, U.S. Pat. 2239277.
56. Birmingham Aluminium Casting Co. Ltd., E.P. 543319, 542886.
57. ZEERLEDER, v. A., *Technology of Light Metals,* Elsevier, Amsterdam, 11 (1949).

E. TREATMENT OF MAGNESIUM ALLOYS

CHAPTER 8

THE TREATMENT OF MAGNESIUM ALLOYS

K. STRAUSS

Pure magnesium is little used for casting production because its mechanical properties are too low for most commercial requirements. Its only use is in form of powder or ribbon for photographic purposes and for fireworks and pyrotechnics. Castings and wrought products are made from magnesium alloys, in which the magnesium content is usually about 90%, the other constituents being aluminium 5–10%, zinc 0·3–6%, manganese 0·15–1·5%, with additions of rare-earth metals, thorium and zirconium.

The melting point of pure magnesium is 651°C and that of the alloys is generally lower so that the casting temperature of most magnesium alloys is between 700 and 780°C. Because magnesium does not dissolve iron and there is no iron pick-up at the casting temperature, the metal can be melted in iron or steel crucibles and even after several remelts there is no appreciable increase in the iron content. But magnesium is a very reactive metal and it oxidizes readily in contact with air. It even decomposes boiling water and forms magnesium hydroxide and hydrogen, and at 700°C oxidation is far more rapid and the metal burns fiercely if no precautions are taken to prevent access of air.

Unlike aluminium, which forms a tough, continuous film of aluminium oxide which protects the metal underneath the film from further oxidation, magnesium forms a granular and porous oxide which will do very little towards stopping further oxidation; a thick white layer of magnesium oxide covers the molten metal underneath which oxidation continues.

First the prevention of magnesium oxide, and, if it has formed, its removal is therefore one of the most important tasks in the treatment of magnesium alloys. This can be carried out by means of fluxes and whilst in the case of aluminium and other metals and alloys it has been shown that fluxes are desirable and useful for producing best results in castings and wrought products, in the case of magnesium they are an essential part of melting practice and reasonably sound and usable castings cannot be obtained without them.

The purpose of the flux is to protect the metal from oxidation; to do this the flux is charged with the ingot into the pot so that by the time the metal is molten it will be covered by a layer of molten flux. But because the metal is so reactive and easily oxidized, the flux cannot completely prevent some magnesium oxide being formed; this must be removed if a satisfactory casting is to be made.

The usual basis for magnesium fluxes is anhydrous magnesium chloride; the melting point of which is 710°C and in order to depress it additions of other chlorides and also of fluorides are made, so that the melting point of the flux is well below that of the metal.

Anhydrous magnesium chloride is a very hygroscopic substance; it absorbs water quickly if left standing in a humid atmosphere and this water cannot be expelled again simply by heating, because the chloride decomposes to form hydrochloric acid.

1. $MgCl_2H_2O \xrightarrow{heat} Mg(OH)Cl + HCl$ 2. $Mg(OH)Cl \xrightarrow{heat} MgO + HCl$

To dehydrate magnesium chloride completely without decomposition, the water-containing salt must be heated in an atmosphere of hydrochloric acid or by adding ammonium chloride and driving off the water together with the ammonium chloride vapour. There are numerous processes available for producing anhydrous magnesium chloride; one which was carried out on a large scale for the production of magnesium metal by the electrolytic method used magnesium oxide as the raw material. This was mixed with carbon, pressed into briquettes and heated with chlorine in specially constructed towers when the following reaction takes place:

$$MgO + C + Cl_2 \rightarrow MgCl_2 + CO$$

The reaction is strongly exothermic and the fused anhydrous salt accumulates at the bottom of the tower and is withdrawn from time to time. During the last 20 years a growing amount of anhydrous magnesium chloride was obtained as a by-product of the production of titanium metal:

$$TiCl_4 + 2 Mg \rightarrow 2 MgCl_2 + Ti$$

and this has served as the basis of most of the magnesium fluxes used in the industry. For each ton of titanium approximately 4 tons of magnesium chloride are produced.

In the early days of magnesium melting, carnallite was used as the base for the flux; this is a double salt of potassium and magnesium chloride $KClMgCl_2 \cdot 6H_2O$. Carnallite is a mineral which is found together with potassium chloride and other salts in some parts of the world and forms a high percentage of the salt content of the Dead Sea.

According to the formula $KClMgCl_2$, the anhydrous carnallite consists roughly of 60% magnesium- and 40% potassium-chloride. Its melting point is 380°C and if used as a flux it becomes very fluid, so that it would be difficult to remove it without leaving some of it on the metal. In absorbing the magnesium oxide it thickens and if molten magnesium chloride is left standing for any length of time it will thicken and become more viscous, but not sufficiently so in most cases for a satisfactory removal of the flux cover. The thickening of the flux is therefore carried out by adding thickening or "inspissating" agents to the flux which will increase the viscosity and facilitate complete removal. Compounds used as "inspissators" are magnesium oxide and calcium and magnesium fluoride, and by their use the viscosity of the molten flux can be so adjusted that it can be skimmed satisfactorily, independently of the amount of magnesium oxide which it has taken up from the melt. By the use of these inspissated fluxes, the difficulties with flux inclusions in magnesium castings, which plagued the industry initially, are now largely overcome. Castings with flux inclusions deteriorate rapidly, have low mechanical strength and corrosion resistance. The reason for this is found in the behaviour of magnesium chloride in contact with the humid atmosphere. The magnesium chloride, being a very hygroscopic material, absorbs moisture quickly and becomes wet. The magnesium chloride solution, so formed, together with the other chlorides of the flux, corrodes the metal and forms cavities and this reaction continues as long as there is still some magnesium chloride left.

Another possible reaction of flux inclusions with the atmosphere is the formation of hydrochloric acid:

1. $MgCl_2 + 2 H_2O \rightarrow Mg(OH)_2 + 2 HCl$
2. $Mg + 2 HCl \rightarrow MgCl_2 + H_2$

The acid will attack and dissolve the surrounding metal until no more magnesium chloride is left.

Magnesium fluoride as an inspissating agent has been very largely replaced by calcium fluoride which is very much cheaper and which will form magnesium fluoride in any case by reacting with an excess of magnesium chloride:

$$CaF_2 + MgCl_2 \rightarrow CaCl_2 + MgF_2$$

The calcium chloride formed during the reaction will first tend to lower the melting point and viscosity but soon the flux will start to get thicker.

The problem of non-metallic inclusions in magnesium base alloys and the flux refining process has been thoroughly investigated by E. F. Emley.[1] Until this paper was published it was generally assumed that in order to protect the metal from oxidation and to remove magnesium oxide from magnesium melts, it was necessary to employ first a very fluid flux which formed a protective cover able to readily absorb magnesium oxide and other non-metallic inclusions. However, due to the low viscosity and to the similarity in density between flux and metal, some of the flux remained suspended in the melt. This was removed by means of a second, inspissated refining flux which is much more viscous than the first and could be skimmed easily.

Typical examples of the two types of fluxes are shown in Table 1.

Emley, after many and exhaustive tests, came to the conclusion that flux inclusions with or without oxides and nitrides readily settle out on standing, that oxide inclusions settle out especially at rising temperature and that flux inclusions in castings must therefore have a different origin than that assumed hitherto. He attributed most of the trouble caused by flux inclusions to carelessness in skimming, to too fluid a flux cover, to pouring flux with the metal at the end of the pour when the crucible is nearly empty and to failure to remove flux from the pouring lip of the crucible. Emley also proved that the viscosity of the flux and its surface tension is of great importance and that the composition of the inspissating fluxes given before can fulfil most of the requirements. These

TABLE 1.

1. Fluid fluxes for covering and cleaning

	(a)	(b) Dow 230
Magnesium chloride	64%	34%
Potassium chloride	28%	55%
Sodium fluoride	5%	
Calcium fluoride	3%	2%
Calcium chloride		9%

2. Fluxes for refining

	(a) Magn. Elektron	(b) Dow 310
Magnesium chloride	40%	50%
Calcium chloride	14%	
Sodium chloride	7%	
Potassium chloride	7%	20%
Magnesium oxide	12%	15%
Calcium fluoride	20%	15%

requirements are that the flux must form a cover sufficiently viscous that it can be held back during pouring so that no loose particles will run into the mould, that the flux cover should be homogeneous, that the flux should absorb magnesium oxide without considerable change in viscosity and that the flux should maintain its viscosity over a fairly wide range of temperature.

The density of the inspissating flux is adjusted so that after absorption of magnesium oxide inclusions it is heavier than the metal and sinks to the bottom. Sometimes, mainly in the case of holding furnaces for diecasting, fluxes are deliberately made heavier by the addition of barium chloride, which ensures that the flux will sink to the bottom of the crucible after refining the melt. To prevent oxidation of the surface, an atmosphere of sulphur dioxide is maintained over the metal by providing a hood over the furnace which is kept free from air by passing SO_2 gas from a cylinder.

Before pouring the metal into the sand mould or die, it must be skimmed thoroughly so that all traces of the flux are removed from the surface. This leaves the metal unprotected and it will oxidize and start burning immediately if no precautions are taken. The most common way of protecting the metal surface is by liberally sprinkling it with flowers of sulphur, usually from a muslin bag. The metal is also sprinkled whilst it flows into the mould. Even if the metal is still covered with flux some magnesium starts to burn and this can usually be quickly suppressed by adding a little sulphur to the burning spot.

The reaction which takes place is as follows:

$$S + O_2 \rightarrow SO_2$$

The sulphur burns and forms a gas, sulphur dioxide, which replaces the air around the metal surface and prevents further oxidation. The heat of formation of sulphur dioxide is higher than that of magnesium oxide so that the former forms in preference to the latter.

There are of course disadvantages in connection with the use of sulphur; sulphur dioxide has a pungent smell, it is an obnoxious gas which makes breathing difficult and is toxic in higher concentrations. For this reason alternative agents for suppressing oxidation were developed, mainly in America. They consist of either volatile or partly volatile fluorides on their own or in mixture with boric acid. The fluorides are ammonium bifluoride NH_4HF_2 or ammonium borofluoride NH_4BF_4 or ammonium fluoride NH_4F, and they are applied through a muslin bag. In contact with the hot metal they decompose and form a film of inert fluoride which prevents further oxidation; some of the decomposition products are gaseous and provide a protective atmosphere and the fumes evolved are far less disagreeable than from sulphur.

DEGASSING OF MAGNESIUM ALLOYS

Gas defects in magnesium castings are far less common than in aluminium castings. The solid solubility of hydrogen in magnesium is high and pinhole porosity does not occur (Fig. 1).

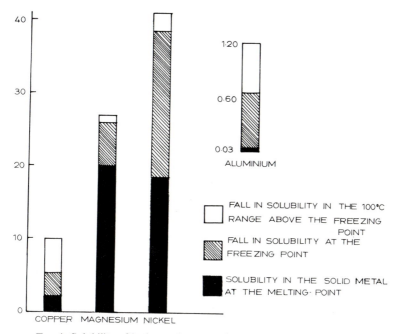

FIG. 1. Solubility of hydrogen in magnesium, copper and aluminium.

However, it has been shown[2] that some micro-porosity which may very frequently be found in magnesium castings is caused by hydrogen gas. To eliminate this defect it is, therefore, advisable to remove the gas from the molten metal.

This can be done similarly to degassing aluminium either with chlorine or with fully chlorinated hydrocarbons.

Degassing with chlorine is carried out either in the melting furnace or in the pouring ladle. Chlorine is bubbled through the melt from a cylinder and magnesium chloride is formed:

$$Mg + Cl_2 \rightarrow MgCl_2$$

The magnesium chloride is in small droplets which are thrown up through the metal by the agitation and will bring some suspended impurities to the top.

Degassing can also be carried out with hexachloroethane C_2Cl_6 and its decomposition in magnesium goes further than with aluminium and more chlorine is developed. This reacts with magnesium to form magnesium chloride.

The great advantage of hexachloroethane degassing over chlorine degassing lies in the fact that in addition to degassing a very good grain refining effect is obtained. This aspect will be dealt with in the next chapter. The application of the compressed hexachloroethane tablets is very simple; the hard-pressed tablet has a density of about 2·0 and if thrown onto the metal surface sinks on its own; decomposition takes place well inside the melt.

GRAIN REFINING OF MAGNESIUM ALLOYS

In order to obtain maximum physical properties, magnesium alloys must be grain refined; without grain refinement they show a coarse structure with low strength. It was known for many years that superheating the magnesium alloy to 850–900°C before pouring would produce a fine grain, but the mechanism of it was not certain. It was thought that nuclei were responsible and that these nuclei were aluminium carbide (the carbon coming from the iron or steel crucible) or iron and manganese compounds. Fox and Lardner[3] have shown that grain refinement by superheating was confined to alloys containing aluminium and that aluminium carbide has a lattice structure similar to magnesium so that it will form the nucleus for crystallization. As in most of the work carried out it appeared that carbon either on its own or in the form of carbide, played an important role. It was logical, therefore, to try elemental carbon or carbon compounds for their grain refining effect, and they have proved to be a satisfactory and more convenient way to promote grain refinement. Graphite powder, paraffin wax, naphthaline and acetylene have all been used with some success; the most satisfactory and convenient way is to use hexachloroethane[4] and hexachlorobenzine.[5] These can be applied in tablet form and as a strongly compressed hexachloroethane tablet is heavier than the magnesium alloy it will sink on its own and at the same time function as a degasser. It must be understood, however, that grain refinement with these tablets is confined to magnesium alloys containing aluminium and that for those without aluminium other ways have to be used.

One of them is superheating to about 900°C, holding there for a few minutes and then cooling rapidly to pouring temperature—this practice differs from most other metals where superheating is undesirable.

Fig. 2. Grain structure of pure magnesium. × 25. (Courtesy of Magnesium Elektron Ltd.)

Fig. 3. Grain structure of Mg-Zr alloy (ZA). × 100. (Courtesy of Magnesium Elektron Ltd.)

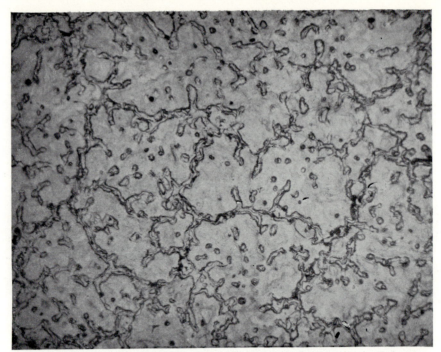

Fig. 4. Mg-Al-Zn-Mn (A8). × 100 as cast. (Courtesy of Magnesium Elektron Ltd.)

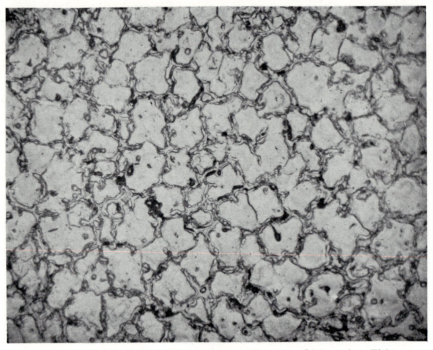

Fig. 5. Mg-Al-Zn-Mn (A8). × 100 superheated. (Courtesy of Magnesium Elektron Ltd.)

Zirconium additions to magnesium-zinc alloys, produce a very fine grain if the zirconium content is *ca.* 0·5%. A large number of patents cover the alloys and the way in which the zirconium is introduced. It is assumed that grain refinement occurs because of the precipitation on cooling of fine zirconium particles. In alloys containing aluminium the refining effect with zirconium is far less pronounced.

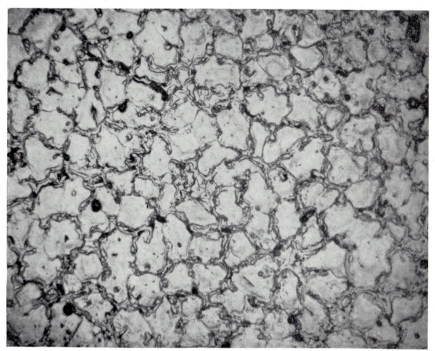

FIG. 6. Mg-Al-Zn-Mn (A8). × 100 grain refined with C_2Cl_6. (Courtesy of Magnesium Elektron Ltd.)

ALLOYING OF MAGNESIUM ALLOYS

The main alloy constituents of magnesium are aluminium, manganese and zinc, but during the postwar years a new series of alloys has been developed which contain zirconium, thorium and a number of rare-earth elements.

The alloying of aluminium and zinc does not present any difficulties; these metals are usually added in form of virgin metal as soon as the magnesium is molten and they go into solution immediately.

Manganese can be added as an aluminium-manganese hardener with 20% manganese but more often in form of anhydrous manganous chloride $MnCl_2$; the reaction with magnesium takes place according to the equation:

$$MnCl_2 + Mg \rightarrow MgCl_2 + Mn$$

Manganous chloride crystallizes with 4 molecules of water, but this can be easily removed without decomposition, by heating to 150°C. The density of the anhydrous salt is 2·4 so that tablets made from it will sink in the melt; to allow for losses it is advisable to use a surplus of manganous chloride.

Zirconium can be added either as a hardener alloy or in form of a salt mixture. The hardener is prepared by reacting molten magnesium with zirconium chloride and it is usually added as a magnesium-zirconium alloy with 30% Zr.

The salt mixture consists of zirconium fluoride and alkali and alkaline earth fluorides; it is added to the molten magnesium alloy and allowed to react at a temperature of about 800°C.[6]

Cerium is added as the metal; it alloys well with magnesium but it oxidizes even more easily than magnesium and reacts preferentially with chlorine so that the alloy should not be degassed with chlorine or with hexachloroethane, as the cerium will be lost.

Thorium is generally introduced as a master alloy with 20% thorium. Like cerium it is very reactive and must be protected from oxidation and no chlorine should be used.

TREATMENT OF SAND FOR MOULDS AND CORES

Because of the special requirements for mould and core sands for magnesium castings, it is more appropriate to deal with these here rather than in the general chapter on moulding sands. Only very few natural moulding sands in the world are suitable for casting magnesium; they are usually a mixture of silica grains of varying size and of clay which makes the sand plastic if mixed with water but at the same time can lower the gas permeability, which would make it unacceptable for magnesium castings. For magnesium alloy castings a synthetic sand is required; this is usually pure silica sand of fairly uniform grain size and clay, in the form of bentonite, is added which will impart the required plasticity. This has the advantage that permeability can be controlled within close limits and that the water content can be kept fairly constant. Both these requirements are very important for success in making sound castings. Permeability must be rather high, so that the metal filling the mould can not build up any back pressure. As the density of molten magnesium is only about 1·5 any back pressure in the mould may throw some of the metal out with unfortunate results, or some air may be trapped in the casting.

The water content must be strictly controlled; it is, of course, necessary to have some water present to impart the required plasticity but it should be as low as possible. As mentioned before, magnesium reacts violently with water at high temperature and additions of special chemicals have to be made to the sand to prevent this reaction, without interfering too much with the moulding properties of the sand. These substances are called "inhibitors" and amongst the many which have been tried the choice is limited to a few, namely, sulphur, boric acid and some fluorides. Sulphur and boric acid or borates are usually used together; they are added in amounts of 2–4% each to the sand and they do not interfere to any extent with the moulding properties. Sulphur in contact with the molten metal evaporates and forms sulphur dioxide, a gas, which displaces the air in the mould so that the entering metal finds little oxygen and can not therefore oxidize to any extent. At the same time boric acid or borates form a film on the surface of the metal which will prevent oxidation, so that the two substances work together in preventing oxidation.

Fluorides are added as inhibitors instead of sulphur and boric acid but sometimes they are used in addition. The type of fluorides to be used are usually double fluorides which on decomposition under heat develop gaseous fluorine compounds; these react with the molten metal and form a tough skin of magnesium fluoride which protects the underlying metal from oxidation. In fact, this fluoride film is a better protection against oxidation than a magnesium oxide film, which is not as continuous and tough. Ammonium bifluoride is the compound most frequently used; it decomposes on sublimation and forms hydrofluoric acid which reacts with the metal as follows:

$$NH_4HF_2 \rightarrow NH_4F + HF$$
$$Mg + 2 HF \rightarrow MgF_2 + H_2$$

Other double fluorides which have been and are being used are ammonium silicofluoride $(NH_4)_2SiF_6$, ammonium borofluoride NH_4BF_4 and ammonium fluoride NH_4F.

If sulphur is used as an inhibitor, the foundry atmosphere is contaminated with sulphur dioxide which makes breathing difficult and very good ventilation is required to overcome this drawback; fluorides on the other hand create a much less objectionable atmosphere.

The composition of some typical moulding sands for magnesium sand casting is as follows (Table 2).[7]

Strict control of the water content is of importance, if too little water is added, the mould is fragile and of low strength, if too much water is present, the

TABLE 2.

(1)	Silica sand	87%
	Bentonite	3%
	Boric acid	2%
	Sulphur	4%
	Water	4%
(2)	Silica sand	85%
	Bentonite	4%
	Ammonium bifluoride	6%
	Water	5%
(3)	Silica sand	86·5%
	Bentonite	4%
	Ammonium silicofluoride	3%
	Sulphur	1·5%
	Boric acid	1%
	Water	4%

reaction with the liquid metal can become violent and scrap castings will result.

To prevent the drying out of the sand, small quantities (0·5–2·0 %) of ethylene glycol are sometimes added; this is a hygroscopic compound which slowly absorbs water and holds it for longer periods than sand alone.

Water formed by chemical reaction is also held more tenaciously than water added mechanically. For this reason boric acid and ammonium bifluoride are sometimes added to produce the following reaction:

$$H_3BO_3 + 2 NH_4HF_2 \rightarrow NH_4BF_4 + NH_3 + 3 H_2O$$

Ammonium borofluoride acts as an inhibitor and the water formed during the reaction is more intimately mixed with the sand and dries out slower than when water from a can is added; the ammonia, which also forms, is a gas and will escape.

To core sand, too, inhibitors are added, so that a typical mixture would be as follows (Table 3).

TABLE 3.

Silica sand	90%
Sulphur	1·0%
Boric acid	0·5%
Cereal	1·0%
Core oil	2·0%
Water	5·5%

Cores for magnesium castings must break down easily because immediately after solidification the metal has very little mechanical strength and it would tear if the sand would not readily give way.

THE REMELTING OF MAGNESIUM SCRAP

Unlike the secondary aluminium industry the recovery of magnesium metal from scrap is of far less importance. This is not only due to the much smaller quantities of magnesium scrap which have to be remelted but also to difficulties in dealing with fine magnesium swarf.

Clean and solid scrap in the form of feeder heads, scrap castings and similar massive pieces can be remelted without difficulty in the same way as magnesium ingots are melted. The scrap should be as free as possible from oil, sand and silicates generally, in order to keep the silicon pick-up to a minimum. The fluxing procedure is the same as with ingots, but it will be found that due to higher oxide contents somewhat larger quantities of flux are necessary.

For the remelting of fine magnesium scrap in form of shavings, sawings, drillings and powder, special methods and great caution have to be employed.

The largest crucible furnaces which are in use have a capacity of 2 tons of metal; they are of the tilting type and the steel crucible (usually cast steel) weighs approximately 5 tons. To support this weight the furnace shell and tilting gear must be of heavy construction. In the U.S.A. tilting reverberatory furnaces for large magnesium melts have been developed and because of the attack of the molten metal on ordinary silicate-brick linings, these furnaces are lined with "Tercod"[8] which is a carborundum (SiC) brick faced with a layer of graphite. The capacity of these furnaces can be as high as 20 tons and they have been used successfully for melting magnesium scrap.

Swarf, such as shavings and drillings from the machine shop should be as free as possible from oil and moisture. The melting process is best started with a heel of liquid metal which is covered with a thick layer of a low melting point, low-viscosity flux of Dow 230 type (page 300). The swarf is mixed with some solid flux and then fed carefully into the liquid bath under constant stirring so that the metal is immediately covered with flux; the temperature of the metal should be near the melting point and the swarf should be added at such a rate that the temperature changes very little. This is continued until the crucible is reasonably full but care must be taken that sufficient space is left to allow for vigorous stirring without spilling any metal. The temperature is now raised to that required at which the normal flux refining process is carried out; if the swarf was very dirty, the flux cover should be removed and fresh refining flux should be added. For reasonably dry and non-oily material metal recovery can be as high as 85%.

For very fine material such as sawings and powder, the above procedure is not recommended because of the hazards involved. The dry material should be briquetted with about the same weight of flux and the compact added carefully to the melt through a layer of flux as described above. If the fine scrap is contaminated with oil, moisture and non-metallic materials, remelting should not be attempted but it should be destroyed by burning.

REFERENCES

1. EMLEY, E. F., *J. Inst. Metals*, Feb. p. 431 (1949).
2. BAKER, W. A., *J. Inst. Metals*, **71**, 165 (1945).
3. FOX, F. A., and LARDNER, E., *J. Inst. Metals*, **71**, 1–22 (1945).
4. Brit. Patent 606072 (1948).
5. U.S. Patent 2540366 (1951).
6. MILLWARD, H. J., *Metal Ind.*, **83**, 83–85 (1953).
7. BRACE, A. W., and ALLEN, F. A., *Magnesium Casting Technology*, Chapman and Hall Ltd., London, 1957.
8. NELSON, C. E., *Trans. A.I.M.E.*, **159**, 392 (1944).

F. TREATMENT OF MOULDING SAND
FOR MOULDS AND CORES

L

CHAPTER 9

MOULDING AND CORE SANDS

E. Parkes, G. Westwood and R. Grigg

INTRODUCTION

Moulding and core sands together represent a most important part of the process of metal casting. However good the quality of the metal, the casting produced will be no better than the mould into which it is poured. Whilst the die casting industry is of very great importance, it is of necessity limited by the size and complexity of the casting, and by far the greater proportion of metal castings are produced in sand moulds. This fact has led to developments in the theory and practice of sand technology which have been of far-reaching importance. The importance of this section of foundry technology is emphasized by the fact that of the examples quoted in the *Atlas of Defects in Castings*,† 30% are directly attributable to sand. As a mechanical handling problem, it has been estimated that for each ton of castings produced, at least one ton of sand has to be handled. It must be remembered that waste sand has to be disposed of and this feature alone is causing increasing concern. The object of this section is to outline the methods used for the production of sand moulds and cores.

DEFINITION OF MOULDING AND CORE SAND

Properties Required

The mould forms the outside of the casting; the core forms the internal shape. Because of the difference in the duties performed, there is an obvious difference in the properties required of the two different types of sand. Because cores form the internal shape of the casting, they are usually enclosed by the metal and therefore have to be extracted through sometimes a very small aperture in the casting. To enable this to be effected, the sand must "break down" reasonably quickly, i.e. the binder must be burnt off, thus enabling the sand to run out of the casting in a free flowing condition. It must also be sufficiently highly refractory to avoid fusion of sand to the metal, and also must not produce an excessive volume of gas in order to reduce the risk of gas entering the metal and being trapped in the solidifying metal. If this does happen, then blowholes will occur in the finished casting, thereby producing, in most cases, a "scrapped" casting. The chemical nature of the gases produced is also of great importance, but this will be discussed more fully later.

† *Atlas of Defects in Castings*, Inst. of Brit. Foundrymen (1961).

313

The mould, however, producing the outside shape of the casting is not enclosed by the metal to the same degree and therefore it is not so important for the sand to break down after casting. The main requirements of the moulding sands are that they should be:

(a) easily mouldable so as to produce a perfect mould,
(b) of sufficiently good surface finish to produce a casting possessing the finish required.

There is no doubt that mould and core sand technologies are quite different in most respects, but it is also very apparent that there is an increasing tendency for the two techniques to merge into a common process, with common production techniques.

Economies in Use of Sand

It will be obvious that, in both cores and moulds, the surface of the sand will come into contact with the molten metal. For this reason, it is common practice, in the case of moulds and large cores, to "face" the mould with a high quality sand and to support this "facing" with a lower quality "backing" These sands are termed "facing sand" and "backing sand". It is also common, when large cores are made, to hollow the core out and to fill with material such as coke. This serves the useful purpose of lowering the cost of the core as well as assisting the breakdown of the core after casting and assisting the escape of gases through the core. Obviously the extent to which facing and backing sands can be used will depend upon the relative size of the job being done and the production techniques used. It is worthwhile economically to use a backing and facing sand system on the larger type of work, but this advantage would obviously be lost in the case of the smaller jobs, where speed of production is of greater importance. For the smaller or "lighter" casting, "shell" moulding or core making is very widely used, in which the mould or core is made as a shell of $\frac{1}{4}$ in. to $\frac{3}{8}$ in. thick by means of a technique employing a sand bonded by a particular type of synthetic resin. This will be discussed in greater detail later.

Sand, and the Reasons for Its Use

From the earliest days of founding, sand has been used for producing moulds into which metal is to be poured. The general term "sand" covers a wide range of granular materials consisting basically of quartz, SiO_2, derived from the break-down of rocks, but containing also small amounts of other materials such as felspars and micas. The term "sand" generally refers to aggregates of fine to medium grain size and by far the most important characteristic of a sand intended for foundry use is the size of the constituent sand grain and its shape. Chemical composition is also of considerable importance in deciding the type of work for which the sand can be used. The higher the pouring temperature of the metal, the greater the purity of the sand to be used, although efficient dressing can reduce the importance of purity. Purity of sand also becomes of great importance in some of the recently developed techniques of core production, where the presence of excessive alkalies in the sand interferes with the chemical reactions of the binder systems.

Where the sand contains natural clay materials, the sand is "self-hardening" when tempered with the correct amount of water and is then termed "naturally bonded sand". This type of sand plays an important part in moulding and to some extent, in core sands.

Sands are described by two methods: according to the method of deposition or formation such as sea, river, lake, glacial or wind-blown, or according to the geological age or rock system in which the deposit occurs, such as the Lower Greensand formation of the Cretaceous system; Millstone Grit of the Carboniferous system; Bunter sand stone of the Triassic system, etc.

The reasons for the use of sand are:

(a) It is readily available in the quantities required.
(b) Because of its availability in the right form, it is cheap.
(c) It is sufficiently highly refractory (fusion point, depending upon purity 1550–1650°C).
(d) It is readily bonded by clays or other inorganic and organic materials.

Silica is by no means the only material used for mould and core making. It has certain disadvantages, such as the mineralogical phase changes which occur when silica is heated from ambient to 1000°C (which occurs during casting) and which produce volume changes in the sand mass which causes casting defects ("scabs", "fins") and also the health hazard resulting from the presence of fine silica particles in the atmosphere.

TABLE 1. *Quartz Phase Changes occurring upon Heating*[1]

Temperature	Product	Approximate volume change
Heating to 200°C	Quartz	
Heating to 575°C	Quartz	1·4%
Heating to 870°C	Tridymite	14–16%
Heating to 1470°C	Christobalite	14%

The other materials used are all free from these two disadvantages, but are either in short supply and hence expensive, or plentiful but of less desirable grain shape and distribution. These materials are:

(1) Zircon–zirconium silicate (ZrO_2SiO_2) which occurs extensively in Australia and Florida. It is produced in the same grain size as silica sand but its density is twice that of silica. Other advantages include high fusion

point (2200°C), high conductivity, low expansion ($\frac{1}{3}$ that of silica) and resistance to wetting by molten metal.

(2) Olivine–magnesium–iron–ortho silicate ($Mg_2SiO_4 + Fe_2SiO_4$), found extensively in Scandinavia.

(3) Chromite ($FeO \cdot Cr_2O_3$), from South Africa, being developed considerably in recent years, mainly as a result of the rise in price of zircon.

The Assessment of Sands

Sands can be assessed by the following tests, all of which have now been standardized and the methods for which are readily available.

Clay content—the clay is removed by repeated washing with water under controlled conditions.

Grading test—after removal of the clay and drying the sand is passed through a nest of screens

usually BSI   22,   30,   44,   60, 100, 150, 200

microns   710, 500, 355, 250, 150, 105,   75

shaken for a standard time.

Grain shape—determined by either microscopic examination or coefficient of angularity determination.

Specific surface area.

Chemical analysis, the important determinations being:

SiO_2, Fe_2O_3, Al_2O_3, CaO, Na_2O, K_2O.

Loss on ignition.

TABLE 2. *Silica Sands used in the Foundry Industry. Typical figures (not specifications)*

	B.S.S. Sieve No. (microns)							
Sand	22	30	44	60	100	150	200	−200
Aperture microns	(710)	(500)	(355)	(250)	(150)	(105)	(75)	
Congleton	—	2–4	14–16	34–36	42–45	2–3	0–1	—
Chelford Washed & Sieved	—	4–6	18–22	42–46	24–28	2–4	—	—
Messingham 45	—	4–6	15–18	54–56	22–24	1–2	—	—
Redhill 65	—	4–6	10–12	22–24	40–42	10–12	6–8	1–2
Kings Lynn 95	—	—	—	4–6	24–26	60–65	6–8	1–2

The importance of the effect of sand characteristics upon the efficiency of the moulding operations cannot be overstated, and it is important that the technically correct choice of sand should be made, even though this may not be the cheapest. The wrong choice of sand can lead to casting defects such as metal penetration, scabbing, porosity and inclusions, as well as to the use of excessive binding materials which can materially increase the cost of the mould or core.

TECHNOLOGY OF PREPARED FOUNDRY SANDS

Introduction

The technology of moulding and core sands can be divided into three distinct sections:

The chemistry and technology of binding materials.
The mechanics and technology of sand preparations plant.
Applications of the various processes to the particular type of work to be done.
It must be appreciated that all three sections are interdependent upon each other and that each job should be appraised according to the quality and quantity of the finished product required and to the production conditions existing. The important points to consider, both in mould making and core making are as follows:

Metal to be poured.
Dimensional accuracy required.
Quantity to be made.

Choice of Process

The choice of process lies between the following (Tables 3, 4). The length of the list illustrates the difficulty of choosing the correct moulding technique.

TABLE 3. *Moulding Processes*

Green sand	Sand–clay–water, without drying. Low pressure up to 30 psi (2·1 kg/cm²). High pressure, 100–250 psi (7–17·5 kg/cm²).
Skin dried	Low-pressure moulds skin dried to a depth of 0·5 in. by portable driers, radiant heat lamps, etc.
Dry sand	Entire mould dried.
Sodium silicate processes	CO_2 hardened. Nishyama process. Self-setting. Fluid sand.
Cold-setting (as facing or block mould)	Sodium perborate hardened oils. Stoved or green.
Cold-curing	Resin. Isocyanate hardened oil.
Cement	Air hardened.
Shell	Dumped or blown.
Hot-box	Blown.

TABLE 4. *Core making Processes*

Stoving process	Oil/cereal sand.
	Resin/cereal sand.
Sodium silicate	CO_2 hardened.
processes	Nishyama process.
	Self-setting.
	Fluid sand.
Cold-setting	Sodium perborate hardened.
	Oils—stoved.
Cold-curing	Resin.
	Isocyanate hardened oil.
Cement	Air hardened.
Shell	Dumped or blown.
Hot-box	Blown.

Essential Requirements for Process

The object of the exercise of making a casting is to produce a dimensionally accurate reproduction of the pattern. In both moulding and core making this can only be done by:

Producing a mould or core that is accurate to the pattern prior to casting.
Producing a mould or core that will have sufficient resistance to metal pressure to retain its shape.

In all cases, the ability of the sand mass to satisfy these conditions will depend upon the strength of the sand immediately upon removal of the mass from the pattern and upon its strength at the time of casting.

In the first case, a certain minimum strength is necessary to enable the core to withstand the tendency to deform during the operation of removing it from the box, handling to the next stage, which may be that of stoving, and the stoving and other finishing operations given to the core or mould.

In the second case, sufficient strength is necessary to withstand deformation due to metal pressure and erosion of the surface due to the flow of metal over the surface.

The method normally used for determining the strength of the sand is either the compression test, which can be done either in the "green" state or "dry" state or the tensile strength.

Reference should be made to standard methods of testing foundry sands to understand the method of determining these properties. Table 5 shows the relative values for the processes in Tables 3 and 4.

It is quite obvious, from the range of compression strengths quoted, that a considerable choice of processes presents itself to the foundryman. This choice in some cases will be resolved by the requirements of the casting being produced, whether in design or in type of metal, in others by the quantity required or again by the type of operation which the particular foundry is equipped to carry out.

TABLE 5. *Compression Strengths*

Process	Green		Dry or set	
	psi	kg/cm²	psi	kg/cm²
Green sand moulding	5–10	0·35–0·7	80–120	5·6–8·4
Skin dried	5–10	0·35–0·7	100–200	7·0–14·0
Dry sand	6–15	0·42–1·05	150–300	10·5–21·0
CO_2 process	0–2	0–0·14	200–400	14·0–28·0
Nishyama process	0–2	0–0·14	up to 1000	70·0
Self-hardening silicates	0–2	0–0·14	100–200	7·0–14·0
Fluid sand	nil	nil	50–150	3·5–10·5
Cement	1–2	0·07–0·14	250–700	17·5–49·0
Oil bonded stoved	0·5–2·9	0·035–0·203	700–1000	49·0–70·0
Oil bonded cold-set	0–1·5	0–0·105	> 1000	> 70·0
Resin bonded	nil	nil	600–1000	42·0–70·0
Shell process	nil	nil	> 1000	> 70·0
Hot-box	nil	nil	> 1000	> 70·0

Technology of Application

Green sand

This is the most widely used moulding process for small to medium sized castings in all metals. Providing the properties of the moulding sand are correctly controlled, and the pattern equipment satisfactory, castings of high quality can be produced at the lowest cost of all processes. The advantage is that the process can be of very short time cycle, i.e. sand preparation, mould making, closing, pouring and shake-out are ideally suited to a mechanized, continuous process. Machines can vary from simple, hand operated pattern draw machines to the sophisticated jolt squeeze machines available today.

Green sand moulds must be of uniform compaction to present high resistance to metal penetration, erosion and deformation and accurate control of sand properties is necessary. These properties are: moisture content, green compression strength, plasticity and permeability, but again the level of these properties will vary with the conditions existing in a particular foundry.

The sand may be either naturally bonded or synthetic, i.e. silica sand mixed with a bonding clay and water. An example of each of these is given below (Tables 6, 7).

TABLE 6. *Typical Synthetic Sand Mixture (for steel)*

Mixture %		Properties
Medium silica sand	90–92·5	Green permeability No. 150–250
Wyoming Bentonite	4–5	Green compression 6–9 psi
Cereal	0·5–1·0	(0·42–0·63 kg/cm²)
Water	2·7–4·0	

L *

TABLE 7. *Typical Naturally Bonded Sand (for grey iron)*

Mixture %		Properties
Old sand	70	Green permeability 16–53
New sand	20	Green compression 3–6 psi
Coal dust	4	(0·21–0·42 kg/cm²)
Water	5–7	

High-pressure Moulding

Considerable mention has been made in recent years of high-pressure moulding, i.e. moulding pressures of 100–250 psi (7·0–14·0 kg/cm²). The advantages claimed for the process are:[2]

Dimensional accuracy. This applies particularly to grey iron castings, when the effect of mould wall movement is of great importance. Mould wall movement has been shown to be markedly affected by the water content of the sand and the hardness and density of the moulds. High-pressure moulding techniques enable lower water contents to be run and high densities to be achieved and hence castings of consistently greater accuracy are produced. This is of particular importance in the modern high production factories such as automobile machine shops where automatic machining operations demand a very close consistency of dimension.

Surface finish. This is generally good, being dependent upon permeability of mould and degree of compaction.

Because of the need to obtain uniform density under the high pressures used, it becomes very necessary to utilize squeeze heads which follow the contour of the pattern. This is a further reason for the limitations of this method of moulding to long-run jobs, and it appears that the majority of high-pressure moulding equipment is integrated into automatic moulding systems, these having high rates of production of the order of several hundred moulds per hour.

Because of the need to produce optimum packing density, the characteristics of the moulding sand used become more critical and greater care is necessary in their control. It has been found that the following composition produces a sand of suitable characteristics (Table 8).

TABLE 8. *Composition of Sana suitable for High-pressure Moulding*[2]

Mixture %		Properties
Medium silica sand	85–81·5	Shatter index—85–95
Clay	8–10	Permeability—50–70
(such as Bentonite)		Green compression—14–20 psi
Water	2·8–3·5	(0·98–1·4 kg/cm²)
Coal dust	4–5	
Dextrine	$\frac{1}{4}$–$\frac{1}{2}$	

Skin Dried and Dry Sand Moulding

This is used where greater mould rigidity is required than with green sand moulding and is applied to the heavier sectioned castings. Because of the necessity to dry the moulds, production cannot be mechanized to the same extent as green sand moulding, but here again, because of the type of work with which dry sand moulding is used, the actual drying time need be a comparatively small proportion of the total moulding time.

In principle, the same type of mixture is used as with green sand moulding, but there is a tendency to make additions of fine silica flour to the mixture, which enables an even finer surface finish to be obtained, as well as an increased hot strength. Additions of starch or particularly dextrine, will increase the surface hardness.

This section completes the moulding processes based upon the use of clay bonded sands. The next section will deal with the methods which have been increasingly used over the last few years. The mechanism of the bonding process is the same, whether used for cores or moulds.

HARDENING IN THE BOX TECHNIQUES—GENERAL PRINCIPLES

The principle upon which the modern processes are based are:

The use of very free flowing sand which renders the making of the mould or core very easy.

The chemical hardening of the sand whilst the mould or core is still in the pattern or box, whether by heat (shell process or hot-box) or by reactions in the cold (cold-setting).

The effect of these later processes is shown in several ways:

They have taken much of the personal skill out of moulding, because, in effect, the sand has only to be poured on to the pattern and "tucked" around it. It becomes virtually impossible to "ram" the sand as the skilled moulder would ram a clay bonded sand. Defects such as soft ramming are greatly reduced. Because the sand is hardened around the patterns before the pattern is removed, inaccuracies due to careless stripping or sagging after stripping are removed, and, in effect, a very accurate reproduction of the pattern is achieved. This has enabled machining allowances to be cut, saving considerably in weight of metal and machining time. This increased accuracy has enabled the engineering designer to design for minimum weight of metal and maximum performance of his part. In fact, many of the advances in design of turbine blades, automobile engines, to mention only two, would have been impossible without such advances, core making in particular.

Production per unit of floor area has risen considerably. This has been a vital factor in the development of the foundry industry since the war. This applies particularly to shell moulding.

In some cases, the saving in quantity of sand handled and hence handling costs, have been considerable.

All these advantages far outweigh the fact that the processes are without exception, more costly than the use of clay bonded sand.

A brief account of each process will be given, together with its relative advantage and disadvantage.

Cement Moulding

The bonding medium is Portland Cement, together with the correct amount of water. In certain cases, an addition may be made to the sand to accelerate the hardening process and to improve the "break-down" of the sand after casting. This process dates from 1930 and is sometimes referred to as the "Randupson" Process.

The advantages are:[3]

Ease of ramming of the sand (typical of cold-setting process).

High dry strength of the sand when set—prevention of warping.

Moulds can withstand high ferrostatic pressure.

Permeability is high.

Accuracy of moulds is high.

Repairs (in case of damage due to storage) can be easily made.

It has been suggested that the drying off of combined water from the cement sand limits metal penetration and provides a light chilling effect.

Elimination of scabbing.

Its disadvantages are said to be:

Its refractoriness is not very high and is therefore not very suitable for thick sections of high melting point metals such as steel.

Somewhat liable to erosion.

Its high retained strength after casting retards the shrinkage of the casting.

The process is not very suitable to small castings.

The patterns (if wooden) are sometimes subjected to the high moisture effect of the sand and sometimes to considerable pressure on stripping.

Long air drying times—from 24 to 48 hours.

TABLE 9. *Typical Cement Sand Mixture facing Mixture*

Sand mixture %		Properties
Dry clean silica sand 84–79 (AFS 60)		Setting time—12–24 hours
Portland cement	7–10	Compression strength—700 psi
Additive	4	(49·0 kg/cm^2)
Water	5–7	Full hardened

This latter is possibly the most serious disadvantage, although it can be alleviated by means of additions such as molasses. Using latest techniques of preparing sand in continuous mixers, both the setting time and addition of cement can be reduced.

Sodium Silicate Processes

Mould and core making processes based upon sodium silicate represent a high proportion of the total chemically hardened sands used in the foundry industry at the present day. What are the advantages of sodium silicate as a bonding material?

Cheap and in plentiful supply.

Inorganic and does not evolve gases upon heating when the metal is poured.

Also, it does not collapse or burn out after casting, hence it gives a thermally stable sand.

It is capable of being hardened in several ways without the use of heat.

It is not critical to the quality of the sand.

The disadvantages are:

For core making, the fact that it does not burn out or collapse after casting implies that the core sand will not be easily removed from the casting. This is a serious disadvantage. For mould making, when the mould surrounds the metal and not vice-versa, this feature is not so important. Additives or break-down agents, usually organic in nature, can be used to improve this property, but often at the expense of strength. Because of its low fusion point, the effect of surface dressings becomes of great importance when good surface finish is required. Fettling, i.e. the cleaning of the castings, is more difficult with silicate bonded sands than with the organic bonded sand.

Again, because the binder does not burn out upon casting, the sand becomes difficult to "reclaim", i.e. to use again. Moreover, the silicate adheres very efficiently to the sand grains and is difficult to remove by mechanical or chemical means. Sand reclamation units are available which will enable a certain proportion of used sand to be recycled.

Low Strengths

Consideration of these points indicate that the sodium silicate processes are somewhat more suited to mould making than to core making and this is found in practice.

Hardening by carbon dioxide. This was the original method used and has the great advantage of giving an almost instantaneous "set". The following points are important: The gassing techniques are critical. Sufficient gas must be allowed to enable the optimum results to be obtained. Excessive gassing is undesirable for the following reasons:

The strength decreases rapidly with over-gassing and formation of sodium bicarbonate and sodium carbonate.

Storage properties become inferior.

CO_2 gas is expensive.

It becomes difficult, in complex shapes, to ensure hardening of the sand right through the core or mould, because of the difficulty of ensuring that the gas reaches all of the sand. "Bench life" of the sand, i.e. the length of time during which the sand remains usable, is not critical and hardening can be effected where and when required.

Hardening techniques can be modified easily.

The correct choice and additions of "break-down" additive is of great importance.

Setting by chemical hardener included in the mixed sand. This process is in direct comparison with the various organic cold-setting processes and possesses some of the same advantages and disadvantages, i.e. limitations of the "bench life" of the sand—because the chemical hardening takes place as soon as the hardening chemical is introduced into the sand mixture. The certainty, providing the correct addition has been made, of the sand compact hardening right through within a predictable time.

The two variations of the processes used are:

The "N" process, or Nishyama process. This originated in Japan and uses additions of powdered ferro silicon to effect the hardening. The final result can be a very high strength sand, having excellent hot rigidity. Certain inherent disadvantages of the process, however, require careful control: the reaction can be highly exothermic and a great deal of heat is produced as well as considerable evolution of hydrogen gas, and as is common with such reactive processes, some difficulty in controlling the setting is experienced.

Additions of water can be used to effect this control; the grading and purity of the ferro silicon used is also of importance, consistency being essential if good results are to be obtained.

The dicalcium silicate process. Originally, this was used as a slag produced in the preparation of magnesium by the Pidgeon process, but upon the supply from this source drying up, a foreign source has been found. The use of this hardener produces a gentler reaction and a more controllable rate of setting than the "N" process, although the final strength is appreciably lower. The mechanics of the hardening can be achieved not by variations in the amount of hardener used, as is the case in the organic cold setting processes, but by variations of the chemical nature of the binding material. Thompson[4] quotes the following relationships (see Fig. 1 and Table 10, page 325).

Fluid sand. This process was developed in Russia and has aroused considerable interest in that the sand is produced in a high state of fluidity which is obtained by the use of a foaming agent (a surface active chemical). The result is that the sand can be poured into the box exactly as a fluid, thereby eliminating the need for any mechanical packing. The hardening agent is again dicalcium silicate. Because of the foamed condition of the binder system, physical strengths tend to be appreciably lower than any of the other silicate processes. Also the permeability of the sand mass is high, which creates the need for an effective barrier, in the form of a thick coating of paint, to prevent metal penetration. The characteristics of the sand as outlined above indicates that the process is suitable essentially for making large simple jobs, where the maximum advantage is derived from the lack of necessity of ramming and where its lack of strength is not likely to be a serious handicap. The strengths produced are of the same order as those of dry moulding sand and obviously such jobs as ingot mould manufacture are suited to the process, where conventional methods present severe plant maintenance problems. As yet, it is little used in this country.

A further advantage which is claimed for this process is that the break-down is somewhat easier (it is necessary to relate this to the low strengths) and that it is easier to reclaim the sand after use.

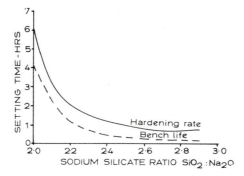

FIG. 1. Hardening rate vs. $SiO_2 : Na_2O$ ratio.

TABLE 10. *Typical Mixture and Properties*

Mixture %		Properties
Coarse–medium grained sand	89	Setting time—about 1 hour
Sodium silicate (2·2 ratio)	6	Compression strength—200–250 psi (14·0–17·5 kg/cm²)
Dicalcium silicate	4	
Water	1	

General Summary

In general, the silicate processes can be used to make the moulds as well as cores for most types of cast metals. Some applications are more suited than others to the particular characteristics of the sodium silicate binder, e.g. because the temperature obtained in aluminium alloys is lower than the vitrification temperature of the silicate, break-down after casting is considerably easier.

Again, because of the high temperatures used in steel castings, there is a tendency for the surface condition of the resultant casting to be good, due to the high degree of vitrification that takes place at steel temperatures. It seems that results are generally less satisfactory with grey iron than with the other metals, because of the pouring temperature and freezing characteristics of the metal.

The particular version of the process to use will obviously depend greatly upon the job to be done and sufficient indication has been given above to assist in the choice. Quite obviously, if small cores or moulds are to be made, then there is no point in using the cold-setting processes because of the difficulty in using up the prepared sand within the limits of its bench life. In this case the CO_2 gassing techniques should be used. Likewise, the heavy jobs are best done using one of the cold-setting techniques.

ORGANIC BINDERS

Processes utilizing organic binders follow the same general pattern as those using the various inorganic binders described above.

The essential difference between the two lies in the behaviour of the binding agent upon and after casting. Whereas the inorganic materials are thermally stable and show little or no change of state upon heating, the organic binders, whatever their composition, burn out upon casting, producing gas and leaving a carbonaceous residue on the sand grains. Moreover, the rate at which they "break-down" after casting depends entirely upon their chemical composition and a tremendously wide choice exists, enabling the founder to obtain the break-down required. This characteristic of the organic binders ensures these important practical advantages:

Easy removal of the sand from the casting.

The evolution of gases ensures a reducing atmosphere in the mould and the formation of a gas barrier between metal and mould leading to excellent surface finish. It must be said that gases could produce mould metal reaction.

Possibility of reclaiming the sand for further use.

All processes give a flowable sand and cores have excellent strength and storage stability.

In effect, a correctly chosen and applied organic binder will satisfy perfectly the requirements of a mould for metal casting, i.e. the sand must remain stable for sufficient time to allow the metal to form a solid skin and then must collapse so as to prevent any hindrance to the contraction of the metal during cooling and to enable the sand to be removed easily from the solidified casting.

The performance of the organic binders can be divided into two general classes:

Cold-curing—achieved by the use of a suitable catalyst.

Heat-cured—with the aid of a catalyst, the curing preferably taking place in the box.

They are divided into natural or derivations from natural products, such as vegetable oils, starches and dextrines or entirely synthetic materials such as phenolic and amino resins.

There have been attempts made to induce hardening by means of an acid gas, a method entirely analogous to the carbon dioxide–sodium silicate process, but there are obviously practical difficulties and this process has not yet achieved production significance.

General Properties of Organic Binders

Certain common properties exist whatever the particular process used:
Ease of break-down—the following order in decreasing rate of break-down
can be generally applied:

Urea–formaldehyde resin.
Urea–phenol–formaldehyde resins.
Oil-based products (the isocyanate hardened oils breaking down easier than the sodium perborate hardened oils).
Phenol–formaldehyde.
Furfuryl alcohol modified resins (the higher the furfuryl alcohol content the slower the break-down).
Furfuryl alcohol polymers.

Constitution of Gas and Effects upon Mould–Metal Reactions

Much has been written upon the subject of porosity in castings, resulting from gas, chiefly nitrogen and hydrogen. From the point of view of choice of binder, nitrogen is the most important and it appears that, with certain metals, the evolution of nitrogen from the mould or core can produce "pinholes", the metals most seriously affected are low carbon steels, spheroidal graphite irons and high duty grey irons. Where freedom from the risk of nitrogen pinholing is desired, two remedies should be sought:

Use of a nitrogen-free binder, which usually means the avoidance of urea containing resins or the use of a hexamine-free system in shell moulding; or use of iron oxide in one of its several forms—red iron oxide, hematite, known as Sierra Leone concentrate, millscale. Hydrogen pinholing from sand is of less frequent occurrence but may be caused by the presence of retained water in, for instance, a hot-box process core, or failure to dry off a water-based paint properly.
"Finning" in a casting results from the cracking of the mould or core face before the metal has had time to solidify, hence resulting in the metal entering the crack, producing a thin line across the face of the casting. Whilst not, by any means, being the entire fault of the binder, some improvement can be effected by choosing a different binder or by taking care that

the mould or core has been cured to the right degree. This applies particularly to shell process moulds or cores, where over-curing produces an increased tendency to crack upon casting.

Cold-setting Process

The various cold-setting processes can be divided into two distinct groups, those using synthetic resins, which are all cold-curing and those using processed oils. This latter can be divided into entirely cold-curing and cold-setting—oven baked.

Cold-curing synthetic resins

Briefly, the history of the use of this resin in the foundry trade dates back to a matter of only 6–7 years, and is connected almost entirely with the introduction of the furfuryl alcohol resins. The advantages, whatever type of resin is used, are based upon the controllability of the setting times by means of the addition of the catalyst, combined with a high final strength, with the control over the other properties mentioned previously by means of the choice of particular resin used. Four main classes are used:

Urea–formaldehyde types, using, for instance, ferric chloride as catalyst used mainly for light alloy work because of its superlative break-down characteristics.

Phenol–formaldehyde resins, using a sulphonic acid catalyst (such as para-toluene sulphonic acid) which were introduced during the recent world shortage of furfuryl alcohol. This type is not widely used because of difficulties in obtaining a good "through-cure", i.e. an even rate of hardening all through the sand, upon which a satisfactory cold-setting technique depends.

Furfuryl alcohol modified resins, which are urea– or phenol–formaldehydes reacted with an amount of furfuryl alcohol, sufficient to give the properties required, the type depending upon whether a nitrogen free resin is required or not. These are used with either phosphoric acid (for U.F. modification) or sulphonic acid (for P.F. modification). The U.F. modifications, in particular, are very efficient cold-setting resins.

Furfuryl alcohol polymers, using phosphoric acid or sulphonic acid. These are nitrogen free and possess very good through-hardening characteristics.

There can be little doubt that the furfuryl alcohol resins represent possibly the most efficient type of cold-setting process, for the following reasons:

A very good balance exists between bench life, setting time and stripping time, i.e. the time after which the sand can be removed from the pattern. Effective control over these factors is possible.

A low gas evolution on casting.

A very hard core, which withstands even the roughest handling in the foundry.

High storage stability.
High hot strengths.
Extreme flowability of mixed sand.

This type of resin has replaced dry sand as the moulding material even in the production of castings up to 40 tons in weight.

There are certain precautions which must be taken, however, to ensure maximum consistency of results, these being:

The use of a clean silica sand.
The avoidance of unusual variations in temperature of incoming sand—particularly the avoidance of warm sand (e.g. above 25°C). Warm sand greatly accelerates the rate of setting and can lead to bench life difficulties.
Efficient mixing and distribution of sand to ensure the use of the sand within a minimum time after mixing. The newly developed continuous mixers are very effective with this process.

Control of properties

Assuming the use of a urea–furane resin and phosphoric acid (1·65 sp.gr.) the rate of setting dependent upon catalyst additions is represented by Fig. 2.

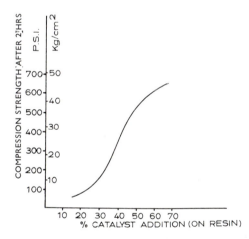

FIG. 2. Effect of catalyst additions upon rate of development of strength of cold-curing resin sand (2% resin addition).

The effect of temperature upon the setting time and the effects of "bench life" of the sand are shown in Figs. 3 and 4.

In practice, the method used is to add the catalyst to the sand first, then add the resin and mix for the minimum time necessary.

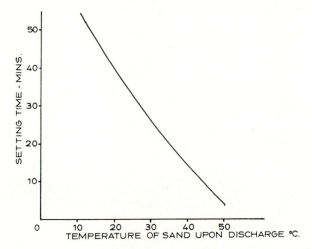

Fig. 3. Effect of temperature upon setting time of cold-curing resin sand (2% resin + 40% catalyst of resin content).

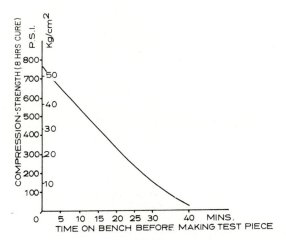

Fig. 4. Effect of bench life upon compression strength of cold-curing resin sand (2% resin, 60% catalyst of resin content).

TABLE 11. *Typical Mixture and Properties—Cold-curing Oil*

Mixture %		Properties			
Sand—clean dry silica	97·5	Setting time—30–40 min			
Oil	2	Compression strengths psi			
Hardener	0·5	(kg/cm²)			
(25% of oil content)		1 hr	2 hr	3 hr	24 hr
		90	214	315	720
		(6·3)	(15·0)	(22·06)	(50·42)

Cold-curing oils

The development of the cold-curing, non-stoving oil system dates back to 1955 but has recently come into prominence. This method utilizes a processed drying oil possessing a certain "hydroxyl value" (i.e. containing some free OH groups) with isocyanate hardeners. The result is a sand mixture which will cold-set within 30–60 min and which rapidly develops strength to a figure comparable with the cold-curing resins. The features of this method are:

A sand possessing high flowability because of its oily nature and lack of the "smell" usually associated with synthetic resins.

A very good strip from the pattern, again in contrast to the resin-bonded sands, the strips of which can be difficult.

The ease of stripping is due to the oily characteristics of the bond, which allows some lubrication and some flexibility in the partially cured sand.

Insensitivity to purity of sand, again in contrast to resins, where the quality of the sand is of prime importance.

Less sensitive to temperature than the resins. This particularly applies where the sand is rammed into a metal box, when the sand is used for moulding. This is of importance when the process is used for moulding and quick strip is required.

Very good build-up of strength and good cured strength.

The hot strength is lower than that of the furane resins and this limits its application, particularly when the high melting temperature metals are used, such as steel. It can also produce "scabbing" in grey iron, particularly from large flat areas of mould or core. This can be alleviated by:

Incorporating a fine powder such as iron oxide.

Torching the surface of the mould in order to partially dry it.

Stove it completely. This obviously eliminates the main advantage of the process.

Because of the use of isocyanates as the hardener, there must be some question of health hazards involved. The isocyanate used is diphenyl methane diiso-cyanate (OCN〈O〉—CH_2—〈O〉—NCO), which is a so-called "blocked" isocyanate. Providing the material is not allowed to come into contact with the skin before mixing the sand, little hazard exists until the metal is poured, when the high temperatures reached may cause some volatilization of unreacted iso-cyanate and also the evolution of isocyanate vapour due to the "reversal" of the chemical reaction between oil and hardener. At this point considerable care should be taken to avoid excessive contact with fumes and ample ventilation should be provided. The maximum allowable limit for isocyanate vapour in the atmosphere is only 0·02 ppm and it has been shown that the maximum concentration at the "knock-out" stage can reach 10 ppm or above. Hence the need for care in the application of this type of binder and particularly care in the use of correct proportions of hardener and binder. The rate of hardening depends upon the free hydroxyl groups available in the oil, not upon the amount of concentration of the isocyanate. The correct amount of hardener must always be used but more than this amount will only exist as free isocyanate in the sand mass, which must be avoided for the reason given. Therefore, any variation in the rate of setting must be achieved by variations in the oil, and the choice of oil does exist to enable the sand technician to achieve the results he desires.

TABLE 12. *Typical Mixture and Properties—Cold-curing Resin*

Mixture %		Properties
Sand—	97·2	Setting time 20–30 min
clean dry silica		Compression strengths psi
Resin	2	(kg/cm^2)
Catalyst	0·8	1 hr 2 hr 3 hr 24 hr
(40% of resin)		250 400 650 900
		(17·5) (28·0) (45·52) (63·05)

This process is quite suitable for use with zircon or chromite when these are desirable for casting results. It is also more suitable for facings backed up by moist backing sand because of the reaction between water and isocyanate.

Processed oil-stoved type

The forerunner to this process, and to all organic cold-set processes, was the processed oil-stoving method. This depends upon the auto-oxidation of a specially prepared drying oil by means of sodium perborate, acting as an oxygen donator. This process was developed in Switzerland prior to 1945 and was introduced into this country in the early 1950's.

The main points are:

Mixed sand of excellent flowability. This proved to be in marked contrast to the heavily bonded sands used at the time to make large cores and moulds.

A fairly quick setting period, producing a shape very true to the dimensions of the box.

A very easy strip from the pattern.

After baking, a very good hard core. Considered in relation to the then common baking practice, baking times were short—about 50% of the normal times. During baking, little or no distortion should take place, but this is dependent upon the characteristics of the oil. Again considered in relation to the existing practice, this accuracy was well in excess of the then accepted standards.

Under some conditions, the core or mould, particularly the mould could be used without baking, in fact about 20% of the work done with this binder is so used.

The conditions for using this system without stoving are:[5]

The core or mould should stand at least 24 hr.

The metal section should be limited to 4 in. (10·16 cm). This refers
principally to grey iron.

Due consideration should be given to the fact that the set strength of a
typical oil mix is of the order of 200–300 psi (14·0–21·0 kg/cm²);
hence, care should be taken in handling the core.

Coatings should be dried.

A small addition of clay helps to increase hot strength.

The finished core, particularly when stoved, shows low gas evolution and a
minimum of mould–metal reaction.

Insensitive to quality of sand.

Some measure of control can be effected by variations in the quantity of
sodium perborate used.

TABLE 13. *Typical Mixture and Properties—Cold-setting
Stoving oil*

Mixture %		Properties
Clean, dry,	97·9	Setting time 2–4 hr
silica sand		Compression strengths psi
Oil	2·0	(kg/cm²)
Accelerator	0·10	1 hr 2 hr 3 hr Baked
(5% of oil)		25 150 200 1400
		(1·75) (10·5) (14·0) (98·0)

Heat Curing Processes

These include the most automated and most rapid of all core or mould
making processes. There are two processes in widespread use: the shell process
(Croning or "C" process, described as such after the developer of the process)
and the hot-box or "wet-mix" process. The shell process is the older of the two,
being developed in Germany before 1945 and greatly developed in the U.S.A.
in the following years; the hot-box process was introduced in the early 1960's
and, likewise, has undergone considerable development.

Both processes have certain features in common:

The sand is cured at temperatures in the order of 200–300°C.

Both are of interest mainly in the mass production of small to medium
castings, but there are quite wide differences in application techniques.

Shell Process

Sand for this process was originally prepared by mixing a powdered phenol
formaldehyde resin (of the type known as "Novolac", i.e. containing an excess
of phenol) which contained the hardening agent necessary to enable the resin

to cure, together with a fine sand, the result being an intimate mixture of resin and sand which, when heated on the pattern plate, would fuse together and cure to a hard, rigid shape following in detail the contour of the pattern. The simple mixture, however, had certain disadvantages:

The resin would segregate from the sand during the moulding process, leaving resin-rich areas, usually against the face of the pattern. This could result in "blown" castings, sticking to the pattern plate, etc.

The handling of the mixture constituted a health hazard because of the fine silica particles in the atmosphere.

It proved difficult to utilize "blowing techniques" in making the core or mould, largely because of the segregation of resin from sand.

Because of the inefficiency of the "bonding" it was found that an excessive amount of resin had to be used, which limited the application of the process because of raw material cost.

These difficulties were overcome in the late 1950's by the development of "precoated" or "resin-coated" sand. This is made by one of two methods:

Using an alcohol solution of the resin to coat the sand grains, then driving the solvent off by continuing the mixing with the application of heat and finally screening the product and passing it through a cyclone to cool it and to break up any aggregates of sand which might have passed the screen. The hexamethylenetetramine (hexamine, in short) curing agent is incorporated as a separate addition at the start of the process together with the addition of a wax such as "Acrawax C" or calcium stearate which serves as a release agent and also to improve the flowability of the sand. This process is called the "Warm Process".

Using a solid resin, in the form of coarse powder, flake or "pearls", which is added to sand heated above the melting point of the resin (i.e. about 130°C) the resin melts and coats the grains. The temperature of the sand is then dropped below the melting point of the resin and the necessary hardener introduced by adding a water solution of hexamine. This causes the mixture to harden and to break down into free flowing sand, in which condition it is discharged from the mixer, sieved and passed through a cyclone. Wax is either incorporated in the resin or added as a separate addition together with the resin, either as powder or water emulsion. This process is called the "Hot Process".

Resin-coated sand prepared in either of these two ways has certain common points but there are also minor differences between sands produced by the two methods.

The common properties, compared with the powder mixture are:

Very much more efficient coating, resulting in the need for lower resin additions. This saving can be as much as 50%, which renders the shell process economically worth while, as well as technically more advantageous because of the lower gas evolution from the moulds.

No segregation of resin from sand. This enables the coated sand to be used in core blowers in the application of automated production methods.

Easy stripping from the pattern or core box, again important in saving the use of expensive release agents and vital to automatic operations.

Greatly improved flowability.

Elimination of health hazards—no dust is evolved from the sand.

The two individual processes diverge in the following ways: in general, the hot process tends to produce a more brittle sand which is more prone to cracking upon casting resulting in "run outs" or finned castings. The warm process will give more satisfactory results in the casting of heavier sections. Hot process sand will cure quicker and will give a more even "shell", and because of this, is of greater advantage in the manufacture of cores.

The advantages of precoated sand are such that very little resin/sand mixtures are used now.

A wide range of resin contents is available, to suit all types of castings, from 1·5% to 6·5%. Additions can be made to suit specific requirements, such as the use of "mill scale" to eliminate "orange peel" defect in low carbon steels; starch to produce break-down properties essential to light alloys; iron oxide to assist resistance to metal penetration. Other granular materials such as zircon and chromite can be coated when their use is required to obtain specific casting properties. Very latest developments include the production of nitrogen-free coated sands to avoid porosity in castings due to nitrogen.

The method of use of resin-coated sand is twofold.

The dump box method—whereby the hot pattern (200–250°C) is placed face downwards upon a "dump box" or bucket containing the sand. This is then inverted, thereby causing the sand to drop on to the pattern. After a certain time in this position, the "investment" time depending upon the thickness of shell required, the dump box is re-inverted, the unused sand falling back into the dump box and the pattern holding the sand removed. At this stage, the mould consists of a hardened skin backed up by a semi-hardened layer, totally being 0·25–0·37 in. (0·63–0·95 cm) thick. The pattern, complete with mould is then cured by either being moved into an oven or by having a heating hood pulled over it and when cured, is removed by means of ejector pins set in the plate. After removal, the two half shells are fixed together, usually by synthetic resin glue and the mould can then be stored or poured immediately.

The blowing method—in which the sand is blown into a box, which can be either the normal type of core box or a contoured mould box. The sand is blown into the box at low pressures (20–40 psi, 1·4–2·8 kg/cm²), allowed to invest for the necessary time (from 5 sec to 30 sec is typical) and then the surplus sand is allowed to drain from the core, leaving a shell. This is allowed to cure in the box before being removed again for storage or immediate use.

The advantages of the shell process are obvious:

It uses a very minimum of sand, as compared with the large amount of sand used by traditional methods of moulding.

The maximum use can be made of floor space because of the small volume of sand used and the possibility of a rapid throughput.

Comparatively unskilled labour can be used, and by automating the process as much as possible, a minimum amount of labour is required.

A great degree of dimensional accuracy can be obtained. This was originally the main claim for the process, but now the process is economically sound as a means for the rapid production of small to medium sized castings.

The automobile foundries have made great use of it, for parts such as manifolds and crankshafts.

Hot-box or Wet-mix Process

Preparation of the hot-box sand consists of the mixing of sand, resin and catalyst, without heat and in the normal type of mixer.

The essential differences between this process and the shell process are:

1. *Nature of the mixed sand.* Because a liquid thermosetting resin and usually a liquid catalyst is used, the sand as prepared will be wet and will possess some degree of cohesion or "green bond". This green bond will tend to increase upon standing. This feature contrasts with the free-flowing dry sand characteristics of the shell process with its unlimited bench life and immediately pinpoints two of the most important features of hot-box mixtures:

The flowability of the sand is not as good as that of shell sand. Higher blowing pressures have to be used (80–100 psi, 5·4–6·8 atm) and it is more difficult to blow the complicated shapes which are easy to blow with shell process sand.

The life of the sand is limited where that of shell sand is unlimited. The bench life will depend upon a number of conditions, such as temperature of the sand, degree of mixing, atmospheric conditions and the chemical characteristics of the resins and catalysts used.

2. *Resin chemistry.* The chemical characteristics of the resins used are quite different. In fact, there is a considerable choice of resins for the hot-box process and one can certainly choose a resin which will give the properties required, whether high or low strengths, freedom from nitrogen, easy break-down, etc. This makes hot-box cores more suited to castings demanding specific properties, i.e. light alloys.

3. *Rate of cure.* This is undoubtedly the *raison d'etre* of the hot-box process. Very rapid rates of cure at normal temperatures can be achieved with the facility for curing at lower temperatures than can be used for shell cores. Using a reactive furane resin, rates of cure as low as 5 sec are possible, an average figure being 10–20 sec. This is cure time, i.e. the length of time before the core is removed from the box, and is only a fraction of the time required to remove a shell process core, i.e. total cycle time including investment time and cure

time. The facility for easy stripping is a result of the very hard skin produced in a very short time, sufficient to bear the weight of the core upon ejection from the box. After ejection, the process of through-curing continues and the core progressively hardens, becoming hard and rigid throughout.

4. *Hollow shell or solid core.* Because of the rapid cure which takes place it is impossible to evacuate any of the uncured sand after ejection, so the only means of making a hollow core is to use a former inside the core box around which the sand must be blown. This is usually difficult to arrange and is often more expensive than the saving of sand justifies. This necessity to produce a solid core implies a major difference in the relative weight of sand to produce cores by shell or hot-box, thus obviously affecting the economic balance between the two processes. Under normal circumstances an indication of relative costs is that resin-coated sand is three times dearer than hot-box sand mixtures; hence, if the weight of a hot-box core is less than three times the weight of a shell core, then the hot-box core will show the advantage; if it is more than this the reverse will be true.

5. *Application.* Obviously, there is no possibility of "dumping" hot-box sand as is done with shell sand, all hot-box cores having to be blown. If a mould is made, this has to be blown. This does impose a severe restriction upon the use of hot-box mixtures for moulding and, in effect, the process is used almost entirely for core making.

Practical Details of Hot-box Method

The type of resin used will depend upon the casting to be made (i.e. section, type of metal). The catalysts are usually liquid and consist of ammonium salts, except where completely nitrogen-free systems are used.

Resin types. Urea–formaldehyde-furfuryl alcohol resins suitable for most types of castings except steel. The furfuryl alcohol content will control the ease of break-down, this being faster as the alcohol content decreases. The urea–phenol–formaldehyde resins, because of equivalent performance at lower cost, have largely replaced the furfuryl alcohol resins.

TABLE 14.

Resin type	General application
Urea–formaldehyde–furfuryl alcohol (urea furane)	Where nitrogen can be tolerated Properties dependent upon furfuryl alcohol content
Phenol–formaldehyde	Nitrogen-free or lower nitrogen for steels
Urea–phenol–formaldehyde	Low nitrogen and suitable for most types of grey iron castings
Urea–formaldehyde	Good break-down. For light alloys

TABLE 15. *Typical Properties of Hot-box Mixture*

Mixture %		Properties					
Clean medium		Tensile strength					
grained sand	97·5	Cured at 220°C					
Resin	2·0	10	20	30	40	60	sec
Catalyst	0·5	440	540	560	570	450	psi
(25% of resin content)		(30·8)	(37·7)	(39·1)	(39·9)	(31·5)	kg/cm^2

Additions to sand. Although it is more usual to select the resin to donate particular properties to the sand, it may be desired to add iron oxide in one of its forms to reduce a tendency to pinholing and metal penetration.

Methods of Testing for Shell Process and Hot-box

Both these methods demand specific tests. The following are suggested:

Shell process:

Loss on ignition.

Investment rate.

Tensile strength $\begin{cases} \text{hot, upon ejection} \\ \text{cold.} \end{cases}$

Hot-box process:

Bench life.

Tensile strengths: 10 to 60 sec cure $\begin{cases} \text{hot, upon ejection} \\ \text{cold.} \end{cases}$

OVEN-BAKED CORE SAND MIXTURES

The application of these mixtures diminishes as the other processes rapidly develop. In principle, the method is the same as has been used for many years, but recent developments have been made to increase the rate of baking of the core. This has largely been based upon the use of vegetable oils such as tung oil and oiticica oil and mineral oils produced during the catalytic distillation of petroleum oils. An indication of the rate of baking is shown in Fig. 5.

A further improvement of recent years has been the replacement of dextrine–oil–water mixtures by pregelatinized starch–oil–water mixtures, these latter giving better flowability and bench life characteristics.

Whilst this method still constitutes the cheapest way of making cores, it suffers from the disadvantage of inaccuracies due to removal of the core from the box in an unhardened state and transfer to a core stove, where further distortion may take place.

Constant attempts have been made to replace oils by synthetic resins of the phenol–formaldehyde or urea–formaldehyde types, but their inherent stickiness has limited their application. Recent developments in phenolic formulations show that this defect can be overcome to a considerable degree. The very important feature of these types of resin is their very rapid curing properties, enabling stoving cycles to be cut by 50% compared with oil binders.

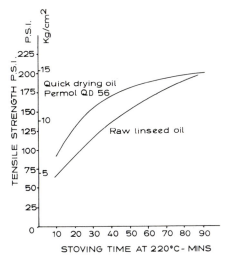

FIG. 5. Comparison of rates of development of tensile strength raw linseed oil and quick drying oil Permol QD56 (0·75% oil, 1·2% starch, 2·0% water).

TABLE 16A. *Typical Mixture and Properties of Oil-bonded Sand*

Mixture %		Properties
Clean, dry silica sand		Green compression 1·0 psi
AFS No. 55	95	(0·07 kg/cm²)
Oil	1·0	Tensile strength—baked at 220°C
Cereal	1·5	30　45　60　90　min
Water	2·5	260　295　303　320　psi
		(18·20)　(20·64)　(21·20)　(22·40 kg/cm²)

TABLE 16B. *Typical Mixture and Properties of Resin-bonded Sand*

Mixture %		Properties
		Green compression 0·6 psi
Clean, dry silica sand		(0·042 kg/cm²)
AFS No. 55	95	Tensile strength—baked at 220°C
Resin	1·25	10　20　30　45　min
Cereal	1·5	250　420　435　380　psi
Water	2·25	(17·5)　(29·4)　(30·4)　(26·6　kg/cm²)

TECHNOLOGY OF SAND PREPARATION

Sand preparation techniques are a vital link in the chain of sand technology. The function of the "mill" or "mixer" is to distribute the bonding media around the sand grains and to discharge the sand in the condition necessary for the particular process being carried out. Obviously the duty will depend upon the characteristics of the sand which it is desired to produce and these can be divided into two main classes:

(i) Medium to heavy bonded sand, such as moulding sand, using clays as the principle bonding material.

(ii) Light to medium bonded sand, including sand for cores and moulds utilizing the chemically hardened or gassing processes.

Medium to Heavy Bonded Sand

This type is invariably of rugged construction because of the heavy duty performed, both in quality and quantity of sand produced. They can be of batch or continuous types and utilize wheels and ploughs to "knead" and agitate the sand. The pans can be stationary or rotating. The principle of the operation has not changed greatly over the years. The energy input is a measure of the ability of the mill to work upon the clay bond and a figure of 1·5 to 3·0 h.p. per ton of sand has been quoted as being necessary.[6] The mill must be chosen to give ample power for the work to be done—a mill of insufficient energy input will result in poor moulding sand.

The various types of mill can be described as stationary pan/vertical wheels; rotating pan/vertical wheels; horizontal wheels; continuous milling pan type; blade type or horizontal rotating drum type.

Light to Medium Bonded Sands

This includes oil-bonded sands and all the recent developments in cold-setting and hot-box mixtures. It is important to consider the main requirements:

Efficiency of coating of the sand grains because of the low percentage of binders used.

Accurate measurement of the additions.

Efficient mixing of the sand without the development of excessive heat. This is important, particularly with resin-bonded sands, because of the very great effect of even a small amount of heat upon setting times and bench life.

Absence of any tendency to crush the sand grains, thereby creating fines in the sand.

These conditions indicate the use of less energy input into the mixing and much more of a kneading or rubbing action than the heavy mulling which is necessary with clay-bonded sands. Hence the description of sand "mixer" rather than "mill" or "muller" commonly given to moulding sand preparation machines.

Types of Sand Mixers

Batch. A typical machine of this type is the "S"-blade type shown in Fig. 6. The action is to rub the sand between the blade and the "rubbing plates", causing the sand to turn over as it rotates. Capacities of this type of mixer range from 2 lb (laboratory mixer) to 1 ton. Mixing times vary from 2 min to 5 min, depending upon the type of binder used. Sands of green strength from nil to 5 psi can be efficiently mixed in this machine.

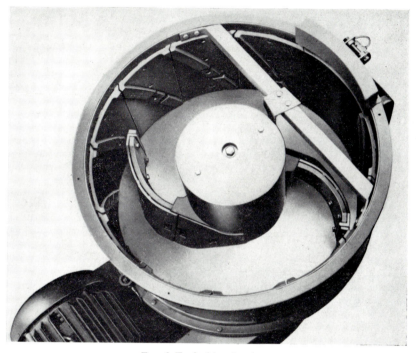

FIG. 6. Typical batch mixer.

Continuous. This machine was developed essentially to mix sand for the modern cold-setting and hot-box techniques and constitutes not only a mixing machine but also a means of conveying the sand into the box or, in the case of hot-box operations, into the storage hoppers above the machines. Liquid binder, oil or resin together with the appropriate catalyst, whether liquid or solid, are fed into the sand stream and blended together and conveyed along the trough by means of a screw which is arranged to give the required combination of mixing and conveying blades. The sand is then discharged into the box or hopper. All the materials are accurately measured into the mixer, the sand through a variable orifice and the liquids through positive displacement metering pumps controlled from the panel, and powders by means of a calibrated screw feed. The whole operation is controlled by one man pressing the stop/start buttons and this man can also be responsible for filling the box. The sand is then literally available when required and avoids wastage of expensive sand, which can be severe in the case of batch mixed, cold-setting material. The combination of this type of machine with the newer processes of core and mould making has had a remarkable effect upon production times and use of floor space.

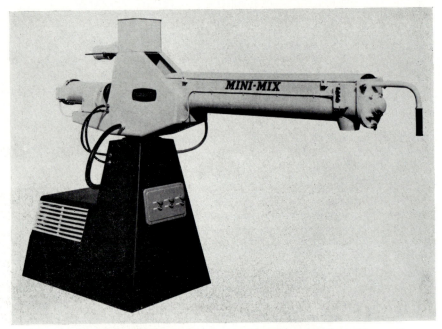

FIG. 7. Continuous mixing and conveying machine for core and moulding sands.

Capacities range from 4 tons per hour of mixed sand to 10 tons per hour and even higher capacity machines are available for specialized application, such as ingot mould production.

TABLE 17.

Type of binder	Type of mixer	Time to make core (min)	Stoving time (hr)
Oil dextrine	Batch	90	9
Cold-set oil	Batch	25	6
followed by			
stoving	Continuous	5	6
Cold-cure resin	Batch	25	No stoving
	Continuous	5	required

Resin-coated Sand Production

This process again can be divided into two sections, batch and continuous for the two processes.

Warm coating. Because of the need to evaporate off the solvent and the time factor involved in this, the warm coaters are invariably batch type machines. The essential parts of the equipment are:

A means of weighing sand and dispensing it into the mixer.
A means of dispensing resin, hexamine and wax or calcium stearate.
A means of heating the sand in the mixer, this being done by passing hot air through the mixer.
A vibratory screen.
Means of cooling the sand, either by cyclone or by fluid bed cooler.

The mixing machine, in which the coating is effected can be either "S"-blade type or horizontal wheel type. Possibly the most important condition in sand coating is that the sand should not be subjected to abrasive action either during mixing or after discharge from the mixer. Loss of resin from the sand by abrasion can easily occur because of the brittle nature of the resin film and this loss can amount to more than 10%. This can cause trouble with resin segregation as well as resulting in loss of strength. The gentler action of the "S"-type blade is of benefit in this respect and excessive pipe runs to the cyclone should be avoided, since any change in direction of the sand will result in loss of resin.

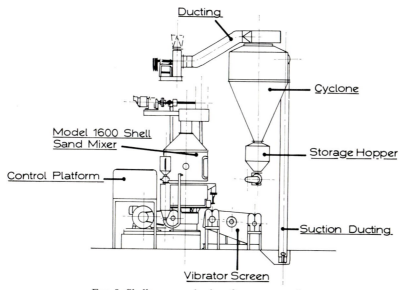

FIG. 8. Shell-core sand mixer for warm coating.

M

Hot coating. The essential conditions are somewhat different from those for warm coating and are as follows:

A means of measuring the necessary amount of sand and heating it to the coating temperature (about 130°C). A means of dispensing resin, aqueous solution of hexamine and possibly extra water, to effect the cooling of the sand to below the melting point of the resin.

A vibratory screen.

A means of cooling the sand. Because of the greater heat content of the sand, more efficient cooling is required and conditions favour a fluid bed cooler rather than a cyclone. In general, because of the faster cycle time and the faster rate of chemical reactions occasioned by the high operating temperatures, it is necessary for the process to be well instrumented and controlled.

Again, because of the higher temperature, it is desirable to use either a solid resin or a water borne resin to avoid the risk of explosion in the coating chamber.

Plant can be either batch, batch/continuous or continuous.

Figure 9 shows a typical machine.

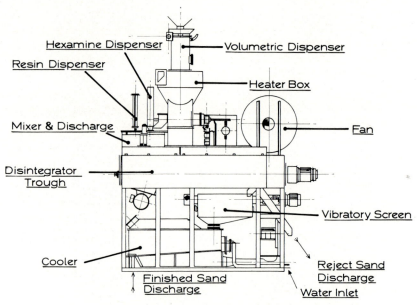

FIG. 9. Continuous shell-core sand mixer for hot coating.

CHEMICAL TECHNOLOGY

Sodium Silicate

Sodium silicate is produced by the fusion of silica sand with sodium carbonate at temperatures around 1650°C.

$$SiO_2 + Na_2CO_3 \rightarrow Na_2O \cdot SiO_2 + CO_2$$

The product is a glass which is cooled, broken up and dissolved in water to give a solution of the specific gravity required.

The relative weights of SiO_2 and Na_2CO_3 will control the type of silicate produced, the type being characterized by the molecular ratio $SiO_2 : Na_2O$.

TABLE 18. *Typical Sodium Silicates*

Molecular ratio $SiO_2 : Na_2O$	Specific gravity	Soda content %	Silica content %	Water content %
2·00:1	1·56	15·3	30·6	53·1
2·50:1	1·50	12·5	31·1	54·4
3·20:1	1·37	8·6	27·6	63·8

The type of silicate will depend upon the hardening mechanism used, usually being the 2·00:1 silicate for CO_2 hardening and preferably a high ratio for the other chemical hardening processes, where the hardening rate is governed by the molecular ratio. In the latter case a balance has to be maintained between faster setting properties and unfavourable sand characteristics caused by excessively fast rate of water loss leading to short bench life.

Hardening by CO_2 Gas

The chemical reactions between sodium silicate and CO_2 are:

$$Na_2O \cdot 2SiO_2 + CO_2 = Na_2CO_3 + 2SiO_2$$
$$Na_2O \cdot 2SiO_2 + 2CO_2 + H_2O = 2NaHCO_3 + 2SiO_2$$

In this reaction, the SiO_2 separates as a silica gel, which causes an increase in viscosity of the bond and hence an increase in strengths of the sands.

In addition to these chemical reactions, some loss of water from the silicate occurs when the CO_2 is passed, which also causes an increase in viscosity of the bond and an increase in strength.

Thus, the final result is a combination of chemical and physical reactions, the relationship between which governs the practical results achieved. This relationship will be determined by such factors as the rate of gassing and the length of time and condition during which the core stands exposed to the atmosphere.

R. Worthington[7] suggests that the idea that is known as "over-gassing", i.e. the production of friable surfaces, is not due to excessive rates of flow of gas but due to long flows, where there is high carbon dioxide absorption leading to excessive gel formation and friability upon drying out. He considers that the balance between CO_2 entering the silicate film to form hydrogel, and H_2O leaving the film to produce the physical "glass" bond lies around 15% of chemical bond for normal foundry practice. He expresses the conditions existing in the bond as follows:

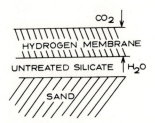

FIG. 10. Conditions existing at the surface of the sand grain during gassing with CO_2.

The direction of the movement of molecules, i.e. CO_2 inwards and H_2O outwards, will depend upon the flow of gas.

Amount of CO_2 gas required. Theoretical considerations show that the ratio of gas required to complete the bicarbonate stage is 0·22 per unit weight of silicate. However, practical results show that this ratio approaches 1·0 to 1·5 or even higher. This is a measure of the inefficiency of the use of CO_2 gas.

Chemical Hardening

Chemical hardening is the result of the reaction of the hardener with the water in the silicate bond to produce the silica hydrogel by loss of water. The efficiency of the reaction will govern to a large extent the strengths produced.

In the case of ferro silicon, the reaction is strongly exothermic, which results in:

The development of the hydrogel bond.
Physical dehydration resulting in the formation of a glass.

If we consider that the "glass" bond gives the strongest bond possible with silicate, we can understand the very high strengths which are obtained with this process. The reaction, however, is so strong that physical variations in the ferro silicon can materially affect the relationship between hydrogel bond and "glass" bond and hence the properties of the sand.

When, however, dicalcium silicate is used, the absorption of water by the hardener is considerably more gentle and little or no dehydration of the bond takes place. The formation of hydrogel takes place around each particle of dicalcium silicate, the continuity of the hydrogel bond being dependent upon the distribution of the particles of dicalcium silicate. Again, the efficiency of the reaction will depend upon the chemical composition and particle size of the dicalcium silicate.

TABLE 19. *Chemical Analysis*
of a Dicalcium Silicate Slag

SiO_2	31·3%
Fe_2O_3	5·4%
Al_2O_3	1·9%
CaO	58·1%
MgO	3·8%
Loss on ignition	Nil

It is obvious from the above that any chemical which will react with or absorb water at a controlled rate will favour the formation of the hydrogel bond from sodium silicates and thus enable "cold-setting" of the sand compact to take place. Portland Cement has also been used successfully.

Clays

The general term "clay" refers to a group of minerals in which the chief constituent is hydrated silicate of alumina. They occur as aggregates of minute flakes, the chemical structures consist of sheets of atoms, the nature and detailed arrangement of which determines their bonding characteristics.

The different classes of interest to the bonding of foundry sand are:

Kaolinite or China clay, including refractory clays of low plasticity. Its composition corresponds to the formula: $Al_2O_3 \cdot 2SiO_2 \cdot 2H_2O$. Its low plasticity makes it of little use for bonding sand, but it is of great use as a refractory stabilizing clay in core washes.

Ball clays. Similar in mineralogical constitution to the China clays but very much finer and therefore produce a more plastic mass. Used to donate increased hot strength and some degree of green strength to cold-setting mixtures.

Bentonites. The chief mineral constituent is montmorillonite (general formula $Al_2O_3 \cdot 4SiO_2 \times H_2O$). The best deposits are found in America and this clay is characterized by very great swelling and moisture absorption properties which makes it of immense value in bonding foundry sand. It finds great favour in steel foundries. The clays are plastic over a wide range of moisture, the sodium clays being more plastic than the calcium clays. Western Bentonite or Wyoming consists mainly of sodium montmorillonite, Southern Bentonite consists of calcium montmorillonite.

Fullers Earth. Found in Britain and developed as an alternative to Bentonite. The plasticity and swelling properties can be increased considerably by treatment with soda ash to replace the calcium by sodium. Calcite is the principle impurity which markedly reduces green and dry strengths.

Secondary Mica, occurs as a constituent of naturally bonded sands. The

Illite group is invaluable as extenders of Bentonite. The properties are mid-way between those of the Montmorillonite and Kaolinite groups.

Limonite. Hydrated iron oxides present in natural moulding sands.

The plasticity of clays. The structure of clays are characterized by layer lattices of various types. The "flakes" of clay minerals are held together by forces acting through the flat surfaces of the flake, the intensity of these forces depending upon the type of clay mineral. Clay becomes plastic when a film of water separates the flakes and the ability of the clay to absorb water, upon which its plasticity depends, will depend upon the strength of the forces between the layers of clay minerals and also upon the strength of the forces between the clay and water layers.[8]

Cement

The general term "cement" has been associated with a wide range of calcareous earths which have been used since pre-Roman days to produce adhesion between stones, bricks, etc. "Portland" cement—so called because of the similarity between the appearance of the hardened cement and Portland stone—originates from the early 1820's (the first patent was taken out in 1824). Portland cement is defined as a product obtained by intimately mixing together calcareous and argilaceous, or other silica, alumina and iron oxide-bearing materials, burning them at a clinkering temperature and grinding the resultant clinker. A typical analysis is as follows:

$CaO = 64 \cdot 1\%$, $Al_2O_3 = 5 \cdot 5\%$, $Fe_2O_3 = 3 \cdot 0\%$, $SiO_2 = 22 \cdot 0\%$,
$MgO = 1 \cdot 4\%$, $SO_3 = 2 \cdot 1\%$, $Na_2O + K_2O = 1 \cdot 9\%$.

These oxides can be combined to form various binary and tertiary compounds such as tricalcium silicate $3CaO \cdot SiO_2$.

The mechanics of the setting and hardening of cement can be visualized as such:[9] the cement grains are acted upon by water to form a supersaturated solution from which the gel-like mass of crystals precipitates. Diffusion of water molecules to the surface or even into the crystal lattice to react *in situ*, will play a part in the hydration. Whilst still in the plastic condition, the cement will show a small contraction. Once the mass becomes rigid, a small expansion occurs because the gel mass deposits around the cement particles and causes them to swell. The quantity of the gel progressively increases and spreads into the intergranular spaces. Upon further drying, the "cement" undergoes an irreversible contraction as the gel changes into its more stable form.

The sequence can be illustrated as such:

Unhydrated cement → plastic cement-water mass → gel of crystals of colloidal dimensions together with crystalline products above colloidal dimensions, (CaO, hydrated calcium aluminates and sulpho-aluminates) → stable gel → crystalline products of coarser dimensions. This last change infrequently occurs under normal circumstances.

The effect of substances such as molasses is not easily understood. Molasses was used for hundreds of years in India to increase the hardness of concrete. Sugar, which is an important constituent of molasses, is said to accelerate the initial set but to retard final set. There is also evidence that the type of molasses, i.e. cane or beet molasses, has considerably different effects. It seems likely that the addition of molasses affects the formation of the gel structure, which is shown by the behaviour of the sand when molasses is added. Green strength drops appreciably and the sand becomes very free flowing.

Mesh size of the cement is of great importance, the finer the particle the greater the strength produced. The surface area of a normal cement is of the order of 2250 cm²/g. Quick drying cements are of the order of 3250 cm²/g.

Oils

The chemistry of the natural and processed oils and the mechanism of their curing is complex and the latter is still open to some discussion. Generally speaking, the raw oils, such as raw linseed oil, are used for the stoving processes whilst processed oils are used for the cold-setting and cold-curing binders, because of the greater degree of control over properties such as rate of cure demanded by these processes.

TABLE 20. *Oils Used in the Foundry Trade*

Material	Source
Linseed oil	Flax plant, Argentina, India, Canada.
Tung oil or China Wood oil	America, China.
Oiticica oil	Brazil.
Marine oils	Oils extracted from herring, pilchard and whale.

Mineral oils—lubricating oils
    unsaturated oils
    (so-called petroleum polymers)
	Crude oil

Esters of fatty acids derived from the above plus Tall oil, etc.	Paper pulping industry.
Alkyd resins	Surface-coating industry.

Chemical Composition

Vegetable and marine oils are composed of the glycerides of long chained fatty acids (usually containing 18 carbon atoms). The type of fatty acid present will control the behaviour of the oil, which may be drying, non-drying or semi-drying.

The drying power of the oil depends upon the content of fatty acids containing two or more double bonds ($-CH=CH-$) and this degree of "unsaturation" is measured by means of the iodine value.

Thus, stearic acid:

$$CH_3(CH_2)_{16}COOH \text{ Iodine value} = 0.$$

Linolenic acid:

$$CH_3CH_2CH = CH—CH_2CH=CH—CH_2—CH=CH—(CH_2)_7COOH.$$
$$\text{Iodine value} = 274.$$

Natural oils consist of a mixture of such acids. For instance, linseed oil contains typically:

22% of oleic acid
17% of linoleic acid
51% of linolenic acid

and has an iodine value of about 170.

Certain other oils, such as Tung oil, consist of a high concentration of oleo-stearic acid which contains "conjugated double bonds" and behaves somewhat differently during the drying process.

Esters. Just as oils are natural glycerol esters of fatty acids, so these may be made synthetically by reacting at high temperatures (200–230°C) the appropriate fatty acid with glycerol or pentaerythritol. This is a convenient way to use low priced, naturally occurring fatty acids, such as those occurring in Tall oils, and esters of this type are of wide application in foundry oil preparations.

Alkyd resins. These are used very widely in the surface coating industry and find some application in foundry core oils, mainly as a constituent of cold-setting and cold-curing oils.

Essentially, they are the product of the esterification reaction between di-hydric alcohols and di-basic acids. The result is termed as "polyester". The complexity of the reaction product increases greatly as the "functionality" of the reactants increases. A commonly used reaction is that between glycerol and phthalic anhydride, the product of which is made compatable with drying oils by cooking the drying oil constituent (either as fatty acid or oil) in with the reaction.

Alkyd resins are usually classified according to "oil length", i.e. the proportion of oil, expressed as fatty acid, contained in the resin. They are termed:

Short oil alkyds (20–45% fatty acids).
Medium oil alkyds (45–60% fatty acids).
Long oil alkyds (60–80% fatty acids).

Their properties will depend upon the type of oil used and upon the "oil length".

Process of Curing

Linseed oil will be taken as a typical example.
The two processes are involved, depending upon the conditions of curing.

Autoxidative polymerization, occurring at room temperature (i.e. cold-setting).

Addition polymerization—occurring with heat curing (oven baking).

Autoxidative polymerization. The following sequence takes place:

Absorption of oxygen from the air (or from oxygen donator, such as sodium perborate), increase in viscosity and final transition to a solid film of linoxyn linkage between two or more molecules producing a polymeric structure.

It has been shown that during the drying process the peroxide value rises to a maximum and then decreases, due to the decomposition of the hydro peroxides, but the viscosity increases rapidly to a gel, following the decrease in the peroxide value.

Thermal Polymerization of Drying Oils

It is believed that the fatty acid radicals present in partially polymerized oils are joined together by a carbon–carbon bond. This is brought about by the following two reactions:

A rearrangement or isomerization of the molecule, in which the double bonds occur in such a way that the system becomes conjugated.

Linking of the molecules to form larger molecules by a Diels-Alder type addition reaction; the formation of a dimer occurs first followed by the trimer and so on to give a complex polymer.

It has been found in air drying oils that the amount of conjugation present reaches a maximum before the oil sets and decreases when the oil sets, but it must be remembered that this polymerization of oil is not simply an addition process but involves condensation and other complex reactions to produce low molecular weight compounds.

The Effect of Metal "Driers"

Metals, introduced in the form of compounds such as naphthenates, materially accelerate the drying mechanism at both room temperatures and elevated temperatures. The metals used are cobalt, vanadium, lead, manganese, zirconium and mixtures of two or more of these.

Their effect is two-fold:

Acceleration of the activation by oxygen of the methylene groups.

Acceleration of the decomposition of the hydro peroxide thereby increasing the rate of cross-linking.

M*

Stages in the Drying of an Oil

The drying of an oil can be followed by means of the changes in viscosity.

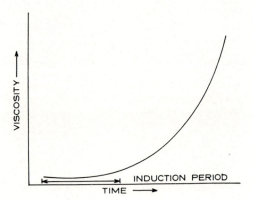

FIG. 11. Diagrammatic representation of drying of oil film at room temperature.

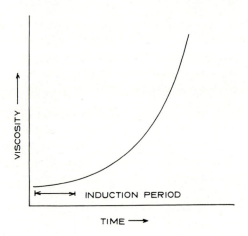

FIG. 12. Diagrammatic representation of drying of oil film at elevated temperature.

The length of the induction period is important to the foundry application of oils because it represents "dead time", either in oven time or "box time", and depends upon:

The presence of antioxidants in the oil. Raw oils contain substances, such as phosphatides which are present as "footy material", i.e. particles of links carried over in the extraction process. The presence of these materials considerably lengthen the oxidation process because they have to be destroyed before oxidation can take place. Careful selection and purification of the oil by filtration is necessary to reduce the content of antioxidants to a minimum. The use of an "air-blown" oil (i.e. one that has been partially polymerized) will considerably shorten curing times.

The addition of metal driers will shorten the induction period.

Unsaturated Mineral Oils

These oils provide useful additions to quick-drying stoving oils. They are produced during the cracking process and consist of various unsaturated hydrocarbons such as:

$$
\begin{array}{cc}
HC\!\!-\!\!-\!\!CH \\
\| \quad \| \\
\text{Cyclopentadiene } HC \quad CH \quad \text{and butadiene } CH_2\!=\!CH\!-\!CH\!=\!CH_2 \\
\diagdown \diagup \\
CH_2
\end{array}
$$

and can be removed by such processes as clay absorption. They provide oils with iodine values varying from 100 to 300. They are highly reactive and are useful for quick-drying oils.

Mode of polymerization—Diels-Alder type addition across conjugated double bond.

Isocyanate Cold-cured Oils

The principle of this cold-cure system is the cross-linking of long chain unsaturated ethylene double bond type of compounds using aromatic diisocyanate derivatives.

The oils used for this process consist mainly of the natural unsaturated oils and the synthetic modified oil as outlined above. The mechanism of cure for these cold-set oils depends on two types of reactions:

Oxidation at room temperature with the use of metal driers (usually cobalt or lead naphthenate) to increase the rate of this oxidation polymerization (as set out above, under normal drying of oils).

Reaction with diisocyanate derivatives (usually diisocyanate-diphenyl methane composition) to give a number of complex reactions resulting in a high molecular weight polymer.

Starches

Starch is used in two forms:

Pregelatinized starch.
Dextrine.

Practically, the difference lies in the fact that dextrine is soluble in cold water and has superior adhesive powers but absorbs less water and therefore demands lower water contents in the sand mixtures.

The feed stock for both materials is corn, wheat or (less important) potato, from which the starch is removed by either a dry milling or a wet milling process. The separated starch is then cooked or gelatinized on steam-heated drum dryers, the resulting cake being ground to produce the pregelatinized starch, powders or cereal binders used in foundry sands.

In dextrine manufacture, starch is digested with hydrochloric acid, which converts the starch to dextrine, the properties of this being dependent upon the time and temperature of treatment as well as the acid content. Dextrine manufacture is essentially a batch process, whereas pregelatinized starch can be produced on a continuous basis.

Variations in the feed stock will produce starches of different characteristics, an important factor being the amount of protein that is taken out of the feed stock. The fineness of grinding can also affect the properties of the starch.

Assessment of properties

Pregelatinized starch. The following properties should be determined:

Water absorption.
Green strength produced over a range of water additions.
Dry strength.

TABLE 21. *Typical Properties of Two Pregelatinized Starches A and B*

Properties	Starch A		Starch B	
% moisture	4·83		8·92	
Bulk density	0·446		0·525	
Sieve analysis:				
100 mesh 150 microns	7·80		17·84	
150 mesh 105 microns	14·42		24·88	
200 mesh 75 microns	75·34		55·36	
2·0% starch				
Water tolerance	psi	kg/cm^2	psi	kg/cm^2
1·0%	1·90	0·133	1·25	0·088
2·0%	2·10	0·147	1·35	0·094
3·0%	1·50	0·105	1·15	0·081
4·0%	1·15	0·081	1·05	0·074
5·0%	1·00	0·07	0·95	0·066
6·0%	1·00	0·07	0·75	0·052
7·0%	0·90	0·063	0·65	0·045
8·0%	0·85	0·059	—	—

Dextrine

Viscosity of water solution.
Water content.

The higher the viscosity, the higher the green strength produced.

Usually a dark colour signifies a greater degree of conversion, a more soluble dextrine and lower green strength per unit addition of water.

THE MANUFACTURE OF SYNTHETIC RESINS

The general procedure for the manufacture of these resins follows the same pattern.

Initial reaction between the resin chemicals, i.e. urea or phenol and formaldehyde, under closely controlled conditions of catalysis. Rates of heating, time of reaction, are important conditions to be controlled during the resinification process. Control is effected by such tests as pH value and viscosity. Distillation under vacuum to remove water introduced with the reactants and formed during the resinification by the process of "condensation".

Further modifications to obtain the physical and chemical properties desired, such as viscosity by addition of solvents i.e. ethanol, isopropanol or water, adjustments by means of various chemical additives to obtain specific properties such as resistance to thermal shock, break-down after casting, etc. These include derivatives of natural resins or other thermoplastic materials.

The plant to effect these chemical processes consists of stainless steel reaction kettles, fitted with the necessary heating and cooling devices. Water-cooled condensers to condense the vapours taken off the kettle during boiling under atmospheric or reduced pressures; a receiving vessel to take the condensate and a vacuum pump to enable very nearly complete evacuation of the kettle. The whole plant should be well instrumented to enable close control over the process to be maintained.

The finished resin will be either in the form of a solution, ready for use, or in a molten state, in which case it must be run out of the kettle into trays for crushing or granulating or flaking directly from the molten condition.

Raw Materials

Urea

$$CO \begin{cases} NH_2 \\ NH_2 \end{cases}$$ A white crystalline solid, m.p. 132·6°C.

Due to the introduction of resinous products using urea as a raw material it was necessary to synthetize on a large industrial scale. This is achieved by allowing liquid carbon dioxide and liquid ammonia to react at high temperature and pressure in steel autoclaves, the urea being removed by vacuum distillation or spray drying.

Formaldehyde (HCHO)

Formaldehyde is a gas at normal temperature and is highly irritant to the eyes and nose with an unpleasant odour. Formaldehyde is prepared industrially by the catalytic oxidation of methanol using a heated finely divided silver catalyst. It is available commercially in polymeric form ranging in different concentrations.

Hexamine (Hexamethylene tetramine)

Hexamine is a colourless crystalline solid soluble in water and is prepared industrially by reacting ammonia and formaldehyde solution at ordinary temperature.

The main use of hexamine in the foundry industry is for the hardening of "Novolac" resins in the shell moulding process.

Phenol

Phenol is a colourless crystalline solid, m.p. 42·9°C, and is prepared industrially either by extraction from coal tar or by synthesis by the Cumene process which involves the chemical oxidation of cumene to give phenol and acetone.

Furfuryl alcohol

Commercial furfuryl alcohol is a mobile amber coloured liquid and is manufactured on a large scale by the reduction of furfuryl which is extracted from Pentosan containing agricultural residues which include corn cobs, oat hulls, etc.

Urea-Formaldehyde Condensations, i.e. U.F. Resins

For a detailed review of the chemistry of urea-formaldehyde resins it is suggested that the reader is referred to the standard texts dealing with the above.

In general, urea and formaldehyde are reacted under pH values in the range of 4–8 with a formaldehyde-urea ratio of 1·5:1. Primary condensations occur under neutral or mildly alkaline conditions which favours formation of methylol derivatives. The resinification and final hardening is carried out under acid conditions. The increase in viscosity as the preparation of the resin proceeds is indicative of the rate of interlinkage of the polymer chains and therefore the reaction may be stopped at the desired level.

The addition of hydrogen ions, i.e. acid, to the final viscous resin causes further cross-linking through formaldehyde to give a three-dimensional structure, i.e. a hard infusible solid.

Effect of Sand on the Catalysis of Foundry Resins

The presence of sand has a marked effect on the speed of cure of foundry resin, both with the hot-box and cold-set processes. This is not specific to foundry resins, but is normally found where there is a resin-filled system.

The presence of the sand affects the cure of the resins in one or more ways:

The presence of a large amount of sand tends to absorb the acid catalyst leaving little to catalyse the resin.

The presence of this large amount of sand prevents any build-up of heat produced by an exothermic reaction between resin and catalyst.

The presence of any basic impurities in the sand tends to neutralize the acid catalyst so leaving less to catalyse the resin binder. This can be determined experimentally by means of the "acid demand" of the sand.

An excessive amount of oxides of iron in the sand will lead to the formation of acid iron phosphates which appear to act as buffer salts. The effect of such sands has been studied by Parkes, Andrews and Walker.[5]

PHENOLIC RESINS

"Phenolic" is the general term for a wide range of phenol formaldehyde polymers which have achieved importance in foundry sand binding processes. They can be divided into two main classes—single-stage or "Resole" resins and two-stage "Novolac" resins, each class possessing distinctive properties. A simple distinction is that the single-stage resins are produced by reacting one mole of phenol with more than one mole of formaldehyde and are thermosetting, whereas the Novolac resins are the result of reacting one mole of phenol with less than one mole of formaldehyde, rendering them thermoplastic until heated with a cross-linking agent such as hexamethylenetetramine. The single-stage resins are for any application in cold-setting and hot-box techniques; the

Novolac resins are used in the shell moulding process, using hexamethylene-tetramine as the curing agent. The two classes are of entirely different molecular formation.

Resole or Single-stage Resins

The reaction of phenol and formaldehyde at ratios of greater than one mole of formaldehyde per mole of phenol under alkaline or neutral conditions give rise to a series of resins called resoles. These resoles are a complex mixture of mono- and poly-nuclear molecules to which are attached methylol groups. Resins with very different properties can be made from the same phenol formaldehyde ratio by adjustment of reaction time, temperature, catalyst and catalyst concentration.

The final cure of a single-stage phenolic resin which consists of a multifunctional chain polymer may rapidly undergo condensation to a three-dimensional structure—the final cured product. The final reactions may be brought about by the application of heat or if the reaction is too slow the rate may be increased by the addition of an acid to the resin. In the case of cold-set binders strong acids are added in the cold, and in the case of hot-box binders weak acid and heat (usually acid salts, e.g. NH_4Cl).

Catalysis (i.e. acid hardening of one-stage resins)

One-stage resins, those used for binders in the cold-set methods of production of cores and moulds in the foundry industry, frequently need cure at low temperatures, i.e. about room temperature. This is achieved by the addition of acids to the resin/sand mixture. Acids used for the curing of foundry phenolic resins are usually either phosphoric acid (H_3PO_4) or para-toluene sulphonic acid:

$$CH_3$$

$$SO_3H$$

Cured resins are obtained by reaction of phenol with an excess of formaldehyde under acidic conditions. This rate of reaction is approximately proportional to the concentration of hydrogen ions, i.e. acid concentration.

As difficulty is experienced in the control of reaction rate under acid conditions, the phenol and formaldehyde are best combined under alkaline conditions and the resole obtained hardened by addition of acid.

Although hot-box binders (phenolic) are chemically hardened in a similar way to cold-set binders, it is required that these resins only cure at elevated temperatures (at the temperature of the hot-box mould) and remain stable at ordinary ambient temperature. To achieve this a "latent" catalyst is used, i.e. a catalyst that only becomes active at high temperature. Ammonium salts (usually ammonium chloride) are used for this catalysis of hot-box binder, and the principle of use is as follows:

$$NH_4Cl \xrightarrow{\text{Dissociation heat}} NH_4^+ + Cl^-$$

Rate increases as temperature increases. The NH_4^+ radical produced reacts with formaldehyde present in the binder to form hydrogen ions.

$$NH_4^+(Cl^-) + HCHO \text{ aq.} \rightarrow NH_2 \cdot CH_2OH + H^+ + (Cl^-) \rightarrow$$
$$NH_2:CH_2(H^+ + Cl^-) + H_2O$$

the hydrogen ion so produced catalyses the binder to give the final cured polymer.

Novolac Resin (a two-stage resin)

As already indicated, a resole is a complex mixture of mono- and poly-nuclear phenolic molecules having attached to each one a number of reactive methylol groups. A novolac is completely different in that the phenolic nuclei have no methylol groups attached to them since the acid catalyst favours the formation of methylene bridges between phenolic molecules rather than that of phenol alcohols. Also the ratios of reactants employed leads to a resin which is deficient in formaldehyde, so much so that the resin will not cure without the addition of a compound which will cause adjacent chains of novolac resin to cross-link.

The structure of a novolac resin is linear chains of phenol molecules linked with methylene bridges and may have as many as 40–50 phenol units in each. Commercial novolac resins consist of chains of phenol units of molecular weight varying between 200 and 1200.

As already explained, novolac resin will not cure without the addition of a curing agent, and so it is possible during manufacture to completely remove all the water put in as formalin and that produced during the condensation process by distillation to 130°C at atmospheric pressure or by distillation under reduced pressure. The novolac produced in this way, when cooled down, is a hard glass-like solid and may be ground for use in the manufacture of precoated sand by the hot-coating technique. Alternatively the solid novolac may be dissolved in ethyl alcohol and the solution used to make precoated sand by the warm-coating method.

Cure of Novolac Resins

The curing agent used in the shell sand process is hexamethylenetetramine, commonly known as "Hexamine", the structure of which has already been discussed. The actual role of hexamine as a curing agent for novolac resin is not fully understood, but hexamethylenetetramine is believed to break down under the influence of heat to give methylene-free radicals :CH_2. These free radicals are highly reactive and will link adjacent chains of the novolac via a methylene bridge.

Furfuryl Alcohol Resin

The mechanism of polymerization is complex and the mechanism postulated involves the intermediates formed via intermolecular dehydration to give furane rings linked by methylene bridges. As this reaction and other side reactions proceed, build-up in molecular weight results to give a polymer. The complete resinification to a hard infusible thermally set resin is explained by the additional polymerization involving nuclear double bonds. This is brought about by addition of an acid catalyst.

Furfuryl alcohol is used in the foundry industry mainly for blending with urea-formaldehyde and phenol-formaldehyde resins to give a final resin with enhanced properties. The presence of the furfuryl alcohol gives the final resin/ sand bonded core additional hot strength due to the more extensive cross-linkages and hence a more complex three-dimensional structure. It is believed that the furfuryl alcohol reacts with the urea-formaldehyde or phenol-formaldehyde resins and with itself in the presence of acid catalysts to give this final resistant thermoset product, acid salts being used for the hot-box process and strong organic or mineral acids for the cold-set process, as previously described for the foregoing binder systems.

REFERENCES

1. SEARLE, A. B., *Sands and Crushed Rocks*, Vol. 1, p. 233, table 1, Hodder & Stoughton (1923).
2. ROBERTS, W. R., *The British Foundryman*, **60**, 176–186 (1967).
3. LE BORGNE, J. and JAUMAIN, M., *Fonderie*, **185**, 209–219 (1961).
4. THOMPSON, R. N., *Foundry Trade Journal*, **123**, 69–76 (1967).
5. PARKES, E., ANDREWS, W., and WALKER, J., *The British Foundryman*, **57**, 235–253 (1964).
6. ROBERTS, W. R., *The British Foundryman*, **60**, 176–186 (1967).
7. WORTHINGTON, R., *Iron and Steel*, **39**, 176–180, 297–300 (1966).
8. DAVIES, W., *Foundry Sand Control*, Chapter 5. The United Steel Co. Ltd., (1950).
9. LEA, F. M., and DESCH, C. H., *The Chemistry of Cement and Concrete*, p. 166. Arnold, (1935).

CHAPTER 10

COATINGS FOR MOULDS AND CORES

F. W. PURSALL

Need for Coatings

Among the properties required of sand moulds and cores are that they should leave a good surface finish on the castings, allow easy stripping of the sand and resist metal penetration and burn-on. If these properties could be made inherent in the sand mixtures then there would be no need for paints or dressings applied to the sand surfaces.

The resin-coated sand employed in the shell process approaches this ideal, and coatings are not needed for relatively small castings of light section. For castings made by the lost wax process the first investment slurry applied to the pattern fulfils the requirements mentioned and functions in place of a mould coating.

However, for large and heavy castings, particularly those poured from the higher melting point metals, the sand surfaces in direct contact with the molten metal need a coating unless the normally used silica sand is replaced by zircon or chromite sand.

From the aspects considered here it is the sand surface in contact with molten metal which is important. Thus, coatings applied to it impart the special characteristics required without having to incur the expense of treating the whole mass of the sand. A property not yet mentioned is that of "wetting" or surface tension.

Mercury spilt on a glass plate immediately forms a number of spheres which traverse the glass without wetting it. Water spilt on the same plate tends to spread over it as a uniform layer. With mercury, the surface tension is high and with water the reverse. Where surface tension is high and wetting negligible the tendency for metal penetration from capillarity is reduced. Molten metal does not wet graphite and sand grains coated with graphite resist metal penetration. This explains why graphite, plumbago and carbon are usually used in mould and core coatings except those for steel.

Other Functions and Requirements

Variables exist with sand-bonding agents, casting size and weight, metal-pouring temperatures and running methods, all of which play a part in deciding the most appropriate sand dressing. It is obvious, therefore, that no single formulation can have universal application. Cost must also be considered relative to the work in hand. It is not fair to expect a good coating to overcome all the deficiencies of an inferior sand; the latter must also be made to suit the job as well as possible.

Other functions of a mould or core dressing are to minimize sand defects such as finning, facilitate fettling of the castings, modify the mould atmosphere, confer a densening or chilling action and to inhibit metal/mould reaction.

Coating Materials

Foundry coatings are of two types, those employing an aqueous carrier and those in which organic solvent carriers are used. The former must be dried after application while the latter are self-drying or can be ignited and dried by their own combustion. Both classes of coating make use of the same refractory materials which are also of two categories, flake and filler particles. Examples of flake materials are graphite, talc and mica, whilst filler materials are represented by silica and zircon flour and refractory clays with particle size of 20 microns or less.

Apart from particle size and shape one has to consider:

1. Specific gravity, which should be as low as possible to facilitate suspension.
2. Refractoriness.
3. Thermal expansion.
4. Chemical reactivity with the cast metal or the underlying sand.
5. Availability and cost.

Grading of the refractories as to particle size depends upon type, nature of the suspension agent and binder and also on the nature of the sand surface upon which the coating is to be used.

The ideal would be to use two coatings, the first containing filler particles to plug the surface pores. Secondly, to apply a dressing containing flake materials. These superimpose themselves in roof tile arrangement (see Fig. 1) and allow surface expansion of the sand without the formation of cracks into which molten metal could penetrate. However, to avoid two separate coating operations the filler and flake particles are usually incorporated in a single dressing.

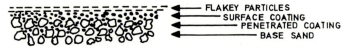

FIG. 1. Ideal application of coatings for sand moulds.

Refractoriness

The higher the pouring temperature of the metal the greater the need for high-grade refractories. They must not fuse or bad casting surfaces will result from "burn-on". On the other hand, theories have been advanced that a certain degree of vitrification helps to resist metal penetration caused by metalostatic pressure.

In the case of some alloy steels, special refractories are used. An example is that of castings made in 14% manganese steel. Here, to prevent the basic manganese oxide from reacting with the silica sand grains to form a slag, a barrier of magnesia is interposed by using a mould coat based on magnesium oxide.

Thermal Expansion

From a theoretical point of view the expansion characteristics of the refractories in a sand coating should be the same as those of the mould or core. Quartz or silica sand is the most widely used core- and mould-making material. Unfortunately, silica undergoes a phase change from alpha to beta quartz at 573°C which causes a rather abrupt expansion of 1·4%. Silica flour is often used in mould and core paints subject to the silicosis regulations. Where it cannot be used the expansion of silica sand must be combatted by other means i.e. the employment of flake-like refractories already mentioned.

Liquid Carrier

Water and organic solvents are the two liquid carriers for the refractory materials. The latter can be combustible, as in flash dressings, or those which evaporate naturally. With water the problems are few and simple such as hardness due to salts in solution. These can affect the functioning of suspension agents adversely. Acidity is also a factor to be avoided especially where clays are used as suspension agents.

Organic solvent carriers offer a wider range of choice which is governed largely by cost and physical properties. Those relating to fire and health hazards are important. With most flash dressings the cost of the carrier is the major item as its proportion, on a weight basis, is generally greater than that of the refractories.

Isopropyl alcohol fulfils most of the requirements of an organic solvent carrier. It has good combustion characteristics with a slow burning front and a moderately hot flame. This reduces the chance of over-heating the sand surface and

TABLE 1.

Solvent	Open flash point	Boiling point	S.G.	Maximum concen- tration	Toxicity	Burning
†IPS 1	12°C	82–83°C	0·79	400 ppm	Very low	Good
†IPS 2	12°C	82–83°C	0·79	400 ppm	Very low	Fair
Methanol	15·6°C	64–65°C	0·79	200 ppm	Low	Too fast
Ethanol	16·0°C	79–80°C	0·79	—	Low	Good

† The difference between the two grades of isopropyl alcohol is in the water content. IPS 1 is practically water free and IPS 2 contains approx. 5% water.

subsequent problems of sand friability. The slow burning properties of isopropyl alcohol help to ensure full curing of the resin binder normally used in these coatings. With too slow burning, curing of the resin may be incomplete, whilst if too fast, the resin tends to char with a resultant loss of bond.

Isopropyl alcohol is also technically acceptable because it is compatible with a wide range of suspension agents and resin binders also used in the formulation of these coatings. Table 1 gives an indication of a number of solvents which may be used, together with other details.

Air-drying carriers include carbon tetrachloride, methylenechloride, chlorothene and chloroform as listed in Table 2. They rely for efficiency on a rapid rate of evaporation which places them in a more hazardous category than isopropyl alcohol. Additionally their specific gravity is higher and they are appreciably more costly, volume for volume. More stringent precautions are necessary during their use in the foundry both from the fire and toxicity points of view. Also these liquids are not so versatile as isopropyl alcohol in the formation of foundry coatings. In many cases they call for specialized forms of gelling media and resin binders.

TABLE 2.

Solvent	Boiling point	S.G.	Evaporation rate (ether scale)	Comments
IPS 1	82–83°C	0·79	0·21	Inflammable and too slow for most uses
IPS 2	82–83°C	0·79	—	
Methylene-chloride	40–41°C	1·33	1·8	Strongly toxic, non-inflammable
Chloroform	61–62°C	1·49	0·56	Very strongly toxic
Chlorothene	74–78°C	1·33	—	Very low toxicity, similar to IPS
Carbon tetrachloride	76–77°C	1·56	—	Strongly toxic

Suspension Agents

There would be no difficulty in keeping solid particles in permanent suspension in a liquid if both had the same specific gravity. This is never the case with foundry sand coatings. Nevertheless, maintenance of the solid particles in suspension is most important and must be achieved as efficiently as possible so as to ensure constant results in practice.

When water is the carrier liquid bentonite type clay is used as a suspension agent. Bentonite swells and forms a gel when mixed with water. Time must be allowed for gelling to proceed to completion. Small differences in the gelling power of bentonites can cause large fluctuations in apparent viscosity of coatings resulting in marked difference in their behaviour and properties when used in the foundry.

Two kinds of bentonite are in common use, one linked with calcium and the other with sodium ions. A base exchange can be effected whereby calcium can be replaced by sodium so as to transform a calcium into a sodium bentonite. As a suspension agent bentonites initially of the sodium type are preferred. Minor variations due to the base exchange treatment with sodium carbonate affect the swelling power of the clay and make control of the viscosity of the coating unpredictable.

Apart from difficulties with quality control of the bentonite it has the disadvantage of tending to induce shrinkage cracks in the coating when dried. Higher proportions of bentonite when employed to suspend heavy refractories, such as zircon, increase the cracking tendency. Even with normal ratios of bentonite, cracks will form if thick coatings of dressing are applied.

In view of the drawbacks associated with bentonite, a search has been made for alternative swelling agents. Of the many tested, two only have come into use, polysaccharide and certain forms of carboxy methyl cellulose. Both of these have disadvantages also. The former require special type mixers, which few foundries possess, to obtain optimum suspension. Subsequently, during storage, the viscosity of the coating increases to the extent that it becomes unusable and cannot be reconstituted.

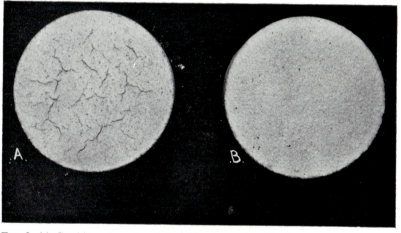

FIG. 2. (a) Cracking of coating induced by bentonite. (b) Cracking eliminated by using carboxy methyl cellulose.

The cellulose type of suspension agent does not suffer in this way, neither does it cause shrinkage cracks as does dried bentonite (see Fig. 2). It has performed well in the foundry industry, particularly in steel foundries where thick coatings are essential. A single thick application can, therefore, replace several thin coatings with their intermediate drying. The proportion of carboxy methyl cellulose introduced into the formulation is much more critical than that of bentonite. Moreover, it does not possess the same flexibility of dilution.

With organic solvent carrier systems different means to obtain suspension of the refractories are necessary. Few materials are available which are able to increase viscosity as do bentonite or carboxy methyl cellulose with water. Modified bentonite known also as organic bentonite or bentone, can fulfil the purpose. Bentones will gel and increase the viscosity of organic liquids such as alcohols and solvents. They have little detectable attraction for water as opposed to the initial unmodified bentonite. Bentones result from a base exchange of the inorganic Ca or Na cation for an organic one which is most frequently quaternary ammonium.

Binding Agents

Numerous materials are available which act to hold the particles of refractory together and attach them to the sand surface. Quantity of binder required for this purpose increases a little as the particle size of the refractory decreases, thereby increasing the surface area for a given ratio in the coating.

It is most important that the minimum quantity of binding agent should be determined. Too little results in poor adhesion but, too much produces a brittle coating which may crack on drying and spall off during casting. Furthermore, resins and similar organic binders evolve gas on heating. Thus, any undispersed binder in the dressing itself or collected in partially dried areas of mould or cores will cause local concentration of gas generation. In this way defects such as porosity and lapping can result.

Most organic binders and many suspension agents used in water suspensions are subject to biological degradation. Precautions must be taken to suppress these reactions especially if lengthy storage of the coating is envisaged. Such changes do not occur with spirit carriers and so precautionary measures to this end are not necessary.

The manner in which binders burn out, when ignited, must be considered together with the kind of residue which remains after combustion. A coating may fail in its purpose simply because the burn-out product is of low refractoriness and allows metal penetration to occur. As the majority of binders are of organic origin they leave carbonaceous residues which are sufficiently refractory. Conversely an inorganic binder, such as sodium silicate, may produce a residue of low fusion point which could give rise to trouble from burn-on.

Rate of decomposition of the binder when heated is equally important because this affects the time period and volume of gas evolution. Resistance to erosion by the stream of molten metal flowing over it must also be adequate and in addition the break-down effect of radiant heat, whilst the mould is filling, must be withstood.

Binders used in water suspensions include sulphite lye, various clays, dextrine, molasses, sugars, sodium silicate and resins soluble or miscible with water. Heat setting or air-drying oils are also used, but under storage conditions separation of the water and oil phases can be a problem; emulsification of the water and oil is then necessary.

Alcohol systems require a natural or synthetic resin compatible with the spirit carrier. They include urea formaldehyde-, phenol formaldehyde-, novolac- and natural wood resins. Cost is always a factor in their selection. If a sand coating is to function satisfactorily in the foundry, all of the factors dealt with must receive careful consideration before it can be compounded.

Practical Considerations

When moulds and cores were all made by hand, casting production depended on craftsmen. Application of coatings was part of the art and each moulder had his own recipe. Coatings were brushed or swabbed on with care and a good skin on castings obtained in this way. Nowadays fast production by mechanical aids has caused the moulder's skill to be built into the machinery. Application of dressings is by spray gun or dipping and no time is available for "touching up" or rectification of faults. Much of the onus is, therefore, on the performance of the coating.

Brushing and Swabbing

Both these methods of applying coatings are used in many foundries. The effort imparted by brushing helps to force the refractory particles into the pores of the sand surface, which is a desirable feature as indicated earlier. The swab is a most useful aid in coating the interior of difficult pockets and re-entrant angles. In both methods the skill of the operator is important.

Spraying

A much faster means of application widely used in foundries of all types. Greater attention is necessary in the selection of coating composition because less mechanical effort is available to force the particles into the pores between the sand grains. This method along with brushing suffers the disability of not being able to coat deep recesses adequately. One reason for this is the back pressure of air which prevents refractory deposition in the cavity. The system of airless spraying provides a means of overcoming this disadvantage.

Capital outlay for good spray-gun equipment can be considerable and it should receive regular maintenance. Selection of the solid constituents and the overall viscosity is more critical for sprayed coatings than for brushing or swabbing. Optimum specifications should be carefully determined and maintained.

Dipping

Used in high production rate foundries as it is the fastest method for coating cores. Carried out manually or mechanically it eliminates much of human variation. The success of the operation depends greatly on the technical merit and "know-how" built into the dressing.

Somewhat better surface penetration than with spraying is obtained because of the head pressure of the coating in the dip tank. Frequently, however, depending upon the kind of binder used in the core sand, additional aids to improve penetration are required.

Even thickness of surface coating is necessary so as to maintain dimensional accuracy and true reproduction of contour. Uneven coating is at its worst when it runs down as tears. This defect can be encouraged by the nature of the surface to be coated but is mainly due to the kind and quality of the suspension agent used in the coating.

Tears and similar coating faults are a source of high gas evolution and casting defects may result. They should be removed, by rubbing, prior to inserting the cores. Rubbing down cores is time consuming, produces loose sand and ought not to be necessary.

The main factors which influence the kind of coating to employ are:

(a) The type of bonding agent used in the sand.
(b) The metal or alloy from which the casting is made.
(c) Whether a drying out of the dressing as a separate storing operation is to be carried out or not.

Selection of Coatings

A new range of core and mould coatings has been developed to suit the advances in sand bonding techniques which have been made in recent years. Where self-hardening sands such as CO_2/silicate, air setting oils and cold setting resins are concerned, combustible alcohol based dressings are the obvious choice. They have a slight disadvantage in that if applied to only partially cured resin or oil bonded sand, softening may occur. However, recovery will take place if sufficient time is allowed after ignition of the coating.

In some cases where cold setting binders are used, particularly air-drying oils, a short heating cycle is performed to speed the setting reaction. In these cases, for reasons of economy, water suspended dressing may be used. With the hot-box method, water based coatings are employed almost exclusively as it is considered that a short heating treatment is desirable to obtain optimum strength for the cores. Where oven drying has to be done as for the older linseed oil based and urea- and phenol-formaldehyde bonds or for conventional dried sand moulds, then water suspended dressings are used almost exclusively because flash dressings lose their chief advantage. It is generally accepted that aqueous dressings are technically superior to spirit based equivalents and they are certainly less expensive.

Light Alloys

Normally little necessity exists for coating cores or moulds for aluminium based alloy castings. Section thicknesses are relatively small and pouring temperatures are at the lower end of the range. Additionally the light alloys readily form a skin which gives a good surface finish without the aid of dressings. Metal penetration is reduced by surface tension of this skin and metal pressures are low.

When a coating for light alloy castings is necessary it need not be highly refractory. Particles of flake morphology such as talc, mica and graphite are common components, used either singly or in combination.

Copper and Its Alloys

Although pouring temperatures are higher than for aluminium alloys, a good casting skin is not difficult to achieve in the majority of cases. Aluminium bronzes and high tensile brasses present little problem. On the other hand, phosphor bronzes are very fluid at casting temperature and tend to wet the sand grains. They are, therefore, prone to metal penetration. Suitable refractories are the same as for light alloy castings and a specially graded molochite is used in certain circumstances.

Steel

Much higher casting temperatures are involved and these coupled with economic considerations limit the choice of coating refractories. The range includes zircon, silica, chamotte and in special cases magnesite and olivine. Zircon is the most popular general-purpose coating for steel. It is relatively inexpensive and can be obtained with constant properties. Its versatility allows of its use on all section thicknesses with equal effectiveness and it fulfils the requirements of the majority of steel casting compositions. Because of its high refractoriness it can be used on surfaces of silica, chromite, olivine and chamotte.

The two steel alloy ranges for which zircon is not completely acceptable are the high chromium–nickel stainless steels and the high manganese austenitic steels. The latter suffer burn-on due to the combination of manganese oxide and silica to form a slag which fuses at the pouring temperature. A coating based on magnesite and free from silica has given the most satisfactory results, particularly on moulds and cores of silica sand. Olivine sands for moulds and cores are an improvement over silica sand.

Coatings for steel castings are often subjected to much higher drying temperatures in the core ovens than those used in the production of castings from lower melting point alloys. This causes problems which are not experienced in the latter case. Also the coatings are applied more thickly in steel casting practice and must have a stronger resistance to cracking and spalling. These contingencies are catered for by modification to the binder system and control of the grading of the refractories.

Cast Iron

It is seen that coatings suitable for use in light alloy, copper base alloy and steel foundries follow a fairly narrow and well-defined path. For a variety of reasons, the opposite state of affairs applies to iron foundries. Choice of sand for cores and moulds covers a wider range and bonding agents of all kinds are used. The size and weight of iron castings also varies greatly. Thus, the number of refractory materials which can be used satisfactorily within this field is considerable.

Before the introduction of the latest type of sand bonding agents, blackings made from plumbago, mineral blacking and clay slurry were in almost universal use. Even today, such dressings find wide usage for general purposes on medium and heavy engineering castings made in dry sand. Their big advantages are cheapness, ease of mixing and the degree of abuse they permit without undue reflection on the surface finish of the castings. The more exacting requirements now demanded underline the need for dressing capable of giving better results and increased reliability. This is particularly apparent in highly mechanized, fast production rate foundries and those producing special castings in high duty and alloyed cast irons.

For these conditions coatings which incorporate more highly refractory materials have been developed. Besides carbon, these include zircon and olivine. Carbon is not wetted by molten iron and clean strips are possible provided that metal penetration has not occurred. Graphite is also a lubricant which aids the passage of the granular refractories through spray gun equipment. For small castings made in green sand no coatings are required because coal dust is an ingredient of the sand.

Special Applications

There are instances where coatings are used to overcome casting defects. Metal–mould reactions occur with magnesium alloys and bronzes and gun-metals. Magnesium alloys, when poured into green sand moulds, require inhibitors in the facing sand, such as sulphur, boric acid and ammonium bifluoride to inhibit the reaction between molten magnesium and the moisture in the sand. These inhibitors are not incorporated into silicate/CO_2 bonded sand which also contains water but at a lower level than in green sand. However, coatings containing inhibiting agents are available for such cases.†

For phosphor bronzes and gun-metals which tend to pick up hydrogen from the moisture in the sand, a mould coating containing finely divided particles of an aluminium–magnesium alloy suspended in a solvent, is employed. This was developed by the British Non-Ferrous Metals Research Association several years ago.

The difficulty in achieving a good cast skin on austenitic manganese steel castings has already been mentioned. Although this is a somewhat indirect metal–mould reaction between silica and manganese oxide, the magnesite fillers in the special dressing advocated help to suppress the reaction and effect a surface finish improvement.

† See chapter on Magnesium, p. 306.

Small steel castings made in green sand moulds may suffer from sand wash or
erosion due to metal flow and the resultant sand inclusions. A solution of resin
in alcohol applied to the surface of the green sands will produce a hardened
skin capable of resisting the eroding action of the molten steel.

The British Cast Iron Research Association has established that the judicious
use of a dressing containing finely divided bismuth metal is effective in suppressing micro shrinkage porosity in some grey iron castings which produce hot
spots during pouring. Examples are the common wall between cores in automobile castings, stud-hole bosses through water jackets and the junctions of
arms with the rims of flywheels. If the sand surfaces adjacent to these trouble
spots are painted with the bismuth containing mould coating the shrinkage
porosity may be overcome.

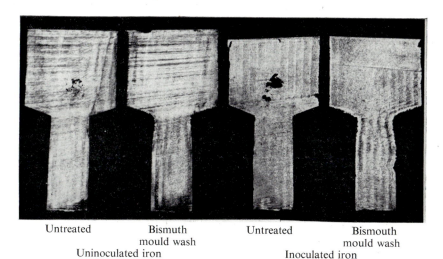

| Untreated | Bismuth mould wash | Untreated | Bismouth mould wash |
| Uninoculated iron | | Inoculated iron | |

FIG. 3. Effect on coatings containing bismuth on shrinkage porosity in uninoculated
and inoculated cast iron (Grade 17).

(By courtesy of British Cast Iron Research Assoc.).

When certain areas of an iron casting require additional hardness this may be
imparted by painting the adjacent mould or core surface with a dressing containing finely divided tellurium. The tellurium is considered to alter the critical
cooling rate of the molten cast iron and increase its crystallization velocity.
In this way a carbide matrix or chilled white iron structure is induced into the
casting for depths up to $\frac{1}{8}$ in. Pouring temperature and the chemical composition
of the iron will affect the degree of chill produced. Unfortunately, tellurium
boils at 1390°C. So, if the flow of metal past the treated area is great, tellurium
vapour may migrate to other parts of the mould which are not required to be
hard and have to be machined. Tellurium containing coatings can also be used
to overcome porosity in a similar manner to those containing bismuth but the
latter do not induce chilled structures.

Resin Defects

The resin used as a bonding agent in flash dressings can be a source of trouble. As the spirit carrier burns away the viscosity of the dressing on the sand surface increases. This may act as a barrier to the escape of the spirit vapour which is thus unable to force its way out. Combustion, therefore, ceases, leaving unburnt solvent. Moulds poured in this condition give off bursts of spirit vapour which may also inject sand grains along with coating materials into the molten metal. Casting defects from this cause can appear as surface cavities, pinholes and cold laps. Thin metal sections are likely to be most affected.

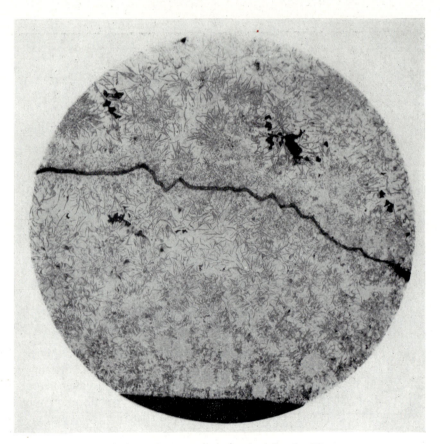

FIG. 4. Micro-structure of resin defect ($\times$ 25).

Resin contents should, therefore, be the minimum which will allow development of a coat hard enough for handling. Even so, non-uniform mixing can cause an undue proportion of resin to be present and result in the same troubles.

Pinhole Porosity

Pinhole porosity on cored faces of castings is sometimes caused by the decomposition products of core binders. Urea-formaldehyde resins are the most notorious for producing this kind of defect.

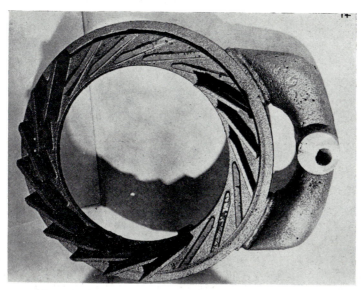

FIG. 5. Example of nitrogen pinholing in spheroidal graphite cast iron.

The use of various forms of iron oxide, notably red iron oxide, goes a long way in reducing pinholing. It can be incorporated in water—or isopropyl alcohol washes together with the normal coating refractories.

Finning

This defect is directly attributable to sand conditions and is particularly troublesome in automobile castings because finning normally occurs in the channels and waterways which are inaccessible to fettling tools. A coating will never completely eliminate severe finning but it can considerably improve it. Where finning is not pronounced, a coating may eliminate it altogether. The coating is based on a combination of refractories which give a fusing range approximately the same as the freezing range of the alloy. One can assume that the coating is effective because the refractories fuse and fill the cavities in the core created by the thermal shock. At the same time, gas is evolved from the coating, and this helps to repel the molten metal. Although effective, because of this gas evolution some discretion is necessary with the use of this type of coating.

These then are a few instances where an improvement in surface finish is of secondary importance but where a coating is of assistance in reducing the number of scrap castings.

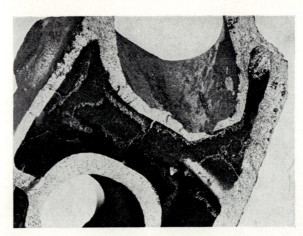

FIG. 6. Finning in a cylinder head casting.
(By courtesy of M. J. Clifford and the Inst. of Brit. Foundrymen).

The concept of foundry coatings has altered little over many years; the basic principle of applying a very fine layer of carbon or refractory material to prevent metal penetration is as pertinent now as it was a long time ago. The major advance has been in the field of binders, suspension media and carrier fluids, coupled with a fuller understanding of the mechanics and technology of coatings. At this moment there is practically no foundry which can operate without coatings. In fact, the largest consumers of coatings are the most modern and up-to-date foundries, e.g. those supplying the automobile industry. The aim for the future, however, is to change moulding and casting techniques in such a way that the need for coatings will be eliminated.

CHAPTER 11

SAND ADDITIVES

J. Beale

In their simplest form clay bonded moulding sands consist of a refractory material, a clay binder and water. However, to produce moulding sands having the required moulding properties, and castings of good quality and surface finish, it is generally necessary to add other materials to the moulding sand mixture. The materials are generally referred to as additives.

MODIFICATION OF BONDING CLAYS BY ADDITIVES

The most widely used clays are the montmorillonites. One of the earliest sources was Fort Benton in the U.S.A., hence the name bentonite. They are the products of weathered volcanic ash. They have no definite chemical composition but the formula $(Al_{1\cdot67}Mg_{0\cdot33}(Na_{0\cdot33})Si_4O_{10}(OH)_2$ in which $Mg_{0\cdot33}$, $Na_{0\cdot33}$, etc., refer to the exchangeable cations represents the main structure of montmorillonite. Individual crystallites in the mineral have a characteristic tri-layer structure in which a basically alumina sheet is sandwiched between two silica layers. In the alumina sheet of montmorillonite some aluminium ions have been substituted by iron and magnesium. As a result three negative charges are formed in the crystal which are balanced by an equivalent number of cations outside the montmorillonite lattice, probably on the surface of the silica sheets. These cations are exchangeable and give rise to the base exchange properties of montmorillonite. The expressions "calcium" or "sodium" montmorillonite indicate that the dominant external cations are calcium or sodium respectively.

Between the triple layer of the mineral several layers of water molecules can usually be bound by Van der Waals' forces and their formation causes the one-dimensional reversible swelling of the lattice which is typical of the montmorillonite.

The monovalent inorganic ions such as sodium and potassium are large and highly hydrolysed thus for a given charge on the clay surface more of these cations can be attracted than, for instance, aluminium or calcium.

As the number of attendant water molecules is much larger with these monovalent cations the zeta potential of the clay will be very large and the force of repulsion between the particles will produce maximum separation. This is the reason why sodium bentonites have a greater water holding and swelling capacity than the calcium types. When hydrogen ions are added to the clay the zeta potential of the particle surfaces may be sufficiently reduced to permit aggregation. A hydrogen ion is small and non-hydrated and thus does not provide an ion atmosphere around the clay particles. These can then easily come together and flocculation results. Because the aggregation is so tight and no attendant water is held, the clay loses its plasticity and a drastic fall-off in properties results. This is the reason why pH control in foundry sand systems pays dividends as the organic additions such as cereal and dextrines break down to give acids and the acid catalysts from cold-set resins, etc., may also find their way into the system.

Sodium carbonate has long been used as the main chemical for maintaining the pH in sand systems and also as the main source of sodium ions in the peptizing of calcium bentonites into sodium bentonites. Thus we have calcium bentonite + sodium carbonate (Na_2CO_3) → sodium bentonite + calcium carbonate ($CaCO_3$).

The effect of additions of sodium carbonate on the rheological properties of various bentonites is shown in Table 1. *Pouring Limits of Bentonite* (i.e. the limiting concentration of an aqueous bentonite suspension which will not flow after 24 hr standing).

TABLE 1.

	%
1. Wyoming natural sodium bentonite	10·5
2. American Southern calcium bentonite (natural)	32·0
3. American Southern calcium with sodium carbonate	17·0
4. Mediterranean calcium bentonite	
Natural	31·0
With 3·7% of sodium carbonate	12·5
With 5·5% of sodium carbonate	14·5

The addition of sodium ions to the clay can be seen to increase the gel strength of calcium clays considerably which in terms of sand properties means increase in wet tensile strength, green strength at high moisture contents and furthermore greater hot strength and resistance to scabbing. It also means that the constitutional or chemical water is more strongly held in the clay lattice. Differential thermal analysis curves confirm that a calcium bentonite loses its chemical water in the range roughly 450–550°C, sodium peptized calcium bentonite from 500–600°C, and finally a natural sodium bentonite from 600–700°C. This latter fact is the reason for the greater thermal durability of natural sodium bentonite.

STARCH MATERIALS

Starch granules are not homogeneous but contain two very similar polysaccharides, namely amylose (10–20% in maize starch), which may be leached out with water at 60–80°C, and amylopectin (80–90%), which is insoluble in cold water, but forms starch pastes with hot water. Although amylose may be dissolved out from the starch granule, when separated from the amylopectin it retrogrades and becomes insoluble in water.

Starches have the following actions upon clay bond moulding sands

1. Reduction of expansion defects; it has been shown that starches increase the wet strength of green sands considerably. The term "wet strength" is associated with the layer of high moisture content in a green sand mould which is produced when the sand surface is heated by molten metal either by contact or radiation. Moisture in the form of steam is driven back from the surface of the mould and recondenses to form a layer of wet sand. This saturated zone acts as a shear plane for the heated and expanding surface shell, allowing buckling and spalling to take place. With an 0·5–1 % addition of starch, however, not only are the free expansion and surface stresses of the sand reduced by 25–30%, but the wet zone strength is increased sufficiently to inhibit any shearing action, thus preventing any rupture of the mould surface due to expansion.

2. Starches increase the green strength and shatter index or toughness of sand mixes and although this improves certain aspects of the moulding properties, i.e. stripping from the pattern, it also reduces the flowability or compactability of the sand.

3. Because of the rapid burn-out of starches, shake-out after casting is improved.

Under this heading dextrines can also be considered. Dextrines act in a very similar manner to starches, the main differences arise from the higher cold water solubility of the majority of dextrines, thus the highest green strengths are obtained with dextrines at very low moisture contents.

Drying out of sand mixes is retarded but migration of the binder to the surface allows non-friable edges to form on moulds which are allowed to air dry.

Because of their lower wet strengths, dextrines are not as effective as starches in inhibiting expansion defects and are not generally used for this specific purpose.

Dextrine is used in the CO_2 process as a breakdown agent for silicate-bonded sand cores.

POLYMER MATERIALS

A new series of additives has recently been investigated by Braddy and Beale.[1] These are organic long-chain polymers which have a very powerful flocculating action with sodium bentonite. The most important of these polymers was shown to be polyacrylamide.

The modifying action of various compounds on clays was also investigated and the type of reactions involved may be summarized from results obtained from an apparatus designed for measuring (a) movement and compaction of sand (see Fig. 1), (b) determination of the shear strength of an aqueous clay paste (see Fig. 2).

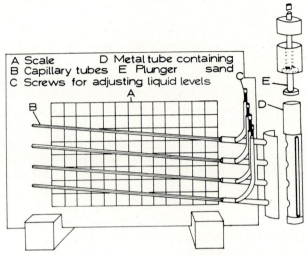

FIG. 1. Apparatus for determining the density gradient of sand mixtures.

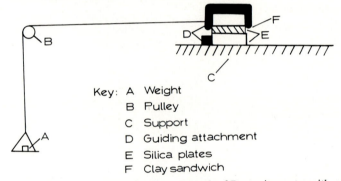

Key: A Weight
     B Pulley
     C Support
     D Guiding attachment
     E Silica plates
     F Clay sandwich

FIG. 2. Apparatus for measuring the shear strength of Bentonite pastes with various organic additions.

CELLULOSE DERIVATIVES

Sodium Carboxy Methyl Cellulose

This material is a fairly good flocculant for clay, showing that it possesses some capability of reacting with clay. However, results in the sand compaction apparatus (Fig. 1) show transmission of energy only over a limited distance of the sand column.

Methyl Cellulose

Methyl cellulose gave even poorer results than sodium carboxy methyl cellulose in the sand compaction apparatus.

Alginates

An extremely impure seaweed extract, containing various alginates and other carbohydrates is used as a foundry sand additive. The recommended addition is 0·05–0·1% of the sand weight except with sand used for facing the mould where 0·5% is suggested. (It is claimed to form a protective colloid with clays and thus increase the binding power, which is especially noticeable with bentonite, where an increase of 25% is instanced. Additional advantages such as improved dispersion, flowability and resistance to some types of expansion defects are also claimed.) Nevertheless, transmission of energy and improvement of the packing in the sand column are low indicating only a slight effect on the clay properties.

This material imparts a slimy texture to a clay paste and consequently shear strength values are low. It is ineffective as a flocculant for clay suspensions.

Sodium Alginate

Pharmaceutical grades of sodium alginate give somewhat improved results over the crude material.

Propylene Glycol Alginate Ester

This material consists of alginic acid which has been esterified with propylene glycol. It reacts similarly to sodium alginate.

Vinyl Polymers

When dissolved in water and neutralized with a suitable base, usually an aliphatic amine, high molecular weight carboxy vinyl polymers produce a highly viscous gel. Thus, in the foundry sand, if sodium carbonate has been added to gel the bentonite, it will also bring the polymer into solution and thus enhance the gelling power of the clay.

This polymer produces interesting results when submitted to the three tests, transmission of energy, shear strength, flocculation. The packing properties of a sand clay mixture show a definite improvement when an 0·2% addition is made, transmission of energy being observed over a considerable distance of the sand column. However, when added to a clay paste, the shear strength values obtained are low. This is due to the material imparting a dry, crumbly consistency to the paste. It is quite an effective flocculant with clay suspensions although aggregation of the particles takes place quite slowly.

Polyvinyl acetate is unsuitable at the pH range existing in normal foundry sands. This was tested in the form of an aqueous emulsion of positively charged polyvinyl acetate in which the polymer particles themselves are insoluble in water, but carry an overall positive charge due to the use of a cationic emulsifying agent in the preparation of the emulsion. This material as supplied has a pH of 5 and as it is stated to be unstable above pH 6·5 it will presumably be precipitated under practical conditions by the sodium carbonate normally added to commercial bentonite. This will bring about flocculation of the clay particles in a similar manner to that which was produced by hydrochloric acid. As with the latter, this material yielded very poor results in these series of tests.

Vinyl Acetate–Maleic Anhydride Copolymer

This material is marketed as a sand conditioning resin for the foundry industry. When an 0·05% addition is made to sand, the workability, flowability and packing characteristics of the sand are claimed to be improved by means of more even clay distribution. Permeability, deformation and compressive green strength are stated to show little or no change.

This material gives a good transmission of energy and even compaction of the sand column in the sand compaction apparatus. It was noticeable that sand treated with this material is drier to the touch and this produces a low density sand with an even density gradient.

The material is quite an effective clay flocculant at high concentrations of 0·25% or above, but at very low concentrations it is quite ineffective. It makes little difference to the texture of a clay and only slight improvements in shear strength values are obtained.

Polyacrylamide

This material gives excellent results in the sand compaction apparatus; transmission of energy occurs all the way down the column and a sand of uniform density is obtained. (When the additions of polymer and bentonite were reduced to 0·1% and 3% respectively from 0·2% and 10% to accord with the amounts normally present in a foundry sand, the improvement was, as expected, less marked though still appreciable.)

Polyacrylamide possesses the remarkable ability of producing extreme plasticity in aqueous bentonite pastes. Thus, if 1% of polymer is incorporated into a 40% paste, the clay can be stretched into thin sheets or pulled into fine strands. Shear strength values on this paste show a considerable improvement over those obtained with a similar clay paste without additions. In addition, even at a concentration of 0·001%, it is found to be an excellent flocculant of a clay suspension.

Polyacrylamide improves the retention of water by a mixture of sand and clay. It has been shown that after standing for 24 hr in humidity chambers at 50% and 80% relative humidity, sands containing 10% of bentonite and 7% of water, with a 0·2% addition of polyacrylamide lose 91% and 80% respectively of the losses sustained by a similar mixture without addition.

The general theoretical considerations on which the above properties are based are as follows:

It may be presumed that, for an organic compound to be capable of reaction with clay particles or sand grains, its molecules must be adsorbed on the solid surfaces. It has been shown that certain organic molecules are adsorbed on the surface of clays by means of hydrogen bonding.

The organic molecules may then either tend to repel the clay particles or bring them together, i.e. the clay may either become dispersed or flocculated. The sand grains are, of course, far too large to be similarly affected, but nevertheless hydrogen bonding will take place between the organic molecule and the oxygen atom on the surface lattice of the sand. Organic molecules may also adsorb on more than one particle and thus bring about flocculation by inter-particle bridging.

Polyacrylamide, which has been shown to possess desirable properties as a sand additive, causes flocculation of clay by this method. It is not completely non-ionic as, in addition to the amide ($CONH_2$ groups which react with clay by hydrogen bonding), it contains a small proportion of amidinium $(CONH_3)^+$ groups whose cationic nature favours adsorption on the negatively charged clay surfaces.

It is this unique combination of two active groups which gives polyacrylamide such marked flocculating ability. It has been shown that it is the only type of flocculation which is desirable in a sand additive. Flocculation of the clay without particle bridging, as caused by hydrochloric acid, does not give a sand with the required properties.

Organic polymers, such as sodium carboxy methyl cellulose, which cause dispersion of the clay due to their high concentration of negative ions, have also been shown to be ineffective as sand conditioners due to mutual repulsion of neighbouring clay particles. However, a material such as a vinyl acetate–maleic anhydride copolymer, which possesses an extended chain with hydrogen atoms available for particle bridging shows quite a strong reaction with clay, though considerably inferior in this respect to polyacrylamide. Its effect on sand–clay mixtures is not, therefore, nearly so marked.

COAL DUST AND BITUMEN COMPOUNDS

Coal dust and pitch compounds are a class of additives widely used in the iron foundry industry to improve surface finish. There are three main theories as to the mode of action of the evolved volatiles from these compounds in preventing burn-on of sand to iron castings.

1. Reducing Atmosphere Theory

Burn-on is caused by the reaction between iron oxide on the surface of the metal with the silica of the mould which produces a fusible iron silicate—fayalite—which promotes adhesion of the sand to the metal. The reducing atmosphere theory assumes that a reducing atmosphere is produced in the mould cavity which can inhibit the oxidation of iron to iron oxide and reduce any oxide that may already have formed. Possible conclusions as to the viability of this theory may be drawn from thermodynamic studies of the mould atmosphere. These indicate that the atmosphere in the mould will not be reducing until the partial pressure of the oxygen is as low as 10^{-3} at a temperature of $1600°K$ ($1327°C$). This conclusion does not take into account speed of reaction but is only the condition existing at equilibrium.

2. Lustrous Carbon Theory

Gilsonite is particularly effective in inhibiting sand burn-on and it can be shown that, using a 2% addition of Gilsonite in a standard foundry mix, a shiny, black, coherent layer of carbon is produced at the metal-mould interface.[4] Chemical analysis of the layer shows it to be carbon whilst the X-ray diffraction pattern shows a number of faint lines corresponding to graphite. It would appear to have a highly faulted graphitic structure.

If an interfacial layer of graphite is laid down, then the sand is not wetted by the metal and iron silicate formation is not possible, due to this physical barrier, and a very smooth casting finish is thereby produced.

The formation of lustrous carbon would appear to be dependent on three main factors:

1. The nature of the hydrocarbon volatiles present.
2. Temperature.
3. The presence and nature of a catalyst.

The formation of lustrous carbon from volatile hydrocarbons takes place via a cracking process at temperatures in excess of $400°C$, in the presence of catalysts. This follows a carbonium ion (i.e. positive ion), mechanism. The first stage of the reaction is the acquisition of a positive charge by the hydrocarbon. Acidic surfaces, including the acidic oxides silica and alumina, can act as catalysts. Aluminosilicates are very effective and in fact until the production of more selective synthetic catalysts became possible, oil refineries used natural clays, particularly of the montmorillonite type, in catalytic crackers. Whilst it is true that the lower the pH of the clay, the greater the catalytic activity, and clays used in moulding sands are commonly alkaline, it is reasonable to suppose that the large surface area of silica and aluminosilicate in a mould is not as Heinze[2] states "inactive". It has been shown that catalytic activity of a western bentonite is eight times that of a silica sand with an A.F.S. fineness number of 55 and that this activity is increased by a factor of 3 on washing with an excess of 0·1 normal hydrochloric acid. However, it was also shown that if the pH of the clay is reduced to below 4, then its swelling characteristics are seriously impaired and its usefulness as a bonding agent lost.

The carbon-forming tendency of hydrocarbons is a function of the tendency of the molecule to acquire a positive charge. An examination of this tendency has been made by Appleby *et al.*[3] He concludes that the aromatic components of the catalytic cracking feed stock are the outstanding source of carbon in catalytic cracking. The carbon is described by him as turbo stratic graphite, and he states that it is formed by intermediate aromatic structures of increasing size and complexity. The carbon-forming tendencies of an aromatic compound correlate well, as one might expect, with this basicity (i.e. tendency to acquire a proton). Figures for the basicity show an increase with the number of fused rings in the system, and also with increasing substitution of alkyl groups on the ring. This correlates well with foundry tests and gas chromatographic analysis which showed that the materials giving the best results, Gilsonite, coal dust and rosin pitch, give rise to high molecular weight aromatic substances on pyrolysis.

The lower temperature limit for carbon deposition in the presence of a catalyst is in the region of 400°C. The fact that substances evolving large amounts of volatiles at temperatures below 400°C tend to fume and be inefficient in their action, may well be dependent on this limit.

GAS CUSHION THEORY

The basis of this theory is that the gas evolved from coal dust or other additives tends to produce a gas cushion between the walls of the mould and the molten metal. Taken in conjunction with the reducing atmosphere theory, this is then said to inhibit the oxidation of the metal and hinder contact of the molten metal with the mould face. The big problem with this idea is that, taken on its own, the efficiency of any addition should be a function simply of the amount of volatiles produced, and be independent of the type of hydrocarbon evolved. The work described indicates the importance of the nature of the volatile constituents. Nevertheless, it must be true that a pressure build-up occurs at the metal-mould interface when a mould is poured and this could serve the function of delaying contact of metal and mould until the sand grains become enveloped and linked by a protective carbon layer. This is obviously particularly important when the bottom face of the mould is considered.

Some correlation of the effect of gas-producing additives has been done from the viewpoints of hot stage microscopy, volatile content and thermogravimetric analysis.

HOT STAGE MICROSCOPE

Coal Dust

Coal dust in an argon atmosphere softens and out-gasses at around 420°C. In air a sintering process begins at around 500°C, which does not occur in argon. This effect would, therefore, appear to be connected with an oxidation process. In air a vigorous burning begins at 650°C and is complete by 700°C. The ash residue melts at about 1320°C and in an argon atmosphere evolution of gas is also noticed at this temperature, presumably caused by the decomposition of inorganic matter. In the inert atmosphere, carbon deposition occurs in the temperature range 650–1000°C.

N*

Coal Tar Pitch

This material is completely liquid at 250°C. Gassing is visible from 400°C leaving a hard, homogeneous carbonaceous residue which does not melt below 1400°C. In the presence of air, the gases enflame at 550°C and at 600°C the carbon is burnt off. The small amount of ash formed fuses at 1300°C.

Gilsonite

Gilsonite is completely liquid at 240°C. Gas evolution is noted at about 450°C. From the change in the colour of the melt it is thought that carbon is produced during this process. The black residue softens at 1200°C due to melting of inorganic matter.

In air three stages of burning are observed:

500–570°C gas burning
570–600°C solid burning
670–700°C solid burning

There would appear to be two forms of carbonaceous material present with different ignition temperatures.

Petroleum Bitumens

These comprise a wide range of materials with varying properties. A particular example would be a material softening at 100–120°C, becoming fully fluid at 150°C and evolving gas in the temperature range 350–450°C with subsequent hardening of the melt.

VOLATILE CONTENTS

Volatile contents of the above-mentioned materials are shown in Table 2.

TABLE 2.

Material	Wt. of volatiles per gram	Volume of volatiles per gram at N.T.P. mls	Carbon residue per gram	Ash residue per gram	Max. rate of evolution mls per min
Coal dust (a)	0·281	289	0·580	0·138	289
Coal dust (b)	0·427	434	0·535	0·038	278
Gilsonite	0·767	697	0·225	0·008	284
Pet. bitumen	0·916	706	0·083	0·001	289
Coal tar pitch	0·734	584	0·259	0·007	424
Rosin pitch	0·971	724	0·009	0·020	396

Thermogravimetric Analysis[4]

Thermogravimetric analysis shows an interesting correlation between the temperature of evolution of volatile materials and the efficiency and fume production of the additive; in general, materials which decompose above 400°C give the best results.

Gas Chromatography[4]

Gas chromatography has shown that coal dust evolves a great range of aromatic compounds including quantities of four ringed compounds having boiling points in excess of 400°C. Coal tar pitch and gilsonite have similar chromatograms, whereas the petroleum bitumens have volatiles which, although of high boiling point, are mainly aliphatic in character and as a consequence somewhat less effective as lustrous carbon producers.

EXPANSION DEFECTS

Coal dust is also used for the purpose of inhibiting expansion defects. It is not particularly effective in this regard but nevertheless it is a useful addition particularly for small, thin section castings. The mechanism by which coal dust works is by relief of the surface stresses of the mould through softening which occurs between 400° and 600°C. However, further hardening of the bond does occur on prolonged heating and thus on larger castings it is possible that expansion defects may in fact be aggravated by the presence of coal dust. It is this subsequent hardening of the plastic decomposition products of coal which is counted responsible for the inhibition of dilation in thick section iron castings.

WOOD FLOUR

Two types of wood flour are normally used as sand additives. Hard wood flour is used in cores to aid breakdown and inhibit veining or finning. Being dense, the hard wood does not absorb the core binder to any great extent and does not, therefore, reduce the baked strength of the core to any marked degree. Soft wood flour, on the other hand, is used in green moulding sand at the rate of 2–3%. It is added mainly as an anti-scab agent, but it will also increase flowability and moisture tolerance. Other effects that wood flour can have on a green sand are to lower the dry strengths, particularly of calcium bentonites, reduce green and dry shear strengths and increase the hot deformation. The mulling action of a sand mill is retarded by wood flour addition.

The chief objections to the wood flour addition are the increase in friability of the sand and the increase in dilation of the casting due to the high hot deformation.

IRON OXIDE

It has been shown that iron oxide is a potent factor in assisting metal penetration by ferrous metals into moulding sand, nevertheless, Fe_2O_3 is added in small quantities to some steel foundry moulding sand mixes in order to increase the hot strength of the sand and prevent erosion. The principal reaction involved is probably formation of the viscous slag fayalite—a low melting point iron silicate—which bonds the surface sand particles together and prevents erosion. In addition, iron oxide in the presence of hydroxyl from the clay would form iron hydroxide $Fe(OH)_3$ which on drying will form a sand bond.

Higher additions, of the order of 3–5%, of iron oxide are used in core sands to inhibit finning and pinholing defects, particularly in S.G. irons, and also to improve hot strength.

In general sand mixtures for copper-rich alloys (brasses and bronzes) do not contain special additives. However, alloys containing phosphorus or lead are more penetrative and additions of coal dust, pitch or blacking (crude graphite) are made to the sand.

Of the light alloys, only magnesium requires special reference.† Inhibitors are incorporated into the moulding sand to prevent reaction between the water vapour from the mould and the molten magnesium. The inhibitor additions for thick section castings are usually 4% sulphur plus 1% boric acid, together with fluoride salts to reduce and strip oxide films. For thinner sections the additions are lower and core sands have generally 1% sulphur and 1% boric acid. It is important that aluminium moulding sands do not become contaminated with magnesium sands as the boric acid particularly can give rise to severe metal mould reaction of the pinholing type with certain aluminium silicon alloys.

REFERENCES

1. BRADDY, D. G., and BEALE, J., *British Foundryman*, **55**, 233, June 1962.
2. HEINZE, G., *Giesserei*, **53**, 449 (1966).
3. APPLEBY, W. G., *et al.*, Am. Chem. Soc. Div. Petrol Chem., Preprints 5/B.71–B.91 (1960).
4. DAVIES, M. J., R. & D. Report No. 248, Foseco International Ltd., August 1967.

† For details see chapter on Magnesium, p. 267.

G. GRAVITY AND PRESSURE
DIE-CASTING

CHAPTER 12

TREATMENT OF DIES WITH COATINGS FOR PRESSURE DIE CASTING

G. KITELEY and M. D. HUNT

INTRODUCTION

The first man to patent a pressure die casting machine was J. J. Sturgis in 1849, and by 1910 pneumatically operated machines were in use. Over the past fifty years control of casting alloys, die design, operating conditions and release agents, have made the process one of the most economical ways of obtaining metal shapes today. The metals most commonly die cast are zinc, aluminium, magnesium and brass, but also small quantities of lead and iron and latterly steel. Depending on the melting point of the metal alloy to be cast and the affinity of the alloy for iron, two basic types of machine are in use.

The "gooseneck" or hot-chamber machine (Fig. 1) is mainly used for zinc-base and lower melting-point alloys and utilizes a metal pump permanently submerged in a holding furnace which is an integral part of the machine.

The cold-chamber machine is used for aluminium and higher melting point alloys and has a piston and sleeve mechanism, whereby molten metal is ladled from the holding furnace into the sleeve (Fig. 2) for each casting cycle.

In the early days the die caster was only concerned with reproducing the metal shape. As the rate of casting production and casting quality became important more thought was given to the ingredients of die casting lubricants or die coatings. Some of the first compounds used were greases and heavy oils, with additions of graphite, where lubrication was considered to be the only property of significance. The lubricant was usually applied to the hot die face by incorporation of a carrying agent, e.g. paraffin or a light oil.

It soon became apparent that in order to obtain consistent results in casting quality, the variables associated with the process would have to be controlled. Some of the variables, namely, die temperature, die surface condition, cycle of operation and lubrication can be greatly influenced by the ingredients incorporated in a die dressing. One major reason why no single die coating can cover the needs of the die casting industry is the wide variation in die temperatures when casting different metals, e.g. lead as compared with brass. Even for a specific alloy, large temperature gradients exist across the die face, especially when casting higher melting point alloys. Oils, fats, etc., which are diluted with paraffin are still more widely used than water dilutable compounds. It is probably true to say that the former coatings give better release and lubrication properties, but in recent years, advances to increase the lubricity of water-based coatings have been made.

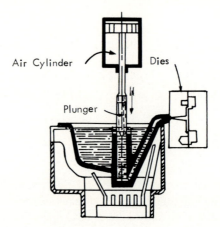

FIG. 1. Gooseneck hot-chamber pressure die casting machine.

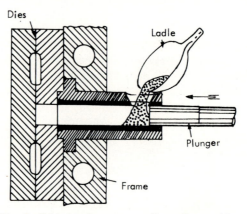

FIG. 2. Cold chamber pressure die casting machine.

REQUIREMENTS OF A DIE COATING

Apart from lubrication the die coat has many other functions to perform. By listing these, and understanding the function of each requirement, a scientific approach can be made in formulating coatings for specific applications.

1. Prevention of soldering.
2. Casting release.
3. Provision of a good casting surface finish.
4. Aid to metal flow and casting soundness.
5. Lubrication of ejector pins and other moving parts.
6. Ease of application.
7. Economy and freedom from fire or health hazard.

1. Prevention of Soldering

This term is applied to either physical interlocking between the casting and the die, or a chemical interaction between the casting metal and the die steel to form an intermetallic compound. Evidence for physical interlocking is given by Roberts and Grobe[1] who showed that soldering occurred on an eroded die surface. When chemical interaction takes place an intermetallic compound forms, e.g:

$$Fe + 3\,Al \rightarrow FeAl_3$$
$$Fe + 7\,Zn \rightarrow FeZn_7$$

Subsequently these compounds become oxidized to give build-up on the die face and galling of the moving parts. Die temperatures in excess of 250°C aggravate this problem. The alloy composition of the die steel is also important.[2] A hard steel, particularly one which contains molybdenum,[3] has less tendency towards soldering than a softer steel. For the higher melting point alloys of Al, Cu, Fe, and if the metal velocity is high, soldering[4] is more likely. Thus a protective film will minimize soldering.

2. Casting Release

In the absence of any soldering, the ease of release is a function of the ability of the coating to provide an impervious film on the die surface.

3. Provision of a Good Casting Finish

Many materials which give excellent lubrication and casting release are rejected because of stains on the finished castings. This is often caused by incomplete carbonization of any organic material prior to the injection of the molten metal, or to the inability of an additive to bond to the die resulting in transfer of the dressing to the casting.

4. Aid to Metal Flow and Casting Soundness

The contribution of a die coating to metal flow and casting soundness is to some extent dependent upon its thermal conductivity. The addition of a filler in order to increase its insulation properties is not common in pressure die casting as opposed to gravity die casting, due to the difficulty of avoiding build-up on the die face. The coating should have low gas emission[5] on impact with the molten metal.

5. Lubrication of Ejector Pins and Moving Parts of the Die

The dressing should perform well as a high-temperature, high-pressure lubricant to prevent any galling or seizure. A major difficulty, especially with aluminium die casting exists here, since adequate die face lubrication is often insufficient for moving parts lubrication, whilst thick greases, etc., are suitable for the latter, but tend to give staining on castings.

6. Ease of Application

This subject will be dealt with in a later paragraph, but essentially the coating should be of a consistency which will be suitable for the various methods of application.

7. Economy and Freedom from Fire or Health Hazard

Certain compounds break down at the temperatures encountered to give toxic or unpleasant fumes. It is also possible for the decomposition products to attack the die surface. Water-based die coatings have the advantage of freedom from fume, combustibility, cleanliness and often economy over an oil or kerosene thinned dressing.

COMPONENTS OF A DIE COATING AND THEIR FUNCTION

A dressing may be split into three components:

1. The carrier fluid.
2. The lubricant.
3. Additives.

1. The Carrier Fluid

Ideally it should possess the following properties:

Non-flammability.
High and wide boiling range.
Viscosity which is unchanged at elevated temperatures.
Low surface energy.
Economy.
Suitability for solution or suspension of the lubricant.

Since the function of the carrier is to deposit the lubricant on the die face, it is preferable for it not to evaporate immediately on contact with the die surface, or penetration of deep draws will be restricted. For this reason water-based coatings have been found inferior to those based on organic solvents. Both aluminium and zinc have a lower surface energy than steel, and thus tend to wet the die. If a carrier fluid and lubricant is applied which has a lower surface energy than either the die steel or the casting alloy, it will wet the steel die and act as a barrier to the casting metal.

Surface energy and the wetting phenomenon may be considered as follows: consider a drop of liquid on a solid with a contact angle, θ (Fig. 3).

At equilibrium $\qquad\qquad\qquad \gamma_{sv} = \gamma_{lv} \cos \theta + \gamma_{ls}$

where γ_{sv}, γ_{lv} and γ_{ls} are the surface tension of the solid/vapour, liquid/vapour and liquid/solid surfaces respectively.

When $\theta = 0$, spreading will occur and

$$\gamma_{sv} \geqslant \gamma_{lv} + \gamma_{ls}$$

If $\gamma_{lv} \ll \gamma_{sv}$, then it can be assumed that $\gamma_{ls} \ll \gamma_{sv}$ and thus liquids of low surface tension will wet solids of higher surface energy.

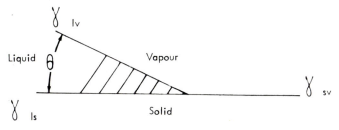

FIG. 3. See text.

2. The Lubricant

Two classes of lubricant may be defined. First, those which remain in the liquid state at the operating die temperature. Second, those which undergo a chemical reaction, and in doing so the new chemical compounds formed provide the lubrication. The "liquid" lubricant is restricted in its usefulness, as high pressures and high temperatures would soon break down the film. The second class of lubricant may be termed extreme pressure lubricants, certain of which can withstand very high loads at elevated temperatures. They may contain chlorides, phosphates, sulphates or organometallic compounds. However, in all cases especially where phosphates are used, the by-products should be checked for any harmful effects to both person and die.

On application of the lubricant to the die face a film is generated to provide a barrier between the cast metal and the die. Initially, there is only sufficient energy to allow a limited chemical reaction. Conducted and radiated heat are liberated, whereupon the carrier fluid evaporates and low thermal stability oils, etc., decompose. On injection of the molten metal the film will prevent soldering, but will undergo changes dependent upon the materials employed. It is important to mention that no one extreme pressure lubricant could cope with all the changing conditions of temperature and pressure in the die casting process. Thus incorporation of a mixed lubricant system will ensure that the film changes to meet the various requirements. Where casting release is not difficult, e.g. zinc castings, low thermal stability oils and waxes can be used which will decompose and be liberated as vapours at low operating temperatures.

At higher operating temperatures, e.g. when casting Al and Cu alloys, or when higher pressures are employed, lubricants which will provide better film strength are required, e.g. higher molecular weight waxes, oils and anhydrous soaps. As has been stated, the changes that take place in the die casting process are numerous and as a result the chemical reactions occurring in the lubricant will be extremely complex. However, simple tests in the laboratory indicate which materials are likely to be suitable as lubricants over a range of conditions. The chemistry of the surface interactions creates the greatest problems. The lubricant, initially, is in equilibrium with the carrier fluid. On application to the hot die surface the carrier evaporates, leaving the lubricant to attain a new set of equilibrium conditions. The lubricant will either spread or roll up in spheres depending on its surface tension relative to the die steel. Water-based die coatings give rise to more problems than those based on organic solvents. In an oil-in-water emulsion, discrete particles of oil are surrounded by chemicals known as emulsifying agents, and these are then distributed through the water which forms the continuous phase. As nearly all these compounds are insoluble in water there exists a balance of forces, namely electrical charges. After application to the hot die, there will be lubricant particles with "free" electrical charges. Provided the lubricant is chosen correctly, these electrical charges will re-orientate themselves on to the die face and form a stable thin film. For this reason and because water is highly "polar" when compared to hydrocarbon solvents (e.g. kerosene), thin films can be produced that have the same lubrication properties as thick oil or grease films.

3. Additives

For lubrication

The incorporation of a finely divided metal powder is fairly common, the most usual additive being powdered aluminium flake. Other metals which have been tried are nickel, bismuth, copper and molybdenum. Several of these have been found to promote both build-up and soldering. Metallic aluminium seems to mask any inclusions and lapping defects, in addition to its function as a release agent.

Among the non-metallic solid lubricant compounds are graphite, mica, molybdenum disulphide, boron nitride and silicon nitride. These materials are refractory and some possess a lamellar structure. For example, the lamellar structure of graphite allows it to slip under pressure, thus providing lubrication whilst the film of graphite forms a barrier at the metal interface preventing welding, etc. Some of these materials, however, suffer from an inability to withstand erosion by the injected metal. Therefore they should be used with caution as they may become impregnated in the surface of the casting and will have a detrimental effect on finishing processes, e.g. plating, painting, anodizing, etc.

For water-based coatings

Additives which are either necessary or desirable include emulsifying agents, surfactants, protective colloids and gelling agents.

The choice of an emulsifying agent should be made with a view to minimizing the staining of castings and build-up of deposits on the die face. The emulsified lubricant should have good "can-stability" over a wide temperature range. The most common emulsifying agents used are the triethanolamine and sodium salts of oleic and stearic acid. A gelling agent may be added in order to maintain a degree of viscosity at higher dilutions. Among these can be included many of the cellulose derivatives, e.g. sodium carboxy methyl cellulose. The addition of surfactants and protective colloids may enhance the emulsion stability, and in the case of a surfactant should also reduce the surface tension of the product in order to improve wetting of the die face.

For oil-based coatings

Oils used in pressure die casting often contain significant proportions of antioxidants, corrosion inhibitors, boundary lubrication improvers (e.g. anti-wear agents), detergent-dispersants, viscosity index improvers and anti-foam agents. If an oil is oxidized by the atmosphere at least two undesirable events can occur. First the oil thickens and second sludge and acidic products are formed which are liable to attack the die surface. Fortunately oxidation can be retarded by oxidation inhibitors.

Usually the viscosity of an oil will decrease with increase in temperature. In die casting it is important that the viscosity of the oil should remain fairly static under varying temperatures. For this reason certain compounds are added to produce oils of a viscostatic nature.

CLASSIFICATION OF DIE COATINGS

Pigmented

Graphitic dressings are still widely used but it cannot be said that they are really popular because (a) the staining which occurs on floor areas around the machines and also on the operator's clothes, (b) build-up of residue on the dies and (c) in some instances, residue on the casting surface which can create problems with respect to finishing operations. Efforts to date to replace graphite have had only limited success. One material which has found universal acceptance is finely divided aluminium flake. Not having the same structure as graphite, its lubrication and release properties are somewhat inferior, but it has the advantage of blending with the cast surface (e.g. aluminium die casting).

When careful additions are made these materials will fill in pits, tool marks, undercuts and other die surface imperfections. Where insufficient film strength is obtained from a release agent, "machining furrows", under sliding pressure (during ejection), tend to lock momentarily resulting in drag, soldering and distortion. The low cleavage point between the lamellae in graphite or mica provides the slip or sliding medium by which "locking" is eliminated.

However, the binders for the pigments must possess sufficient film strength at high temperatures and pressures to withstand the injected metal, otherwise the pigments will be carried away with the flow of molten metal. This is necessary because most pigments have no affinity for die steel. The specific gravity of pigments should be closely related to the medium in which they are to be dispersed, or particular attention should be paid to the possibilities of special suspension systems.

It is well known that cracking of thermally cycled steels can be related to the rate of change of temperature of the steel surface. Thus theoretically the greater the change in temperature, the quicker the die steel will crack. This problem is minimized by the incorporation of an insulating material. However, these will eventually build up on the die face. Mica is often included in a coating because of its insulating properties, as well as for lubrication.

Solvent-based Die Coatings

In any die casting operation it is essential to stabilize the die conditions as much as possible. Even with the best die design and water cooling, there will be some variation in temperature across the die face. Since the thermal conductivity of oils, etc., is relatively poor, the die face temperature will not be greatly affected by the application of an oil. Thus the "shot rate" will primarily depend on the heat sink characteristics of the die, i.e. the efficiency of the water cooling.

Oils should be chosen so that the carbonaceous deposit is as small as possible. However, a small residue is beneficial, as it provides a parting mechanism for the casting. Under sustained high temperatures, certain untreated, or improperly treated, oils (vegetable or mineral) can become acidic and consequently corrosive. For this reason, oils used in die cast release agents are compounded with buffering agents, which are provided in the form of metallic soaps. These, when ignited, decompose to an ash of the corresponding oxide. Other metallic soaps are frequently used as thickeners in die coatings and trouble can result through the formation of excessive ash deposits on the die cavity surfaces.

For optimum results this type of dressing should be sprayed with an "atomizing" jet. In order to facilitate this type of application the die coating is diluted with paraffin or white spirit. The carrier volatilizes on impact with the hot die leaving a film of oil. The additives incorporated accelerate the thickening process, so that when an injection is made the oil is not displaced by the abrasion of the molten metal.

Although in general oily dressings give good lubrication, gas absorption creates a problem. This does not occur to the same extent with water-based dressings as the die is cooled and gas formation from the lubricant will be minimized. Staining from incomplete decomposition also causes a problem. This is almost impossible to foresee since little is known about decomposition with respect to temperature and time of any one oil. As paraffin is often the carrier, health and fire safeguards must be enforced.

Water-based Die Coatings

Few lubricating materials are soluble in water, and thus the lubricant is either dispersed or emulsified in the water phase. As water is a "polar" compound the lubricating particles will be in electrical equilibrium with the water, otherwise separation of the solid matter will occur.

In an oil-in-water emulsion the oil or wax particles are probably $\frac{1}{10}$ to 5 microns in size and are subject to "Brownian" movement. Inversion of an emulsion system can occur (i.e. oil-in-water to water-in-oil) if the emulsion is subjected to a severe temperature variation or a high shear stress. A common cause of spray nozzles becoming blocked is due to this phenomenon.

When an emulsion is sprayed onto the hot die the electrical forces will act to maintain a stable emulsion, but as the water boils steam will be liberated and the rate of loss of water will be rapid. The exceptions occur when the die temperature is too low for all the water to be released before the die closes. In these cases water-based dressings should not be used.

The lubricant in emulsion concentrates should have suitable decomposition and dissociation characteristics with respect to an efficient time/temperature cycle of the die casting machine. No organic or organometallic compound will survive the operating temperatures for any period without decomposition and reaction with the cast metal or die to form new compounds. The ability to control this rate of decomposition is among the most important features of a correctly formulated die coating.

Schafer[6] found that polymers which either possessed a weak bond or produced corrosive elements stained the castings. The functioning of different types of polymers can be explained in terms of their bond strength, configuration and chemical composition. On this basis telomers, polystyrene, polyethylene, silicones, silicates and polyisobutylenes did not stain, whilst polyglycols, polyvinyl chloride and most oxygen- or sulphur-containing organic compounds stained. Other materials often incorporated in water-based emulsions are beeswax, carnauba wax, lanolin, mineral oils and fats.

The advantages of these materials in emulsion form are that they give low fume, no health or fire hazard, low gas emission, and if properly formulated, good "thin-film" strength. As yet there has been no evidence to suggest that water-based dressings cause any thermal shock to the die. Theoretical support is provided by the suggestion that the spray is separated from the die by a layer of steam until the emulsion is dissociated. Also the temperature differential between steam and the die steel is small compared with that between the die steel and the molten metal and this differential increases with increase in the molten metal temperature. It is well known that significant changes in steel structure and properties occur at high and not at low temperatures.

Unless some provision has been made in the formulation of a water-base coating to counteract the effects of hard water salts, chlorides, bromides, iodide, sulphur compounds, etc., it is advisable to use distilled rather than tap water as the carrier fluid.

The water-based dressings are the nearest approach to an ideal coating. They embrace practically all the desirable properties, but although lubricity properties are gradually being improved they are still lacking in this respect where the higher melting point alloys are concerned. However, with good die conditions, i.e. smooth die surface and small temperature gradients, excellent casting results can be obtained.

TREATMENT OF GRAVITY AND LOW-PRESSURE DIES WITH COATINGS

Introduction

The now well-known gravity die casting method of producing castings was, in retrospect, almost a natural progression from the sand moulding method. The immediate advantages were obviously those related to having a permanent, durable metal mould or die from which large numbers of castings could be produced, each being an exact replica of its predecessor. Closer dimensional tolerances could be observed thus eliminating or reducing the need for finishing or machining operations. One disadvantage encountered was that in contrast to a sand mould which is permeable, and permits air or gas to escape during the filling operation, venting systems had to be incorporated into the design of the metal dies at the critical points where air could be trapped. To avoid violent reaction between the molten cast metal and the moisture of condensation, it was found necessary to preheat the metal die before operations began. This also had the effect of bringing the temperatures of die and metal slightly closer together and thus going some way towards minimizing the expansion and contraction stresses developed between die and casting immediately after the casting operation.

The inter-deposition of a suitable dressing between die and cast metal can make all the difference between success and failure since, depending on its formulation, it can perform as an insulator and/or lubricant, an aid to metal flow, a protection for the die metal, a protection against thermal shock and an aid to directional solidification and feeding.

COATINGS FOR THE GRAVITY DIE CASTING OF ALUMINIUM AND MAGNESIUM ALLOYS

The alloys of aluminium are without doubt the most widely used in the gravity die casting process and the most important property that can be conferred by a dressing for aluminium and magnesium alloys is thermal insulation. The degree of insulation of the applied film of dressing depends just as much on its form or texture as on the properties of its ingredients.

The basic constituents of a gravity die dressing are (other than water) fillers and binders. The fillers are refractory powders such as titanium oxide, calcium carbonate, asbestine, talc, mica, kieselguhr, iron oxide, aluminium oxide, etc.

At the normal casting temperatures of aluminium alloys in the gravity process (generally between 650°C and 800°C) the refractoriness of the fillers is sufficiently adequate to ensure that they are not usually subject to any chemical change and their function is largely a physical one. The choice of suitable fillers is controlled not only by their inherent insulation properties but by their effect on the surface finish of the casting, form, ability to interlock with other refractories, particle size and inertness to other refractories and the bonding media. The object is to obtain maximum strength and permeability conducive with the type of finish specified for a particular range of castings. The bonding constituents are usually based on sodium silicate although other materials, such as certain types of clays, starches, aluminium orthophosphate and boric acid are sometimes used.

Most of the dressings are supplied in a concentrate paste form, mainly to avoid separation of the constituents during storage and transit. The dilution rates of the sodium silicate based dressings are not critical, for any excess water in the dressing is driven off on impact with the hot die leaving the fillers and other constituents at their original ratio. Ideally the dressing should be applied to the hot die in the form of a fine mist spray, the object being to allow each individual particle to be cemented firmly and separately to the die until full coverage is achieved. This foundation layer will provide the necessary strong key for the support of the completed film of dressing. When the film has been built up to the desired thickness, the bridging effect of the particles will create voids or air pockets thus considerably enhancing the insulation properties of the layer. In addition the slight bubbling reaction of the sodium silicate on impact with the hot die also creates minute voids in the film further increasing its permeability and insulation properties. In contact with the hot die the sodium silicate is dehydrated producing a film highly resistant to temperature, and to a somewhat lesser degree, abrasion.

The surface condition of the applied dressing is of considerable importance to the real fluidity of the molten metal in the die as well as to the finish on the casting. Although a smooth surface is desirable on the casting this is not usually compatible with the complete and satisfactory filling of the die cavity with molten metal, unless the casting is dimensionally of heavy section. Despite the fact that the natural insulating properties of the dressing are considerably enhanced by the permeability of the applied layer, this layer is of a necessity comparatively thin (usually between 0·005 and 0·020 in.), and with temperature differences of the order of 400–500°C between die and molten metal, the thermal conductivity of the coating film is still relatively high and heat losses from metal to die will be almost instantaneous and comparatively great. If the surface of the applied layer of dressing is relatively smooth, a high percentage of the surface area will be instantaneously wetted by the molten metal and thermal losses, through the dressing to the die, will be both rapid and excessive, probably resulting in underfilling or misrun castings. If, however, the dressing surface is rough, contact between molten metal and coating is considerably reduced because the metal, momentarily supported by surface tension, bridges the "peaks" of the dressing and after an extremely short lapse of time, sinks into the "valleys". As the metal sinks into the "valleys" the air beneath escapes through the channels of the permeable film of dressing. The effect of this phenomenon is to reduce heat losses from the molten metal to a minimum and to maintain its fluidity at the critical moment to allow for the die cavity to be completely filled. Unfortunately the "peaks" on the surface of the dressing are usually progressively snapped off as the casting contracts during solidification and heat losses increase correspondingly with subsequent casts. However, if the "peaks" did not break off it is probable that the casting would be firmly keyed into the die. The dressings must therefore be expendable and on this basis the formulations are a compromise between durability and some consideration for removal and replacement.

As a general rule where maximum insulation properties are required from the dressings, castings are of thin section and large surface area. In these cases the applied film of dressing should be rather thicker and its surface rougher than normally needed for a chunky type of casting. The rougher the surface of the film the more it will be affected by the contraction characteristics of the casting, and consequently, the more frequently will it necessitate "touching up" or replacing. Where a comparatively smooth cast surface finish takes first priority, the problem to a certain extent can be alleviated by super heating the melt to a sufficiently high level above its solidus to counteract the effect of the higher thermal losses of the smoother dressing surface. By this means a more acceptable finish on the casting can be accomplished, but if carried too far, and to a degree dependent upon the characteristics of the alloy, may result in higher gas content, oxidation, coarsening of the grain structure, inclusions and increased losses of some alloying elements. The answer is usually to be found in a compromise between surface finish and metal temperature.

Although, as previously stated, the ratio of the constituents remain constant despite the various dilution rates with water, dilution rates do affect the ultimate form of the applied film. When a well diluted dressing is sprayed on to a hot die, the viscosity of the water/sodium silicate solution is very low, and on impact the refractory particles can orientate more easily into a flat plane before the water is driven off and the bonding effect of the silicate is achieved. With a more concentrated solution refractory particles, surrounded by a more viscous liquid, are cemented or bonded almost instantaneously in random position on meeting the die surface. The result is that some of the particles, with point-contact only, have a weaker attachment to the die; the surface of the applied film is also rougher and, as previously mentioned, reduces thermal conductivity and improves insulation. As a general rule, therefore, a high dilution rate of water to dressing will produce maximum adhesion to the die together with the smoothest possible film surface. A low dilution rate of water to dressing, however, will produce an applied film with less adhesion to the die together with a coarser film surface and lower thermal conductivity and consequently better insulation properties.

The temperature of the die can have a considerable effect on the type of film

COATINGS FOR PRESSURE DIE CASTING 401

produced when the dressing is being applied. If the temperature is too low and the evaporation of the water consequently delayed, the dressing will form an almost impermeable skin which is lifted off the die surface in the form of blisters by the trapped vapour underneath; it may then shrink and crack off when completely dried, leaving a bare patch. If the temperature is too high and the evaporation of the water too violent, the refractory particles, surrounded by a film of water and sodium silicate, will be almost completely detached from one another and also from the die surface by the miniature explosions that occur as the partially trapped steam escapes to atmosphere. The result is a very weak, rough, powdery film which falls away very easily.

The temperature range most suitable for the application of this type of dressing lies between 180° and 220°C, and for the best results an atomizing type spray gun is recommended.

Graphite is sometimes incorporated with refractories in dressings for the gravity die casting of aluminium, but where it is used on the die impressions the advantages are debatable. Theoretically the graphite can only serve to physically weaken the bonding properties and increase the thermal conductivity of the dressing, which may, in some isolated instances, be desirable but in most cases is a reversal of the purpose for which the dressing is normally applied. There is, however, a logical reason for the common practice of applying a graphite dressing as a separate entity on top of the normal refractory one. The formulation for this purpose usually incorporates colloidal or semi-colloidal graphite in water plus a small percentage of a bonding agent such as a special type resin or a starch. This graphite top dressing performs as a lubricant to facilitate the extraction of the castings from the die and also as a lubricant for the protection of the undercoat against the abrasive action created during the solid contraction of the casting. If the adhesive properties of the bonding agent in the graphite dressing result in a hard shell surrounding each particle of graphite, the lubrication properties of the graphite will be impaired or completely nullified. Ideally the bonding agent for this type of dressing should progressively disintegrate at the cast metal temperatures to allow a small percentage of the graphite to fall away or be blown off, after each cast and then to be replaced by a further sprayed application. This may appear illogical but can be explained by the fact that when the dressing is lightly sprayed it orientates into a fluffy film on top of the undercoat. In this form, air pockets are created within the film itself which enhance its insulation properties and aid the flow of molten metal when the casting is being poured. Later, however, when the casting solidifies and contracts, the relatively weak film is compressed flat on to the undercoat, causing the film of graphite to seal the pores of the undercoat and render it impermeable. If the permeability of the undercoat is to be retrieved for the next cast, then the compressed film of graphite must be partly shed or removed by other means.

The Application of Dressings as an Aid to the Control of Solidification

Castings designed for production by the gravity or low-pressure methods can vary widely both in complexity and section thicknesses. Invariably there are comparatively thick sections, isolated from the feed reservoir by thinner sections, which pose difficulties when attempting to produce an internally sound casting. Metal temperature, rate of pouring or filling, etc., can all play some part in overcoming the problem but, so can the dressing. By applying a rougher thicker layer of dressing to a thin section, thereby delaying solidification, and a smoother thinner layer on a thick section, thus accelerating solidification, the freezing characteristics can be brought sufficiently close together to enable a sound casting to be produced. The greater the difference in casting section thicknesses, the greater will have to be the difference between the insulation properties of the applied films of dressing. The desired effect can be achieved by completely removing the dressing from the thick section and applying a thick coarse layer

FIG. 4. Gravity die cast aircraft box and lid in LMG. *Top right and left*, lid and box as cast, *bottom*—finished casting. Length of box 6 in. (casting No. 7500 from same die). (Courtesy of W. T. Crowley & Sons Ltd.)

on the thin. Feeder heads or risers which need to remain molten long enough to supply feed metal to the casting during solidification, usually warrant a comparatively heavy layer of an insulating dressing to assist them in this function.

Magnesium Alloys

The dressings and methods of application for the gravity die casting of magnesium alloys are the same as for the aluminium alloys with the exception of certain types which contain an inhibitor such as boric acid. The normal dressings will be satisfactory if the usual practice of preventing the oxidation of the metal prior to pouring, is carried out by inhibiting the die cavity with SO_2 or dusting the die faces with a mixture of sulphur and boric acid for each cast.

LOW-PRESSURE DIE CASTING

In this process the metal is injected into the die but the pressures involved are relatively low. The die is mounted on top of the holding furnace, which houses a sealed crucible containing the molten metal. Air or nitrogen pressure is applied to the surface of the melt inside the pot, forcing the metal up an attached connecting tube and through a sprue into the base of the die. The pressure required to displace the metal and transfer it to the die cavity is usually only a few pounds above atmospheric pressure; there is no back pressure on the molten metal as it proceeds to fill the die and pressure only builds up where restrictions are encountered or when the die cavity is practically full. The requirements of a dressing for this procedure are exactly the same as those for the gravity die casting and therefore the types, methods and application discussed under that heading apply. The advantages of this method over gravity are the quiet metal entry into the die from the bottom at a predetermined speed creating very little turbulence during transit, and, fettling being reduced to the removal of a small feed sprue.

There is one problem involving dressings which is not common to both processes and this concerns the tube, stalk or venturi, which is the connecting link between the base of the die and the molten metal. This tube is partially submerged in the molten metal at all times and its protection against erosion and attack poses something of a major problem. In efforts to overcome this the tubes have been made in ceramics, plumbago, alloy steel, cast iron, etc. The ceramic or plumbago types are partially successful but usually prove to be too fragile and expensive. The differences in the degree of resistance to attack between the various steel and cast iron tubes is comparatively marginal and a refractory protective coating is necessary to prevent them being dissolved by molten aluminium. Acceptable results have been achieved by coating the tube in exactly the same way as for gravity dies, i.e. spraying a thin film of a highly bonded insulating dressing followed by a graphite/water application. One of the difficulties with this method, however, is to spray the inside of the tube for in some instances the inside diameter is of the order of only 3–4 in. and it can be 2 or 3 ft in length. Another reasonably successful method is to paint or dip the tube into a slurry, and then dry it out very slowly to avoid peeling. The ratio of refractory particles to bonding agent in the slurry is much higher than for the normal sprayable dressing and consequently produces a relatively thick, soft protective film which is often less susceptible to peeling or cracking.

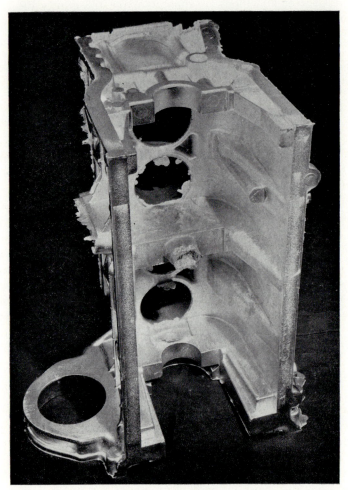

Fig. 5. Low-pressure die casting. Cylinder block in LMG with sprue attached. (Courtesy of Rootes Motors, Scotland Ltd.)

Lubricants for Gravity and Low-pressure Dies

Graphite in water dressings are ideally suited for the lubrication of most moving parts of gravity or low-pressure dies. Where possible the dressing should be sprayed but, there are times when it is desirable to chill a core or core pin in between casts as well as lubricating it, and under these circumstances it will probably be more convenient to make up a well-diluted bath of graphite and water into which the cores or core pins can be plunged.

Gears or ratchets which are not accessible for spraying can be lubricated with a concentrate graphite in oil dressing applied by swab, brush or oil can.

FIG. 6. Same as Fig. 5, fettled.

COATINGS FOR THE GRAVITY DIE CASTING OF COPPER BASE ALLOYS

60/40 brasses. The most widely used range of copper base alloys for gravity die casting are the 60/40 brasses. The zinc, which is volatile at the high temperatures involved, creates one of the most difficult problems associated with the gravity die casting of these alloys, i.e. the deposition of zinc oxide on the die.

This, unless removed, rapidly builds up to such proportions that the surface finish of the castings are adversely affected to a point where they are unacceptable. The method adopted by many existing die casters is to design the die so that it can be taken apart rapidly into comparatively small component units, of handy size and weight. After each cast the components are plunged or partially dipped into a bath or tank of water-based dressing which cools the die slightly, partially removes the residue of oxide and dressing and deposits a fresh film of dressing ready for the next cast.

The dressings in such dipping tanks are usually composed of carbonaceous materials, e.g. combinations of graphite, lamp black, coke dust, etc., suspended in water. The graphite functions as a lubricant and parting agent to facilitate extraction of the casting and ease of operation of the die. The carbonaceous constituents impart insulation properties to the dressing by "fluffing" up the film and thus maintaining the fluidity of the metal during the pouring operation. The dipping process does not completely remove the spent dressing and consequently a build up does take place on the die which necessitates frequent removal by wire brush or some other means.

Aluminium bronze. Casting temperatures for this alloy are usually higher than those of the 60/40 brasses and consequently the working life of the die is somewhat shorter due to the effect of thermal shock. Some die casters employ exactly the same dressings and techniques, i.e. dip tanks, for the production of aluminium bronze gravity die castings as for the 60/40 brasses, despite the fact that there are no oxide build-up problems with this alloy. In fact the normal dressing for the gravity die casting of aluminium alloys can be used quite satisfactorily except that the castings are a little more difficult to extract from the die and therefore a dressing with enhanced lubrication properties is desirable. Just as for the aluminium alloys a film of graphite in water can be sprayed on top of film of refractory/sodium silicate dressing or the two types can be premixed to form a universal film.

Production rates are generally higher for the dip tank method than for the latter.

COATINGS FOR THE GRAVITY DIE CASTING OF IRON

Comparatively few foundries are involved in the gravity die casting of iron, but the process is slowly being extended. Unfortunately not a great deal of detailed information is available concerning die dressings except that a semipermanent refractory type is used as a base and an expendable layer of carbon or soot, deposited by an acetylene or other hydrocarbon flame, on top. It is logical to assume that the function of the base dressing is to insulate and protect the die, and any good refractory/sodium silicate dressing used for the gravity die casting of aluminium could be adapted to perform this function. The flame method of applying the carbon film would ensure that it would be "fluffy" which would enhance the insulation properties and offer some protection to the base dressing.

COATINGS FOR THE PRESSURE DIE CASTING OF
ALUMINIUM ALLOYS

New developments or trends in the design of machines for pressure die casting are usually concerned with faster production, higher quality castings and increased capacity for larger dies. The consequent demands made upon the release and lubricity properties of the dressings for these machines, particularly where aluminium alloys are involved, have increased immeasurably, serving to highlight the inadequacies of the materials and formulations available for this purpose.

Temperature gradients across the dies are usually more acute and at higher levels than in the past. Unfortunately, the discovery or development of materials, new or otherwise, which could be adapted to dressings capable of coping with the prevailing severe conditions, have not kept pace with the development of the machines.

Pretreatment processes designed to protect the dies against the attack of aluminium and to facilitate the release of the casting from the die cavity have, to date, proved only very marginally successful, in most instances hardly worthy of the inconvenience of application.

The emergence of the more sophisticated lubricants such as silicones and molybdenum disulphide, have made very little impact in the sphere of the pressure die casting of aluminium alloys. Few, if any, advantages have been observed when substituting these materials for the cheaper types, such as graphite, in this temperature range.

One of the problems associated with the pressure die casting of aluminium is the natural affinity of aluminium for iron. When molten, and to a degree dependent on temperature, aluminium will readily attack and partially dissolve the die steel, resulting in molecular welding of the two metals.

The increased iron content of certain aluminium alloys such as LM24 has a dampening effect on this tendency for welding, but even under these conditions it is necessary to apply some form of a protective barrier between the two metals and this is a function the dressing is called upon to perform in addition to its other properties.

There are two basic types of dressings in general use, those with pigments and those without pigments. The pigmented types are the most widely used because they offer greater protection against welding, better release and better lubrication properties than the non-pigmented formulations of oils, waxes, greases, etc. Graphite, which is of a lamellar type structure, is the most universally used pigment for these dressings. It imparts better lubricity and release properties than most other available pigments. One of the disadvantages to its use is, that where a high concentration of solids or pigments is necessary, e.g. where die temperature gradients are steep and casting conditions severe, the graphite usually adheres to the casting impairing its appearance and creating problems with surface finishing processes, such as painting.

To counter the effect of too high a concentration of graphite in a formulation,
o

other pigments are usually introduced and substituted for part of the graphite. These may comprise finely dispersed powders of metals or micaceous materials, all of which may possess lubricity properties of a lower order than graphite, but which may impart similar or equal release properties. Aluminium powder is the most universally accepted substitute for graphite basically because it has the advantage of blending invisibly with the casting surfaces. Its leaf-like structure confers a fairly high level of "slip" or lubricity in addition to its good properties of release. The micaceous materials can be expected to impart slightly higher insulation properties to the dressing, but are usually employed as cheap fillers.

The pigments are generally incorporated into a grease or oil base dressing and the ratio of one to the other is fairly critical. If the dressing is too oily through lack of sufficient solids, it will be displaced by injected molten metal and concentrated in isolated areas of the casting resulting in staining, inclusions and substandard finish. Additives for increasing the viscosity of the oils on impact with the hot die are frequently utilized to prevent the film from displacement by the metal stream, and this approach allows of dressings with lower percentages of pigments to be used than would otherwise be possible. The advantage of a dressing with a lower ratio of solids to oils is that build-up on the die is much slower, apart from the economic advantages which usually accrue. It is also necessary for the dressing to initially possess good "wettability" properties to allow it to spread evenly over the die surface and too high a solids content, will not assist in this respect.

The mechanics of the pigmented dressings are that the pigments should be in close contact with the die surfaces, and, bonded *in situ* by the carbonization processes the oil phase passes through before its complete disintegration. The carbon deposit from the oil, plus the pigments, provides a relatively thick strong barrier, between die and casting, facilitating easy ejection. These thicker films fill and smooth out minute defects such as tool marks on the die surfaces, which would otherwise resist the extraction of the casting, and are also more durable and resistant to penetration when subjected to the contraction pressures of the casting.

The method of applying the dressing will depend on its consistency, e.g. by brush, swab, or blow gun if concentrated, and by spray gun if diluted. The brush and swab methods have their advantages in that the pigments are pressed into close proximity with the die surfaces, and a relatively dry film can be achieved, but they are time-consuming. The blow gun is probably the most used, for by this method the blow gun tip is dipped into the dressing and directed initially to areas where a concentrate is necessary. The remainder is then dispersed by air pressure to the less important areas. This method is fast and efficient, but usually creates volumes of smoke and fume and dirty conditions around the machines; it is also the most uneconomical method. For the spray gun applications the dressing is diluted with solvents such as paraffin, white spirits, etc. The solvents volatilize on impact with the die and do evolve smoke and fume, but if the air pressure is reduced to a minimum a film can be quickly

FIG. 7. Pressure die casting LM 24 aluminium alloy, weight 21·5 lb. (Courtesy of Birmingham Aluminium Casting (1903) Co. Ltd.)

FIG. 8. Pressure die castings LM 2 aluminium alloy. (Courtesy of Birmingham Aluminium Casting (1903) Co. Ltd.)

and evenly applied all over the die with less smoke and fume than is evolved by the blow gun method.

The non-pigmented types of dressing are usually cleaner to use but limited in scope. The materials employed are too numerous to list, but the important ones are mineral and vegetable oils, animal fats, natural or synthetic waxes and additives to function as antioxidants, extreme pressure lubricants, etc. Despite endless permutations of the above materials this type of dressing is inadequate in release and lubrication properties for the major proportion of the aluminium alloy pressure die castings which are produced. Their use is restricted to conditions where die operating temperatures are fairly even and low, and casting designs not complex.

Where, however, a pigmented type is applied to the troublesome areas of the die only, the non-pigmented types are frequently employed as auxiliary dressings for the rest of the die.

The bulk of the non-pigmented dressings are extended with solvents such as paraffin, etc., a very much smaller proportion are in emulsion form for dilution with water. Spray gun is the most common method of application.

COATINGS FOR THE PRESSURE DIE CASTING OF ZINC ALLOYS

The lubrication problems associated with the pressure die casting of zinc base alloys are not so great as those for aluminium alloys. The casting takes place, in most cases, in a hot chamber machine where the holding pot for the molten metal is an integral part. The metal temperature ranges between 400° and 450°C. This is closely controlled, and the die temperature is maintained by a constant casting cycle (often automatic). Few major problems are encountered and those that occur are usually concerned with the surface finish.

Most zinc castings are either plated or painted before they go into service and defects which affect these finishes are numerous. To achieve the best surface finish on the casting a "bloom" should be developed on the die impression. This "bloom" takes the form of a very thin oxide film evenly distributed over the surface of the die and is slowly developed by the constant injection of the molten metal as the casting operation proceeds. Until this "bloom" is produced the casting surfaces suffer from minute lap and flow mark defects, but afterwards they are smooth with a mat appearance. Very little information is available concerning this phenomenon, but an opinion is held that the rougher oxidized surface of the die prevents skidding of the molten metal and also imparts a degree of insulation which allows the metal to recover from turbulence before solidifying.

For the production of zinc die castings a highly lubricating dressing is normally unnecessary and the usual procedure is to apply to the moving half of the die, sufficient coating to lubricate the core pins and ejector pins. More than a trace of dressing on the fixed half will almost inevitably result in a reappearance of flow marks and laps on the surface of the casting despite the "bloom".

Briefly the requirements for a pressure die dressing for zinc are that it should impart sufficient lubrication for core and ejector pins and not in any way adversely affect the surface finish of the casting.

Meticulous care is necessary to produce a satisfactory surface on the castings, particularly those which are required to be plated. Oily or greasy dressings, unless applied very sparingly, accentuate the directional flow lines of the metal, the excess oil or grease being displaced and trapped in the folds of the metal stream. The intensified flow lines or folds will usually be reproduced by the plating process to the detriment of the finished castings. Dressings in which pigments such as graphite, mica, etc., are incorporated can sometimes create problems of a different kind, for unless the pigments are securely bonded to the die particles may be detached and flow with the molten metal stream. Being of a lower density than the zinc alloy they are usually transported on the surface of the metal and deposited on or near the metal/die interface and become impregnated into the surface of the castings. If these inclusions are not removed by prefinishing treatments, they can have detrimental effects on the adhesion properties of paints and create minute barriers which weaken the bond between the plating and the casting; also they may become an eventual source of corrosion. Even if the inclusions are removed the resultant pits or cavities left behind may be reproduced by the plating process and impair the appearance of the finished casting. Any lubricating media, therefore, utilized for the purpose of aiding the production of zinc alloy pressure die castings should be of a type which is easily removed or neutralized, by the usual pretreatment cleaning processes and if solid particles or pigments are incorporated in the formulation, they should be preferentially bonded to the die if impregnation of the casting surface is to be avoided.

Materials such as graphite and silicones, etc., particularly silicones, with lubricity and release properties which are practically unimpaired after subjection to the comparatively low temperatures involved in the pressure die casting of zinc alloys, should be used with caution. Their release properties may persist, even after undergoing most cleansing operations and function as parting agents between paint or plating and the casting. The basic materials used in the formulation of dressings for the pressure die casting of zinc base alloys are numerous indeed, and the allowable permutations for blending these to obtain special properties are even greater. Most types of grease, oil, wax, animal fat, etc., at some time or other have been used either in concentrated form or diluted with water or solvent, with varying degrees of success. There is little doubt, however, that water-base dressings have many advantages over the solvent types. Most water-based dressings need to be emulsified, however, and certain types of mineral oils and waxes can be difficult in this respect and produce emulsions which are unstable and break down too easily.

A high proportion of present-day emulsified dressings contain such materials as lanolin, natural or synthetic waxes, oils, petroleum jelly, etc., waxes probably being the most widely used either on their own or in conjunction with one or other of the above materials.

The usually accepted method of applying the dressing is to spray or brush sparingly, at regular intervals, that half of the die which houses the ejector pins. The duration of the interval between sprays depends on the complexity of the casting being produced. Theoretically, however, after every shot, an application of the dressing is the only way to maintain the static die temperature conditions so necessary for consistent quality casting reproduction.

COATINGS FOR THE PRESSURE DIE CASTING OF COPPER-BASE ALLOYS

The pressure die casting of copper-base alloys is usually carried out on "cold chamber" type machines. A feature that precludes the wider acceptance of pressure die casting for these higher melting point alloys is the comparatively short working life of the expensive steel dies. The very high injection temperature of a copper alloy can be 500–600°C greater than the die temperature at the moment of injection. Under these conditions the dies are subjected to severe thermal shock which results in the development of an extensive network of hairline cracks sometimes referred to as "heat checks" or "checking". The rapid progression of this condition towards the disintegration of the die surface is accelerated by the erosive action of the injected metal stream. Intensive research into alloy steel properties and alternative materials for dies is in progress at various institutions in many countries, but to date only partial success has been achieved and the problem still remains acute. If and when this problem is overcome, the economics should allow the pressure die casting of these higher melting point alloys to commercially expand out of all proportion to the present level of production.

Most of the popular dressings for the pressure die casting of copper-base alloys incorporate pigments, chiefly because as well as providing improvement in release properties, the deposition of a film of pigment on the die surface, functions as a buffer against thermal shock and erosion extending, to some degree, the working life of the die. The type of pigment is dictated by the severity of the operating conditions which usually means lubricating properties of a high order at elevated temperatures, good release properties and some protection for the die against thermal shock. Graphite is usually the automatic choice and may be incorporated with a small percentage of micaceous material into a viscous oil for lubrication of the injection ram mechanism, and also for a brush or swab application to the die impression. A similar type formulation can be diluted for spraying. Graphite/water formulations, usually in sprayable form, are used fairly extensively and in these circumstances a colloidal form of graphite should be chosen if suspension problems are to be avoided. Water-based dressings have a greater cooling effect on the dies and while this may be a desirable attribute to higher production rates, it does indirectly reduce the life of the dies by increasing the temperature differential between the cast metal and the dies, thus increasing the severity of thermal shock.

When casting copper-base alloys the dressing needs to be applied for each casting cycle and this frequent application at these elevated die temperatures can result in a relatively rapid build-up of pigment and carbon on the die surface. When the chief alloying element is zinc, as in the 60/40 brass alloys, oxide deposits add considerably to this build-up. At least one of the known companies involved in the pressure die casting of 60/40 brass finds it necessary to clean down the dies at the end of each shift in order to maintain the surface quality of the castings. The cleaning operation is carried out by means of vapour blasting or wire brushing.

COATINGS FOR THE PRESSURE DIE CASTING OF IRON AND STEEL

The pressure die casting of iron and steel is presently in its very early experimental stage, but the associated problems are fairly well known. Most of the experimental work has been carried out on the cold chamber type machines. With metal temperatures in excess of 1000°C the die life is extremely short. To reduce the incidence of thermal shock to the die it is believed that the die will have to be relatively hot, probably in the region of 500°C, and at this temperature oily or greasy dressings will almost certainly be inadequate. The important properties required of a dressing suitable for the pressure die casting of iron or steel will centre round lubricity and insulation; the most likely materials to find favour will be graphite or boron nitride. These may need to be used in conjunction with a finely divided refractory to enhance insulation properties, with water as a carrier and with or without a bonding agent.

REFERENCES

1. ROBERTS, G. A., and GROBE, A. H., *Metal Progress*, **69**, Feb., 58–61 (1956).
2. DAY, E. A., A Visit with "500" Die Cast Plants. American Charcoal Company, Detroit, 1959.
3. *Foundry*, **90** (8), 99 (1962). (Anon).
4. ELIJAH, L. M., *Metal Progress*, **82**, July, 111–112 (1962).
5. KAJOCH, W., *Foundry*, **93** (12), 118–122 (1965).
6. SCHAFER, R. J., Lubrication in die casting. Ph.D. Thesis, Case Inst. of Technology, 1963.

H. FEEDING OF CASTINGS

o*

CHAPTER 13

SOLIDIFICATION OF CASTINGS— GENERAL PRINCIPLES OF FEEDING

R. W. RUDDLE

INTRODUCTION

The purpose of this chapter is to furnish the reader with a general background of the principles of the feeding techniques used in the manufacture of castings. This chapter will not be confined to any one particular alloy group, but will deal with effects which are common to all alloys. Details of special feeding practices relating to specific metal groups will not be considered here, but information regarding them will be found in the appropriate chapters. Furthermore, this chapter deals with the mechanism of freezing in metals or alloys only in so far as this affects feeding practice; other aspects of solidification including topics such as major and minor segregation, the effect and mechanism of grain refinement, dendrite arm spacing and other related phenomena are referred to in chapters relating to particular metal groups.

A necessary prerequisite to an understanding of feeding techniques is knowledge of the way in which metals and alloys freeze. Indeed, it is difficult to see how a foundryman can expect to feed his castings satisfactorily and economically without a thorough understanding of the way in which the metal concerned freezes.

Metals and alloys as a whole exhibit a very wide range of freezing mechanisms, but from the point of view of feeding it is fortunately possible to group the commercially useful metals and alloys into two or perhaps three groups. In most practical cases with which the foundryman is faced, the mode of freezing depends on the freezing range of the alloy, although other factors such as the rate of freezing, the presence of nuclei capable of initiating solidification, etc., also affect the mode of freezing to a significant extent. The picture presented here is, therefore, necessarily rather simplified and for further details of the way in which some of these variables operate, the literature cited in the bibliography should be consulted.[1,2]

It should be emphasized that the mode of freezing of metals and alloys cuts across alloy group lines, and it is therefore desirable to avoid thinking that all copper alloys freeze in this way, that all steels freeze in another way and that all light alloys have yet different modes of freezing.

METALS AND ALLOYS OF SHORT FREEZING RANGE

A. Mode of Freezing

This group consists of the pure metals and alloys which are 100% eutectic, such as aluminium-silicon alloys and many aluminium bronzes. However, many commercial alloys with low alloy additions also behave in a similar manner from the point of view of freezing and are therefore grouped with the pure metals. In addition, alloys of near-eutectic composition and alloys of near-peritectic composition as some of the brasses are also similar in mode of freezing. A few solid solution alloys of narrow freezing range also approximate this mode of freezing. A number of commercially important alloys which fall in this group are listed in Table 1.

TABLE 1. *Some Alloys Freezing by Skin Formation*

Pure metals	Aluminium-silicon	Silicon brass
Low carbon steels	Near-eutectic alloys	Cupro-nickel
Commercial copper	Yellow brasses (some)	
Commercial aluminum	Commercial zinc	
Aluminium bronze	Commercial tin	

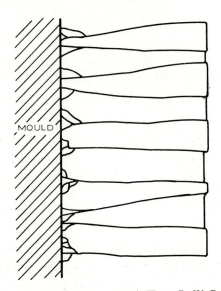

FIG. 1. Wall growth in freezing of a pure metal. (From R. W. Ruddle, *Solidification of Castings,* fig. 9, the Institute of Metals, 1957.)

Freezing begins in these alloys as soon as the temperature of the liquid at the mould walls reaches the freezing point. It is initiated by the formation on the mould walls of a number of tiny crystallites. The more favourably oriented of these grow rapidly into the melt both sideways and forwards, quickly linking up with their neighbours to form a continuous front known as the "solidification front". Thereafter freezing proceeds by the advance of this front into the interior of the casting as shown in Fig. 1.

Separate stages in the freezing of a casting by this mode of freezing are shown in Fig. 2 for a pure metal. It is a particular characteristic of pure metals that the solidification front is almost perfectly smooth as depicted in Fig. 2. The solidification of low alloys such as low carbon steel, commercial copper, etc.,

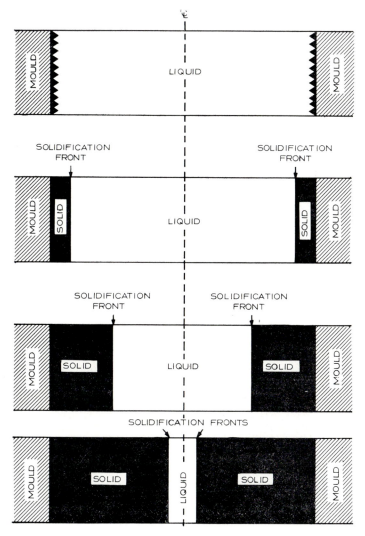

FIG. 2. Stages in the freezing of a pure metal. (From R. W. Ruddle, *Gating and Risering of Copper Alloy Castings,* fig. 18, Fenton Publishing Co. (Foundry), Cleveland, Ohio, 1960.)

is closely similar but in these instances the front is not perfectly plain but is serrated. Stages in the freezing of a casting in this type of alloy are illustrated in Fig. 3. It should be noted that where this group of alloys is concerned, there is little difference in the mode of freezing of sand castings and those cast in chill moulds.

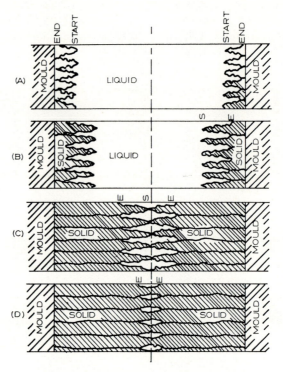

FIG. 3. Stages in the freezing of a low alloy. (From R. W. Ruddle, *Solidification of Castings,* fig. 20, the Institute of Metals, 1957.)

The rate at which any casting freezes is, of course, controlled largely by the rate at which the mould can absorb the heat liberated when solidification takes place. Since the heat absorption by the mould varies as the square root of time,† it follows that the growth of the solid wall in castings which freeze by "skin formation" is also a parabolic function of time.

$$D = q\sqrt{t}$$

In the above equation D is the thickness solidified in inches, t is the time from the start of freezing in minutes and q is the solidification constant in (in./min$^{\frac{1}{2}}$). Table 2 lists values of the solidification constant for castings made in typical sand and chill moulds for a number of different essentially pure metals. It will be noted that in Table 2, the solidification constant for steel is approximately

† See Appendix.

$1 \cdot 0$ in./min$^{\frac{1}{2}}$. It is this fact which accounts for the rule of thumb used in steel works, that the freezing time in minutes of an ingot is approximately equal to the square of half the ingot thickness; this rule yields a solidification rate which is probably within 10% of the true solidification time in the case of slabbing ingots whose width is considerably greater than its thickness, but is likely to be more in error in the case of ingots of more nearly square section owing to the influence of corner and edge effects.†

TABLE 2. *Typical Values of the Solidification Constant of Various Metals*

Metal	Solidification constant (in./min$^{\frac{1}{2}}$)	
	Sand cast	Chill cast†
Pure iron, steel	0·27	1·0
Copper	0·27	(1·1)
Aluminium	0·26	(1·15)
Magnesium	0·50	(2·2)
Zinc	0·23	(1·2)

Figures in brackets are calculated values.
† Steel mould.

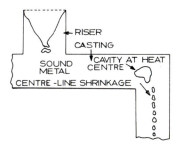

FIG. 4. Forms of shrinkage in alloys of short freezing range. (From R. W. Ruddle, *Gating and Risering of Copper Alloy Castings,* fig. 23, Penton Publishing Co. (Foundry), Cleveland, Ohio, 1960.)

In alloys freezing by skin formation, the outer regions of the casting almost invariably are completely solid. Shrinkage when present occurs as internal cavities which may be of considerable size at the heat centres. In plates and bars especially, centre-line shrinkage tends to occur. Risers tend to pipe deeply; in a pure metal, the pipe is smooth and geometric in shape; in a less pure material, the pipe is less regular in shape and the end may be ragged. These forms of shrinkage are illustrated in Fig. 4.

† See Appendix and p. 423.

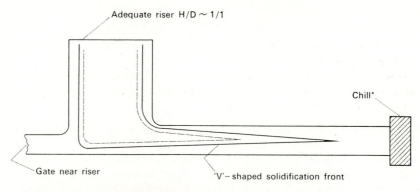

FIG. 5. Directional solidification.

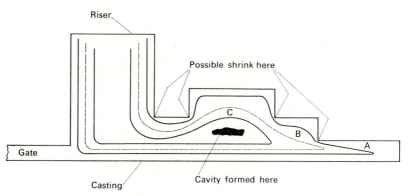

FIG. 6. Effect of local heat centre.

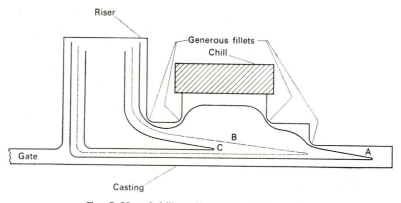

FIG. 7. Use of chill to eliminate local heat centre.

B. Principles of Feeding Practice

The general principles for the avoidance of shrinkage in alloys freezing in this manner may be rather simply stated. The procedure is to attach large reservoirs of liquid metal near the heat centres of the castings (i.e. the thicker sections) and thus further retard their solidification at the same time furnishing liquid feed metal. If necessary, efforts may be made by means of chills or other means to increase the rate of freezing of the more remote sections (see Fig. 5). The influence of *small* local heat centres may be eliminated by speeding up solidification at local heat centres by means of chills as depicted in Fig. 7. If the technique adopted is satisfactory the solidification front will be "V" shaped with the wide end of the "V" towards the riser and the apex of the "V" pointing towards the colder remote areas of the casting as shown in Figs. 5 and 7. Feed metal from the riser is thus easily able to reach the site of final solidification at the apex of the "V".

To enhance the taper of the solidification front towards the riser, it is desirable to keep the gates as close as possible to the risers to insure that the hottest metal enters the risers and the coldest finds its way into the remote regions of the casting where rapid freezing is desired (see Figs. 5 and 7).

One or two other features in freezing in castings in these alloys deserve special mention. In the first place, it should be realized that freezing at edges and corners of the casting takes place considerably faster than at flat surfaces. The reason for this is depicted in Fig. 8, where the casting is imagined to be composed

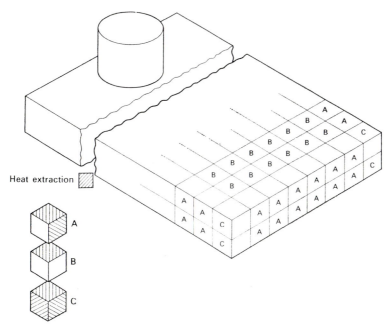

Heat extraction

FIG. 8. Freezing at corners and edges.

of a number of small cubes. It will be seen that the cubes marked A at the edges of the casting have heat removed from two surfaces, whereas those marked B in the interior of the casting lose heat from only one surface. The A cubes thus freeze more quickly. Similarly, the cubes marked C, which are located at the corners of the casting, have heat removed from them by three mould surfaces and thus freeze yet more quickly. These edge and corner effects are most valuable in inducing rapid solidification in the more remote areas of the casting and are thus a considerable aid to directional freezing. The wise rigger of castings makes full use of the corners and edges present and locates his risers away from them so that the corner and edge effects promote the desirable "V" shaped solidification front and thus assist in producing sound castings.

A particular problem arises when the corners and edges are situated some distance from the riser as shown in Fig. 9. In this case, their effect will not persist as far as the riser itself with the result that the solidification front loses its "V" shape at some point located between the edges and the riser.[3,4] This leads to the idea that risers have a limited freezing range, and if this is exceeded, centreline shrinkage will result. The feeding range of risers attached to steel and castings in some other alloys has been investigated at considerable length in recent years and formulae are available which give the freezing range in terms of the dimensions of the casting for several different casting shapes.[3–5]

Figure 10 shows an example of such a formula and chapters dealing with specific alloys may be consulted for more detail. In the event that the feeding range is found to be exceeded by a particular casting and riser configuration, it is generally necessary to either extend the feeding range by the judicious use of chills (Fig. 11) or to increase the number of risers so that feeding range requirements are not exceeded. In the event that multiple risers are found to be necessary their number can be minimized by locating at the mid-point between them, substantial chills which in essence create artificial edge effects as shown in Fig. 11.

The size and shape of the riser are of vital importance. The riser must complete its freezing well after the heat centre to which it is attached. To ensure that this situation obtains, it is necessary that the riser be of sufficient size. For economic reasons, the riser shape should be such that heat losses are minimized, so that the riser can be kept as small as possible, consistent with feeding the casting. From this point of view, a spherical riser is optimum; however, this shape cannot usually be moulded readily unless a Styrafoam pattern is used. For this reason, the next best shape, a cylinder, is generally employed; optimum performance from a cylindrical riser is obtained when the height to diameter ratio (H/D ratio) is about 0·5–1·5. Further economies can often be made when insulating or exothermic riser sleeves are used together with an exothermic top cover; these means usually permit the riser volume to be reduced by a factor of 3 or 4, and the casting yield to be increased proportionately.

Various methods have been proposed from time to time for the calculation of riser sleeves. The most popular of these and that recommended by the S.F.S.A. is the use of the "shape factor" proposed by Pellini et al.[6] According to these

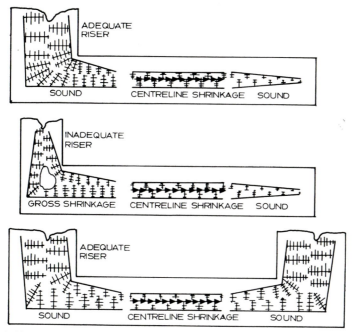

FIG. 9. Shrinkage in casting whose length exceeds feeding range of riser. (From R. W. Ruddle, *Solidification of Castings,* fig. 144.)

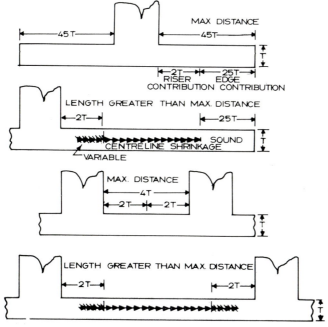

FIG. 10. Feeding range formulas for plate castings. (From R. W. Ruddle, *Solidification of Castings,* fig. 141.)

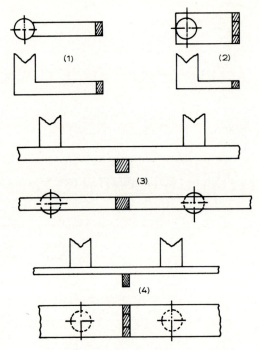

FIG. 11. Use of chills to extend feeding range. (From R. W. Ruddle, *Solidification of Castings,* fig. 157.)

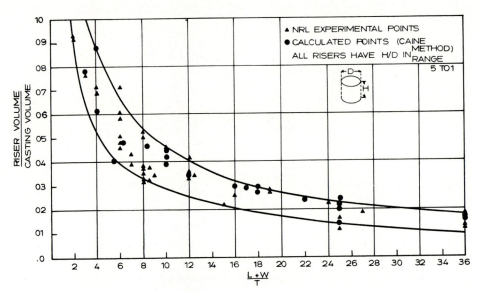

FIG. 12. Risering curve for steel castings. (From R. W. Ruddle, *Solidification of Castings,* fig. 150.)

researchers, the freezing time of a casting (or a part thereof) in an alloy of short freezing range is approximately dependent on the ratio:[2]

$$S = \frac{L + W}{T}$$

where L = length of casting, W = width of casting, T = thickness of casting.

Pellini[7] has presented an experimental curve relating (riser volume/casting volume) to the shape factor for steel castings (Fig. 12). If the riser volume places it above the upper line, the casting is invariably sound; if it plots below the lower line, the casting is always unsound. Riser volumes falling between the lines may or may not be sound. Similar curves are available for some other alloys.[5]

ALLOYS OF LONG FREEZING RANGE

A. Mode of Freezing

This alloy group includes a large number of commercially important materials among them nearly all the light alloys and a number of copper and ferrous alloys. Principal commercial alloys which fall into this group are listed in Table 3.

TABLE 3. *Some Alloys Solidifying in the Mushy Manner*

Aluminium alloys	Gun metals
Magnesium alloys	Leaded bronzes
Tin bronzes	Nickel silver
Red brass	Yellow brass (some)

When alloys of this group solidify under *commercial foundry conditions* freezing commences as before with the formation of crystallites at the mould wall as soon as the liquidus temperature is reached by the metal there. From then on, all similarity with the "skin-forming" alloys is lost. Almost as soon as the crystallites are formed, their growth becomes drastically inhibited. When most alloys of long freezing range freeze, the first-formed crystals are much lower in alloying elements than the liquid metal from which they are formed. Hence many of the atoms of alloying elements present in the liquid from which crystallites are formed are expelled into the liquid surrounding the crystallites. This surrounding liquid thus becomes enriched in alloying elements and this enrichment considerably depresses the freezing point of the liquid (see Fig. 13). If freezing were very slow, so that ample time were available for diffusion of the excess atoms from this enriched layer of liquid into the remainder of the liquid metal, the crystallites would continue to grow and a columnar structure would result. However, in most practical situations there is not nearly sufficient time for these concentrations in the liquid to be dissipated by diffusion so that the liquid metal surrounding the crystallites is unable to freeze until a much later stage during freezing of the casting as a whole.

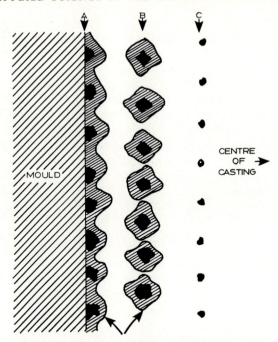

FIG. 13. Retardation of crystal growth in alloys of long freezing range. (From R. W. Ruddle, *Gating and Risering of Copper Base Castings,* fig. 20.)

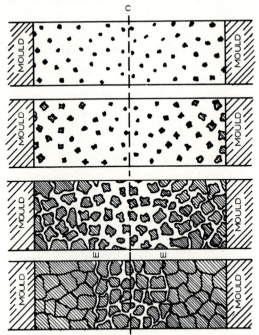

FIG. 14. Stages in the "mushy" solidification of an alloy of long freezing range. (From R. W. Ruddle, *Solidification of Castings,* fig. 21.)

Growth of the crystallites is accordingly temporarily halted. A slight decrease in the temperature of the casting as a whole then takes place and a new crop of crystallites is initiated behind the layer of enriched liquid as shown in Fig. 13. In turn, crystallites of this second crop almost immediately have their growth rate severely limited by the formation of an enriched layer of liquid around them.

A further slight decrease in the temperature of the casting then takes place and a third crop of crystallites is nucleated. The process of nucleation and growth inhibition takes place time and again until small crystallites have been nucleated throughout the entire volume of the casting. Thereafter, solidification of the casting proceeds by the gradual enlargement of all the crystallites, this process taking place simultaneously throughout the entire casting. For obvious reasons, the crystallites closest to the mould wall grow slightly faster than those near the centre of the casting, but the differences are not particularly great.

This sequence of events is illustrated in Fig. 14. The process may be likened to the setting of cement, the material at first being fluid, then mushy and finally rigid. It has been estimated that in many alloys, rigidity is not established until the casting is about 60–70% solid.

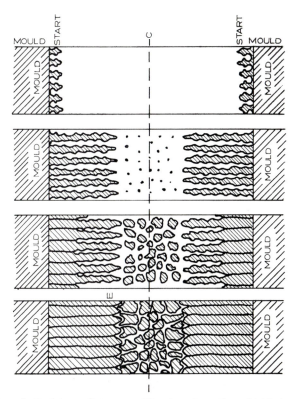

FIG. 15. Stages in the intermediate type of freezing of an alloy of fairly long freezing range. (From R. W. Ruddle, *Solidification of Castings*, fig. 22.)

There is no sharp transition from the "skin-forming" mode of freezing to the "mushy" type, as the alloy content of a metal is increased from 0 to a value which gives rise to the mushy type of freezing described above. Therefore, in practice, intermediate types of freezing are observed in many composition ranges. In an alloy of this type the first crystallites formed have their growth somewhat restricted, but not altogether stopped. Growth takes place in a columnar fashion, long thin dendrites attached to the mould walls being formed

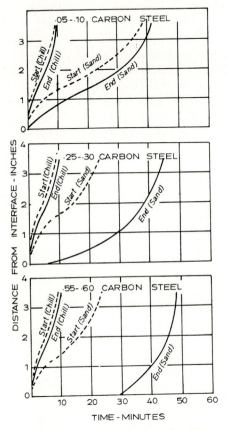

FIG. 16. Effect of carbon content on the freezing of steel castings (From R. W. Ruddle, *Solidification of Castings*, fig. 12.)

as illustrated in Fig. 15. The space between the dendrites is filled with enriched liquid which freezes very late in the freezing process because of its low melting point. Sometimes the growth of the dendrites peters out altogether at a certain stage as illustrated in Fig. 15. In this event, mushy freezing takes place in the central region of the casting and gives rise to an equi-axial structure there. (Both ordinary skin formation and the columnar-dendritic type of freezing give rise to columnar structures.)

The intermediate type of freezing illustrated in Fig. 15 shares an important feature with the mushy type of freezing of Fig. 13 in that freezing proceeds at the same time throughout the entire casting, some of the metal at the mould wall remaining liquid until a *very late* stage of freezing. Thus, no fully solid shell is produced early in the freezing of the casting in contra-distinction from the true skin-forming type of freezing.

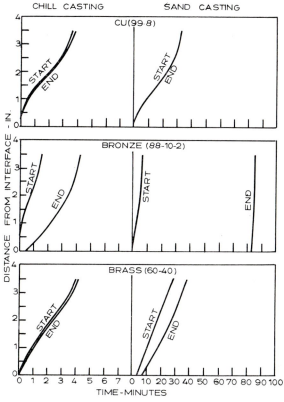

FIG. 17. Freezing of several non-ferrous alloys. (From R. W. Ruddle, *Solidification of Castings*, fig. 18.)

Alloys exhibiting this intermediate type of freezing are quite sensitive to the rate of freezing. Thus, an alloy which shows an intermediate type of freezing in a sand casting may behave almost like a skin-forming alloy when chill cast. High carbon steel (Fig. 16) and some of the yellow brasses (Fig. 17) are a good example of this kind of behaviour. In these figures the time at which freezing begins at a particular location is shown by the "start-of-freeze" curve; the moment at which freezing ends is indicated by the "end-of-freeze" curve. For a pure metal these two curves coincide; for an alloy of long freezing range they are widely separated. Thus the separation of these curves is a measure of the tendency towards mushy freezing.

B. Principles of Feeding Practice

Alloys whose mode of freezing is predominantly of the mushy type exhibit a totally different type of shrinkage from those which freeze by skin-formation. Stages in the freezing of these alloys are shown in Fig. 18. In the initial stages of freezing, the metal both liquid and solid, is quite fluid so that the freezing shrinkage is compensated simply by a fall in the level of the liquid in the risers. This process is known as "mass feeding". As indicated above, the structure begins to become coherent when solidification is about 65% complete. After this happens, freezing shrinkage can no longer be compensated by this movement of liquid and solid. From this stage on, every crystal in the now rigid structure is surrounded by pools or films of liquid and each crystal is competing simultaneously for this liquid. Clearly because of the solidification shrinkage,

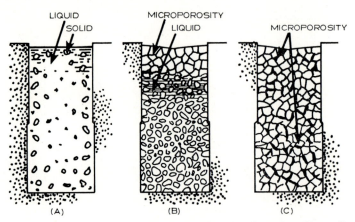

FIG. 18. Development of shrinkage in alloys of long freezing range. (From *Copper-base Alloys Foundry Practice,* 3rd ed., fig. 4.9, American Foundrymen's Society, Chicago, Ill., 1965.)

there is not enough liquid to go round; equally clearly the chances of liquid percolating from the riser to feed crystals in the remote areas of the casting is poor, especially when one bears in mind that the riser itself is in much the same condition as the casting. As the result, small pores are formed in the inter-dendritic areas in the final stages of freezing, and this gives rise to the dispersed microporosity which is so characteristic of alloys of long freezing range.

To ensure soundness in casting in these alloys, solidification must not only be directional towards the risers, it must also be steeply directional. In other words, when the remote areas of the casting are almost solid, solidification in and near the risers must be at an early stage. The establishment of the thermal conditions required to produce this condition presents no particular problem in many castings of thin section (up to about $\frac{1}{2}$ in. in sand cast plates or thicker in chill cast plates) because rapid cooling at the edges and corners produces the desired effect automatically.† However, considerable problems are experienced with castings of greater thickness (say 1 in. and over sand-cast sections), and generally, the only way to produce high soundness in thicker sections in these alloys is to severely chill the greater part of the surface of the casting. The effect of such chilling is shown in Fig. 19. A graduated chilling effect may be produced with the aid of taper chills as in Fig. 20.

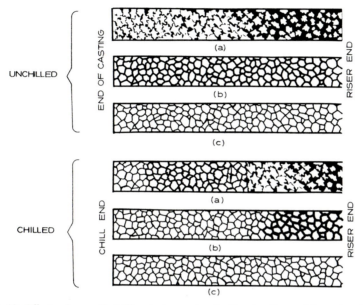

FIG. 19. Effects of overall chilling in improving feeding in alloys of long freezing range. (From *Copper-base Alloys Foundry Practice*, 3rd ed., fig. 4.12.)

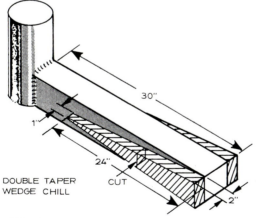

FIG. 20. Taper chills. (From R. W. Ruddle, *Gating and Risering of Copper Base Castings*, fig. 30.)

† This statement does not apply if the length of the casting is excessive.

Typically, the shrinkage in alloys of this kind is general and well dispersed as depicted in Fig. 21. Somewhat coarser shrinkage may be found at the heat centres and the risers, but the particular characteristic of these alloys is that shrinkage cavities are present in almost every region of a poorly fed casting. In certain alloys such as many magnesium alloys, the aluminium–10%–magnesium alloy, and fine-grained tin bronzes and red brasses, the shrinkage cavities tend to form layers which are particularly damaging to mechanical properties. This kind of porosity is sometimes known as "layer porosity".

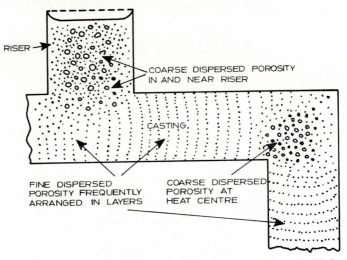

FIG. 21. Forms of shrinkage in alloys of long freezing range. (From R. W. Ruddle, *Gating and Risering of Copper Base Castings,* fig. 24.)

An alternative way of producing the temperature gradients and the steep solidification gradients necessary to produce high soundness in these alloys is to greatly retard freezing of the risers by means of exothermic or insulating sleeves.

The shape and size of the risers are determined by much the same considerations as apply with alloys of short freezing range. The feeding ranges, however, tend to be considerably different from those obtaining with steel and other alloys of short freezing range. Unfortunately, little or no good data are available to guide the foundryman.

The chilling procedures mentioned above are, of course, expensive and generally are economic only for premium-quality castings, such as those used in aerospace work. For general run-of-the-mill castings, different procedures are adopted. In this case a decision is taken to put up with a certain amount of shrinkage and to disperse this as widely as possible and to eliminate local concentrations. This may be done by adopting procedures which are diametrically opposite to those used to promote directional solidification. The risers are located near the thinner sections of the casting and the thicker sections may even be lightly chilled. By these means, solidification is evened out throughout the entire casting and local concentrations of shrinkage are avoided. In this procedure, the size and shape of the risers are also important. The purpose of the riser, when "uniform" solidification is adopted, is simply to pressurize the mushy mass during the stage of mass feeding and thus to prevent surface sinks and draws from forming. The riser does not contribute at all to feeding in the final stages of freezing. Figure 22 shows that in heavy (unchilled) sections in

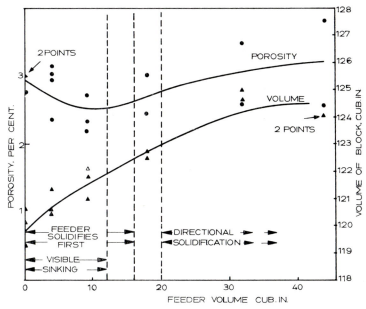

FIG. 22. Effect of riser size on porosity and volume of red brass casting. (From R. W. Ruddle, *Solidification of Castings,* fig. 135.)

these alloys the porosity in the castings is almost independent of riser size but that the casting volume, which indicates the extent of surface sinking is greatly affected by riser size at least up to a certain limiting size. To achieve its purpose, the riser can be quite thin, but is desirably tall (H/D ratio at least 3/1) to give satisfactory pressurization of the fluid mixture of liquid metal and crystals. The riser is thus quite different in shape from that which would be adopted if directional freezing were used, or indeed, the metal had a short freezing range

THE CAST IRONS

The freezing of cast irons needs special mention, first, because this group of alloys exhibits some unusual features during freezing, and secondly, because of the great commercial importance of this alloy group. To make the matter more complicated, the cast irons are themselves divisible into a number of groups, depending on whether the iron is hypo- or hypereutectic, and whether it freezes as a grey iron, as a white iron, or as nodular iron. Most commercially important irons are of hypoeutectic composition, although some of them approach the eutectic composition closely, and for this reason, and to simplify the discussion, only hypoeutectic alloys will be considered here.

The freezing of cast irons is a complicated matter, and in the space available, it is only possible to give a very brief outline of some of the salient features from the point of view of the avoidance of shrinkage. For further details, the reader is referred to the references cited.[8-10]

A. Freezing of White Cast Iron

The white cast irons freeze in a straightforward manner, similar to copper alloys which contain a fair amount of eutectic. The first part of freezing takes place in the "intermediate" manner described above, long dendrites of austenite being formed, and the liquid between them becoming enriched in carbon, silicon and other alloying elements. When, after most of the primary austenite crystals have formed, this liquid reaches the eutectic composition, and if thermal conditions permit, its freezing also begins. Thus, the freezing of the inter-dendritic liquid follows behind that of the austenite dendrites to an extent which depends upon the base composition of the iron. The white irons therefore tend to the formation of somewhat dispersed shrinkage behaving from this point of view somewhat like a high carbon steel or a copper alloy of intermediate freezing range. The feeding techniques are similar to those required for other alloys of moderate freezing range described above.

B. Grey Cast Iron

Freezing of grey cast irons in the initial stages is closely similar to that of white cast irons, beginning with the formation of dendrites of austenite. The formation of the austenite dendrites is accompanied by a decrease in volume, but since this is taking place in the earlier stages of freezing, the decrease in volume is largely counteracted by mass feeding. Towards the end of the formation of the austenite dendrites, solidification of the iron/graphite eutectic begins at nuclei near the mould walls in the spaces between the austenite dendrites.

Very shortly afterwards, eutectic crystallization starts at other nuclei in the interdendritic spaces in the interior of the casting. The eutectic cells thus formed grow contemporaneously throughout the entire casting, in a manner reminiscent of the freezing of an alloy of moderate or long freezing range, the process being most advanced near the mould walls and least in the heat centres (see Fig. 23). Freezing is completed when the eutectic cells touch and coalesce finally occupying all the interdendrite spaces. In low phosphorus irons the eutectic cells coalesce fairly soon near the mould wall; thus a fully solid shell is formed well before the end of freezing (Fig. 24). On the other hand, in high phosphorus irons the eutectic cells do not coalesce until almost the end of freezing of the casting as a whole, even at the mould walls, so that the state of partial freezing continues until very late in the solidification process.

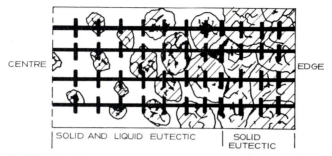

FIG. 23. Solidification of eutectic in grey cast iron. (From J. W. Boyes and A. G. Fuller, *Solidification of cast iron, B.C.I.R.A. Journal* (July 1964), fig. 7.)

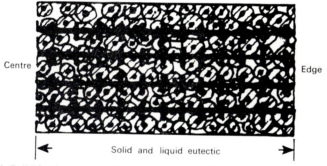

FIG. 24. Solidification of eutectic in nodular iron. (From Boyes and Fuller, *op. cit.,* fig. 10.)

Since the formation of the graphite flakes in the eutectic involves an expansion, grey irons have little or no tendency towards the formation of shrinkage cavities, the net volumetric change being about plus 1%; therefore, in theory, they should not require feeding. This last statement presupposes that the mould is rigid. In practice, this is often not so, at least, where the larger castings are concerned, and mould wall movement frequently gives rise to coarse porosity in the heat centres of grey iron castings, which is closely reminiscent of true shrinkage porosity. It is for this reason that the heat centres of the larger grey iron castings frequently do need some feeding, although the risers required are not generally as large as those needed to feed alloys of short freezing range.

C. Nodular Iron

The first stage of the freezing of nodular iron castings is very similar to that of freezing in grey or white irons, austenite dendrites being formed as before. Freezing in nodular irons, however, differs from that in grey irons in that the eutectic is entirely different in structure, although it forms in discrete cells as in grey iron (Fig. 24). These eutectic cells are degenerated, the core consisting of a graphite nodule and the outer part austenite. Growth proceeds by the deposition of austenite on the outside of the cell and the graphite core enlarges by diffusion of carbon through the austenite layer. The graphite nodules thus grow out of contact with the liquid, whereas in grey iron the graphite flakes are in contact with the liquid during their growth. The result of this is that the expansion involved in the formation of graphite is transmitted directly to the residual liquid in the grey irons, and thus compensates for the freezing shrinkage. In a nodular iron, on the other hand, this is not possible because the graphite nodules are surrounded by a solid phase which prevents transmission of expansion to the liquid. As a result, there is little or no tendency for the normal freezing shrinkage accompanying the formation of the other constituents of the alloy to be compensated by graphite expansion. Nodular irons, therefore, do suffer from genuine shrinkage defects (apart from porosity caused by mould wall movement) and, therefore, need adequate feeding if shrinkage cavities are to be avoided. From this point of view, nodular irons behave like a high carbon steel in that the shrinkage porosity generally tends to be concentrated at centre lines and heat centres, although some dispersion is evident. Substantial feeder heads are necessary if the shrinkage is to be avoided.

APPENDIX

CALCULATION OF SOLIDIFICATION RATES

Many elaborate mathematical treatments have from time to time been proposed for the problem of solidification in moulds. Unfortunately, all of these, despite their complexity, are only of approximate accuracy, or are only accurate under severely restricted conditions.

The results of a fairly simple treatment[1] are described, but it should be noted that these apply only to solidification from flat mould surfaces; in any practical case, corners and edges of a casting are involved and since solidification in these regions is faster than it is from a flat mould surface, computations based on the following treatment will usually be in error, predicting freezing times which are too long. The errors so introduced are greatest with small castings. However, this relatively crude treatment is often satisfactory for many purposes. If really accurate computations of solidification rates in castings are necessary, recourse must be made to the digital computer, which in recent years has shown the capability of handling the complex three-dimensional heat-flow problems involved in the freezing of castings.[11]

Heat Extraction by the Mould

The rate at which a casting freezes is obviously determined by the rate at which the mould can absorb the superheat and latent heat of fusion of the casting, and it is therefore pertinent to examine this factor. (It is here tacitly assumed that the loss of heat from the open surfaces of risers can be ignored.) It should also be again mentioned that the equations given below refer *only to heat flow through a flat mould wall*.

The total heat absorbed in a time t by unit area of a flat mould wall is given by:

$$Q = 1.128b(\theta_i - \theta_0)\sqrt{t} \qquad (1)$$

In the equation above, Q is the total heat absorbed in time t. θ_i is the temperature at the mould interface, assumed to be equal to the freezing temperature of the metal, θ_0 is the initial temperature of the mould and b is the heat diffusivity of the mould material which is itself equal to $\sqrt{(k\rho c_p)}$, where k is the thermal conductivity, ρ is the density and c_p is the specific heat. For simplicity, the above equation is sometimes written:

$$Q = \xi\sqrt{t} \qquad (2)$$

where the quantity ξ is called the mould constant, and is a measure of the chilling power of the mould at the temperature of the metal being cast.

TABLE 4. *Approximate Heat Diffusivities of some Moulding Materials*
0–1200° C

Material	Heat diffusivity (c.g.s. units)
Moulding sand	0·02–0·035
Plaster (0–700°C)	0·014
Magnesite brick	0·079
Silicon carbide (clay bonded)	0·040
Graphite (dense)	0·52
Iron	0·34
Copper	0·87

The heat diffusivity, b, of the mould material is the all-important material property from the point of view of heat extraction; some values of this quantity are given in Table 4. It should be noted that the other important quantity in equation (1) is the freezing temperature of the metal, the heat extraction by the mould being directly proportional to this temperature, referred to the mould's initial temperature. Figure 25 shows total heat and rate of heat removal curves obtained at three different metal freezing temperatures. The significance of the freezing temperature of the metal is clearly evident from this set of curves which were all obtained with the same sand.

P

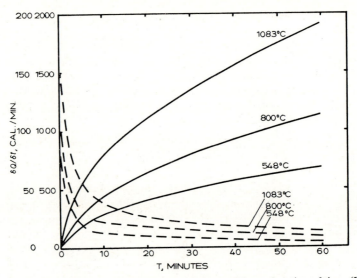

FIG. 25. Total heat and rate of heat absorption by mould as a function of time. (From R. W. Ruddle, *Solidification of Castings*, fig. 56.)

Freezing Rates in Castings

(a) Alloys freezing by "skin-formation"

It can be assumed that the metal solidifies on an advancing front as is the case when skin-formation occurs, then the thickness d of metal solidified in time t must be proportional to the heat removed in that time by the mould wall in contact with the metal. Equating the heat removed from the metal when it freezes with the heat absorbed by the mould time is obtained, the simple relationship mentioned earlier (p. 420).

$$D = q\sqrt{t} \qquad (3)$$

where the solidification constant, q is given by

$$q = \frac{1\cdot128b(\theta_i - \theta_0)}{L\rho'} \qquad (4)$$

In equation (4), L is the latent heat of fusion of the metal and ρ' is its density. Values of the solidification constant for a number of pure metals are given in Table 2, p. 421.

The above relationship assumes that the metal is cast without superheat. In practice this is never so, but allowance for the superheat can be made by including it with the latent heat of fusion. The above equation may be used to predict the thickness of the solid shell of metal at an intermediate stage in the freezing process of a "skin-forming" alloy with reasonable accuracy, say 10–15%.

Similar reasoning may also be used to predict the time of total freezing of the casting. The total heat content liberated during complete solidification is:

$$W[L + S(\theta_c - \theta_i)] \qquad (5)$$

where W = weight of casting, θ_c = casting temperature and S is the specific heat of the liquid metal. If the time of complete solidification is t_c then the above quantity of heat must be equal to that absorbed by the mould in this time. Assuming that the mould extracts heat uniformly over the entire surface of the casting† the heat absorbed by the mould may be obtained by multiplying equation (1) by the surface area A of the casting.

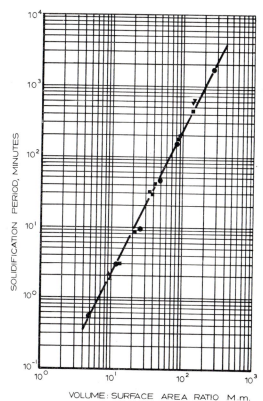

VOLUME : SURFACE AREA RATIO M.m.

FIG. 26. Logarithmic plot of freezing time of castings against $(V/A)^2$ ratio, illustrating validity of Chvorinov's rule. (From R. W. Ruddle, *Solidification of Castings*, fig. 93.)

The following equation is thus obtained for the solidification time:

$$t_c = \frac{W[L + S(\theta_c - \theta_f)]}{1 \cdot 128 Ab(\theta_i - \theta_0)} = \frac{\rho'V[L + S(\theta_c - \theta_f)]}{1 \cdot 128 Ab(\theta_i - \theta_0)} \qquad (6)$$

θ_f = final solidification temperature.
θ_c = casting temperature.

† i.e. corner and edge effects are ignored.

since $W = \rho' V$, where V is the volume of the casting. All the quantities in equation (6) are properties of either the metal or the mould material with the exception of the volume and surface area of the casting. It follows that for any casting in a given alloy, the freezing time is given by:

$$t_c = \text{constant} \times \left(\frac{V}{A}\right)^2 \tag{7}$$

This is the theoretical basis of the well-known Chvorinov's Rule.[12] Although this rule ignores corner and edge effects, it nevertheless applies quite well over a wide range of casting sizes as shown in Fig. 26.

(b) Alloys of long freezing range

Equation (2) which gives the rate of skin formation cannot, of course, be applied to alloys of long freezing range, since there is no actual solid skin formed. It may, however, easily be modified to yield the amount of solid material frozen as a function of time, irrespective of the location of the solid material.

Equations (6) and (7) giving the time of total freezing of the casting, apply equally well to an alloy of long freezing range, which solidifies in the "mushy" manner.

REFERENCES

1. RUDDLE, R. W., *Solidification of Castings*, Inst. of Metals (1957).
2. PELLINI, W. S., *Proc. Elect. Furnace Steel Conf.*, A.I.M.E., **14**, 48–91 (1956).
3. BISHOP, H. F. and PELLINI, W. S., *Trans. AFS* **58**, 185–197 (1950).
4. BISHOP, H. F. and MYSKOWSKI, E. T. and PELLINI, W. S., *Trans. AFS* **59**, 171–180 (1951).
5. FLINN, RICHARD A. and TROJAN, PAUL K., Mold Design and Controlled Solidification of High Integrity Copper Alloy Castings. *AFS Brass & Bronze Committee* (1966).
6. BISHOP, H. F., MYSKOWSKI, E. T. and PELLINI, W. S., *Trans. AFS* **63**, 271 (1955).
7. PELLINI, W. S., *Trans. AFS* **61**, 61 (1953).
8. HUGHES, I. C. H., *J. Austral. Inst. Met.*, **10**, Feb., 4–16 (1965).
9. HUGHES, I. C. H., *The Solidification of Metals*. Iron & Steel Inst., 184–192 (1968).
10. MORROGH, H., *The Solidification of Metals*. Iron & Steel Inst., 238–241 (1968).
11. HENZEL, J. G. and KEVERIAN, J., *Cast Metals Research Journal*, **1**, June, 19–36 (1965).
12. CHVORINOV, N., Giesserei, **27**, 177–180, 201–208, 222–225 (1940).

CHAPTER 14

THE FEEDING OF STEEL CASTINGS

D. R. ALLEN and M. W. BARLOW

It would not be possible in a book such as this to give every aspect of heading practice because of the complexity of casting shapes encountered. General principles can be presented, however, and many investigators favour the "shape factor" as the best and most convenient method of calculating riser dimensions which will achieve a sound casting. It is the intention to give only an approach for feeding castings with exothermic aids, using the shape factor method.

THE DEVELOPMENT OF EXOTHERMIC FEEDING

The modern exothermic material used as a feeding aid derives its heat from the oxidation of aluminium; other easily oxidizable metals and alloys have been tried, such as magnesium, calcium silicide, ferro-silicon and calcium carbide, but aluminium is the best material with regard to heat output, price and ease of handling. The use of aluminium as a fuel for exothermic mixtures goes back to the end of the last century when Hans Goldschmidt developed his process for the production of steel and other metals by igniting mixtures of metal oxide and aluminium, a process which is generally known as the "Thermit process". Chemically it can be summarized as follows:

$$2\,Al + Fe_2O_3 = Al_2O_3 + 2\,Fe + 204\ \text{kcal.}$$

The reaction proceeds at very great speed and is finished within a few seconds and the reaction products consist of liquid iron and slag; the temperature during the reaction can go up to 2500°C, although only 204 kcal are developed. Actually the burning aluminium and the oxidation to Al_2O_3 develop 400 kcal; from this has to be deducted the energy required to reduce Fe_2O_3 to Fe which is 196 kcal which leaves 204 kcal for the total reaction.

The iron readily separates from the molten slag especially if slag forming compounds such as calcium fluoride are added and consequently this "Thermit mixture" is and has been extensively used for welding rails, for welding jobs generally and for all those jobs where liquid steel is required and where no melting equipment is available. Such a material is also used in steelworks and foundries as a late addition to feeder heads as will be described later.

It is reasonable to suppose that if this type of mixture, in which aluminium and iron oxide are present in stoichiometric proportions were mixed with a bonding agent, it could be made into a sleeve and used to surround a feeder head. Such a sleeve, however, would burn very violently on ignition, the

443

reaction would be finished in a short time and the reaction products—molten iron and aluminium oxide slag—would intermingle with the steel and contaminate it. In order to control this reaction an obvious step is to dilute the mixture with a refractory; this lowers the rate of burning and the final temperature achieved so that the residue remains solid and keeps its shape. Such compositions have a few serious disadvantages:

1. They contain a great deal of metallic iron which lowers the insulating value of the burnt-out shape.

2. The temperature necessary to ignite a Thermit mixture is very high; ignition is generally initiated by means of special ignition mixtures. These consist of barium peroxide (BaO_2) and aluminium powder to mention only one of many possible combinations. They can be ignited with a burning magnesium ribbon or a red hot iron poker and they burn off creating high enough temperatures that the Thermit mixture will be ignited. Liquid steel alone sometimes fails to ignite the mixture, especially if it is diluted with refractory material.

There remained the problem to develop a heat-producing mixture which would burn with a higher temperature than liquid steel but low enough that it would not fuse, the burning of which could easily be controlled with regard to the start of the reaction, and the rate of burning and which would be cheap enough to be used economically even on cheap grades of steel.

Such a product was developed in the Foseco laboratories in 1946/7 and as a result a patent was granted to the organization in 1949.[1] It was found that the addition of small amounts of fluorides in the form of alkali fluorides, or cryolite or aluminium fluoride, influence the properties of exothermic mixtures decisively in several ways. The reaction will start at a much lower temperature than ordinary Thermit mixtures; it proceeds evenly and quickly from the spot where it started until the whole of the aluminium is consumed and by slight variations in the amount and the type of oxidizing compounds in the mixture important variations in burning rate can be obtained. These exothermic materials are no longer stoichiometric mixtures of aluminium and iron oxide but consist basically of aluminium powder, refractory fillers, small amounts of fluoride and oxidizing compounds whereby the amount of oxidizing materials is far below the stoichiometric percentage so that there is not enough oxygen present in the mixture itself to burn the aluminium completely. This is the fundamental difference from "Thermit mixtures". It was found that once the reaction is initiated the aluminium will burn using oxygen of the air as the oxidizing medium, whereby the fluoride acts similar to a catalyst. For this reason the total heat output of the new exothermic mixtures is greater than of a similar weight of "Thermit" (providing the aluminium content is the same) because due to a much lower oxide content the energy required to reduce the oxide is much less and the heat of formation of the aluminium oxide can be nearly fully utilized. The addition of a binder will make the exothermic material into a mouldable product which can be moulded into any desired shape similar to a core sand.

The burnt-out shape contains only very few globules of iron (depending on the amount of iron oxide present in the original mixture) and has very good insulation characteristics.

MECHANISM OF ALUMINO-THERMIC REACTIONS

The chemistry of the reactions which take place when these exothermic mixtures are burning and the role which small amounts of fluorides play in the propagating of the reaction has been thoroughly investigated in the Foseco International laboratories, but the results have so far not been published and are available only in internal research and development reports.[2, 3]

According to these reports the reaction takes place in two stages, although in a burning mass of exothermic material, no doubt, both stages are proceeding simultaneously. The first stage predominates at the lower temperature and serves to start the reaction when external heat is applied. Oxidizing compounds such as nitrates (alkali and alkaline earth) or chlorates which are present in the exothermic mixture react with the aluminium and are reduced, whilst the aluminium is oxidized to aluminium oxide. The reaction probably takes place as follows:

$$(1)\ 2\,NaNO_3 \xrightarrow{heat} 2\,NaNO_2 + O_2$$

$$(2)\ 2\,NaNO_2 \xrightarrow{heat} Na_2O + NO_2 + NO$$

$$(3)\ Na_2O + 2\,Al \rightarrow Al_2O_3 + 6\,Na$$

The sodium immediately burns on contact with air and forms Na_2O and sufficient heat that the temperature will rise above $700°C$ when the second reaction is initiated. This consists of a reaction between the fluoride and aluminium:

$$3\,NaF + Al \rightarrow AlF_3 + 3\,Na$$

$$2\,Al + AlF_3 \rightarrow 3\,AlF$$

$$6\,AlF + 3O_2 \rightarrow 2\,Al_2O_3 + 2\,AlF_3$$

In the first place aluminium subfluoride AlF is formed; this is a very reactive, unstable gas which reacts with the oxygen of the air and forms aluminium oxide and aluminium fluoride; the latter is deposited in the pores of the burning shape and is available for further reaction; this continues until all the available aluminium is oxidized. The whole process of the oxidation of aluminium thus occurs in the vapour state and is cyclic in nature and this could explain the relatively small quantities of fluoride needed to ensure the reaction of all the available aluminium.

The temperatures created during the reaction are so high that no doubt a number of other reactions take place including the formation of a considerable amount of nitrides by the reaction of aluminium subfluoride AlF with the nitrogen of the air:

$$3\,AlF + N_2 \rightarrow 2\,AlN + AlF_3$$

The end products of the reaction are all solid and the total amount of gas generated is comparatively small; the fact that not only oxygen but nitrogen from the air takes part, means that there is little tendency for the reaction to choke itself.

After the shape is burnt out, it consists of aluminium oxide, aluminium nitride, refractory material and small amounts of aluminium fluoride; the aluminium oxide can, of course, react with the refractory material when the shape is still very hot and form aluminium silicates of various compositions.

FIG. 1. View of "Amitec".

EXOTHERMIC PERFORMANCE ASSESSMENT

Although many tests are carried out by the manufacturers of exothermics the results tell the user very little about the comparative effectiveness of the different types of products available when used under practical conditions. Consequently several foundries have set out their own empirical procedures for ensuring quality. The best of such tests is undoubtedly one in which metal is poured into a standard sleeve of exothermic compound. For this test the sleeves should be twice as tall as their internal diameter to compensate for end effect heat losses and the rate of cooling measured by thermocouples at the geometric centre of the sleeve.

Briggs, Lock and Ashbrook[4] and Wlodawer[5] have adopted this method of assessment and have elaborated the technique by using hollow spheres of exothermic material with a thermocouple at the centre. Wlodawer compares the time of solidification of a standard sized sphere (usually 150 mm diameter lined with 35 mm of exothermic material) with that of a similar sized sphere in sand. The efficiency of the exothermic material is then expressed:

$$\frac{\text{time of solidification in exothermic material}}{\text{time of solidification in sand}}$$

A good exothermic should give a ratio of 2 or above.

In practice the performance of an exothermic material is determined by the heat flow properties, rather than by its calorific value.

Equipment has, therefore, been developed by the Foseco Group in order to determine the calorific value and also the heat flow through exothermic and insulating materials. This apparatus ("Amitec") is shown in Figs. 1 and 2.

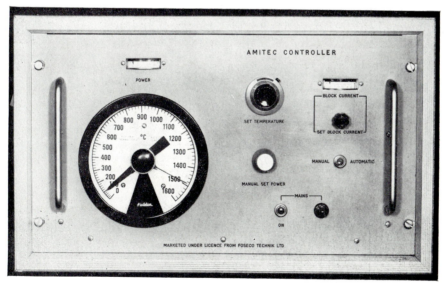

FIG. 2. Control instrument for "Amitec".

P*

"AMITEC" TESTING

The apparatus consists essentially of a thermostatically controlled hot-plate which is capable of operating very precisely at any selected temperature up to 1500°C. A monitoring device records the power input required to maintain the plate at the chosen temperature under all conditions. The cumulative power input is recorded continuously and a print-out counter operates each minute. At steady-state (i.e. in the absence of any test material) a constant rate of input is recorded due to the background loss of heat from the hot-plate. With reference to Fig. 3, consider the sequence of events following the application of a known weight of exothermic material to the plate. The steady-state loss referred to above is shown as the zero value in Fig. 3.

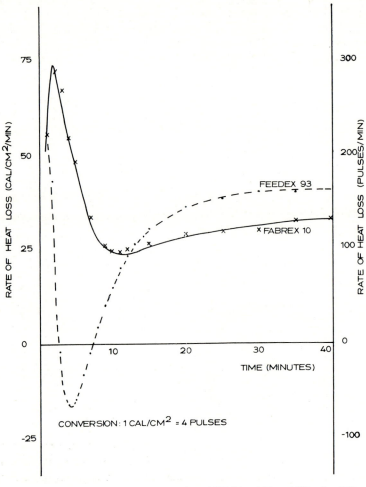

Fig. 3. Comparative thermal properties of "Fabrex 10" and "Feedex 93".

Initially the sample is colder than the plate and extracts heat from the plate. The instrument responds by increasing its power input in order to overcome this "chill". The extent of the response provides a measure of the chilling properties of the test sample. In Fig. 3 the maximum chill is indicated by the first point of inflection.

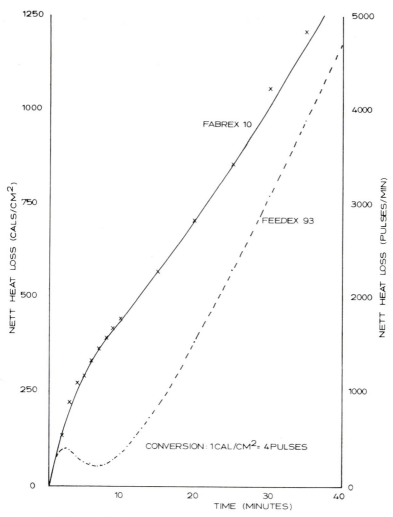

FIG. 4. Comparative thermal properties of "Fabrex 10" and "Feedex 93".

At some time after the chill has been overcome, the exothermic material is ignited by the heat of the plate. Heat is now offered to the plate by the burning exothermic and the input rate of the instrument decreases, and may fall below that value which corresponds to steady-state. The second point of inflection represents the maximum exothermic output of the sample. The distance between the first and second points of inflection on the thermal axis describes the total heat output, whilst that distance on the time axis describes the rate of ignition or sensitivity of the material.

When ignition is complete, the burnt-out residue of the test specimen acts as an insulator, retarding the heat loss from the hot-plate. An equilibrium is achieved where input by the instrument balances heat loss through the sample. This state is represented by the straight line portion of the "rate curve". From the gradient of this portion of the curve, the "K" value of the residue may be deduced.

The interpretation of curves derived from insulators is somewhat simpler. Only one inflection occurs; that between "chill" and equilibrium insulation loss.

Figure 4 shows the cumulative heat loss measured throughout the test and is of less value in the assessment of materials since the information which it provides is less detailed than in the case of the "rate curve". Its value is in its description of total heat loss measured at any selected time.

THE APPLICATION OF EXOTHERMICS TO STEEL CASTINGS

Broadly speaking there are three main types of exothermic material used in steel foundries:

1. Surface additions—these include anti-piping compounds and Thermit materials.

2. Mouldable exothermic materials for lining risers.

3. Mouldable exothermic material to replace metal padding.

SURFACE ADDITIONS

For a gravity riser to be efficient it is essential that the column of steel is kept open to atmospheric pressure for as long as possible—this means that radiant heat losses from the riser surface must be cut down to a minimum to delay the solidification of the steel on the surface. According to most investigators these heat losses are approximately 25% of the useful heat in the riser, when no cover is used, and any saving of this heat would lead to more efficient feeding. Adams and Taylor[6] showed that by using an insulating cover on a 4 in. diameter steel riser the solidification time increased from 5 minutes (with no cover) to 13·4 minutes. For maximum efficiency it is essential that anti-piping compounds are used when exothermic sleeves line the risers, since the percentage heat loss by radiation from the surface will be greater than with a sand riser.

There are three main types of anti-piping compound with which the foundry-man can cover the surface of risers to minimize radiant heat losses:

1. Carbonaceous.
2. Insulating.
3. Exothermic.

Under this last heading one can include Thermit material which can be used as an anti-piping compound, but which is mainly used as a feeding aid.

1. Carbonaceous anti-piping compounds are essentially crushed coke and/or blacking. They are slightly exothermic by virtue of carbon burning and insulating because of their low density. However, with this type of anti-piping compound, carbon is picked up at the steel surface, which helps to reduce the liquidus temperature, but which can also lead to contamination by convection currents carrying the higher carbon percentage into the casting. Gertsman[7] showed in a study of insulating and mildly exothermic compounds that although carbonaceous materials gave a flat contour to the riser surface, the carbon can increase to $1\cdot2\%$ a quarter of an inch below the surface. The British Steel Castings Research Association have shown in a film of the solidification of a wax casting that a dye injected into the surface is carried by convection to nearly all parts of the casting. Therefore, it is possible that a high carbon percentage created under the steel surface of the riser could be carried into the casting.

2. Anti-piping compounds are essentially low density, low heat capacity and low thermal conductivity materials, such as crushed fire-brick, vermiculite, etc. However, although they have a low heat capacity "skulling" (prema-ture solidification of the steel surface) may occur if the steel in the riser has a very low superheat, since the materials take some heat in raising their own temperature to that of the steel. Using special apparatus described previously Foseco International laboratories have shown that although vermiculite has excellent insulating properties at room temperature, these deteriorate at 1450°C where the thermal conductivity is $0\cdot0012$ cal/cm²/sec/°C/cm. This still compares well with dry silica sand, whose thermal conductivity is $0\cdot003$ to $0\cdot005$ cal/cm²/sec/°C/cm. This latter figure should discourage the use of dry sand as an anti-piping compound.

3. Exothermic anti-piping compounds are useful in that the heat emitted by their reaction delays the "skulling" of low superheat steel risers and the sintered residue, which bridges the riser cavity, serves as an insulator by virtue of its low thermal conductivity. This value can be judged from the results of heat losses through equal thicknesses of various materials. Silica sand has heat losses of $1\cdot22$ cal/cm²/sec. Ferrux 101 (a proprietary anti-piping compound) has heat losses of $0\cdot61$ cal/cm²/sec and Ferrux 40 $0\cdot32$ cal/cm²/sec. To achieve the necessary insulation, sufficient anti-piping compound to give a minimum thickness at least equal to one tenth of the

riser diameter is required and it is necessary to add this as soon as pouring ceases. This will generally amount to approximately 4 to 6 lb of anti-piping compound per ton of castings.

THERMIT MATERIALS

Although small amounts of Thermit can be used as anti-piping compounds its main use is as a feeding aid. The essential difference between this exothermic material and the mouldable exothermic has been explained and the technique of application is obviously different.

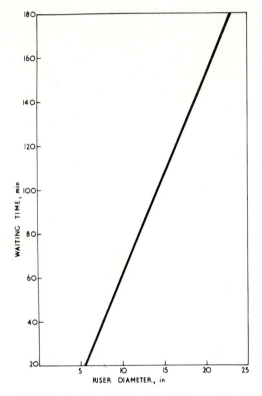

FIG. 5. Waiting time for "Kalmex" or "Thermexo" additions.

The reaction of the Thermit produces approximately 45% of the original weight of the material as a metal. This metal is essentially iron with a certain percentage of silicon. It is, therefore, obvious that one does not want this iron in the steel casting and the technique of using the material is to ensure that the resultant metal remains on top of the parent metal in the riser.

The casting is headed up as for sand risers, with head volumes ranging from 50% to 100% of the casting volume, depending on the casting modulus. These heads are then deliberately short-poured to a height equal to $\frac{1}{3}$ of the riser diameter, with a minimum height of 3 in. If there are different diameter heads on the castings they are poured to the ruling height of the largest diameter. Anti-piping compound is added, and then, depending on the riser diameter a waiting period is necessary to allow the initial feeding to occur. On diameters up to 5 in. the waiting period is negligible and the Thermit can be added within a minute or so of the termination of pouring. The graph shown in Fig. 5 gives an indication of the required waiting time, but because of varying conditions, i.e. pouring temperatures, moulding materials, etc., can only be a guide to the actual time of application in individual foundries.

Practice will lead to the correct time of application for any individual foundry. This will be when the surface metal of the riser head has skulled over except for a liquid pool in the centre equal approximately to one quarter of the original riser diameters.

When this situation is reached the anti-piping compound is removed and the requisite amount of Thermit is added.

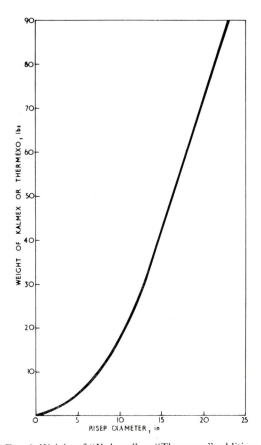

FIG. 6. Weight of "Kalmex" or "Thermexo" additions.

Figure 6 gives a guide to the weight of material required, and this should be adjusted in the light of experience but will be generally found to be approximately equal in pounds to one quarter of the cross-sectional area of the riser in square inches.

It can be seen that the heads are allowed to pipe and the Thermit reaction, which attains a temperature of approximately 2000°C, is sufficient to remelt the parent metal which then fills the pipe cavity and the Thermit metal, being less dense by virtue of its temperature and alloying remains on top.

The residual heads have a very flat contour and yields of up to 80% can be achieved, depending on the original heading practice. The remelting of the residual heads, as scrap, presents no problem, since silicon is generally added in many melting processes, thus the residual silicon percentage in the heads can be useful.

MOULDABLE EXOTHERMICS

Most of the useful heat lost from a riser is through the walls and any reduction in this improves the efficiency of the riser. Exothermics are available in powder form and as prefabricated sleeves so that riser cavities can be lined with exothermic material either by ramming up *in situ* or by insertion of a complete sleeve. There are many grades of exothermic, classified by their sensitiveness (time to react in contact with molten steel) so that the fastest would be used on smaller diameter risers and the slowest on large risers. The suppliers will advise on the correct grade of exothermic to use.

Except on very small diameter risers, the exothermic riser does not raise the temperature of the steel in the riser, but is sufficient to prevent premature freezing of the steel and then the incandescent residue insulates against rapid heat losses. Investigations at Foseco International showed that the heat loss through a thickness of medium sensitivity exothermic was 0.58 cal/cm^2/sec, whereas through a similar thickness of silica sand the heat loss was 1.2 cal/cm^2/sec. The freezing time of an exothermic riser can thus be extended to twice that of a sand riser of equal diameter. Since exothermics liberate quantities of gas when burning, and in turn need atmospheric oxygen to complete the oxidation of aluminium, adequate venting is required at the interface of the sleeves and backing sand and one quarter inch diameter vents every 2 in. should prove adequate for this purpose. Care must be taken to ensure that the steel does not run over the sleeve and block these vents and any that does so should be removed as soon as possible before applying anti-piping compound.

The two most important aspects of a riser are the thermal capacity and thermal conductivity of the surrounding material. During the early part of solidification of a casting when the rate of solidification is most rapid the thermal capacity of the surroundings is more important and can account for the majority of heat taken from the steel in the riser. The advantage of exothermics is due to the fact that the heat generated by the reaction satisfies its own thermal capacity and so there is no drain of useful heat from the steel in the riser.

As the time of solidification increases, as with larger diameter risers, the relative importance of the two properties mentioned changes and the low thermal conductivity plays the major part in improved feeding. Exothermics are still useful up to, say, 36 in. diameter riser since the incandescent residue is insulating, but beyond this diameter it is better to employ a more suitable insulator than exothermic, or resort to a Thermit material. These points are illustrated in Fig. 7, showing results obtained in the Foseco International laboratories of heat losses in the steady-state. It can be seen that no matter what material is added to the

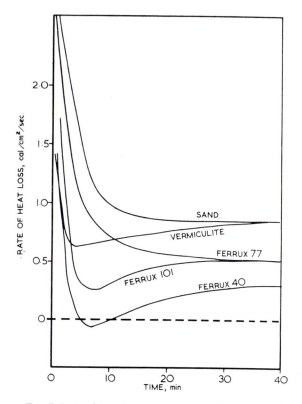

FIG. 7. Rate of heat loss against time (with top covers).

top of the riser, or what material is used to line the riser as shown in Fig. 8 there is an immediate heat loss corresponding to thermal capacity requirements. Then depending upon the characteristics of the material this rate reduces rapidly and eventually heat may be added to the riser metal (where the graph goes below the zero line) and then the rate of heat loss becomes fairly steady. This rate of heat loss depends on the thermal conductivity of the materials used.

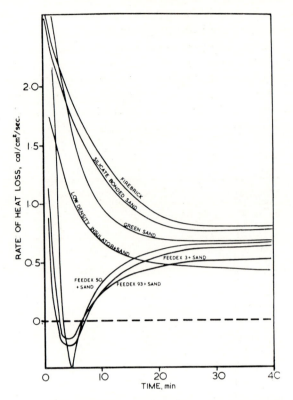

FIG. 8. Rate of heat loss against time (with side linings).

THEORY OF EXOTHERMIC RISER CALCULATIONS

Following earlier work Chvorinov[8] on modulus of castings, Wlodawer[9] devised a mathematical approach to the heading of castings both for sand and exothermic risers. Bishop, Taylor and Powell[10] had previously published graphs relating riser diameter to casting shape factor and at the International Foundry Congress in Liege, Belgium in 1958, Snelson[11] gave graphs relating exothermic riser diameters to casting sections and also showed the effect of increasing diameter and increasing wall thickness on resultant feed. Figure 9 shows the effect of variations in riser percentage, riser diameter and sleeve wall thickness of exothermic lined risers. The series shows that even a 50% exothermic riser can result in shrinkage of the casting if too small a diameter is used, because of the adverse volume to surface area modulus. Increasing the diameter flattens the feed so that a smaller residual head can be tolerated, to ensure a sound and segregate free casting, by reducing the height of the riser; it should also be noted that thickening the wall of the sleeve also flattens the feed and this step can be resorted to if, for any reason, the requisite diameter of riser for the section to be fed cannot be used.

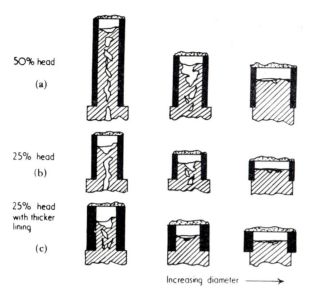

FIG. 9. Effect of variations in riser percentage; diameter and sleeve wall thickness.

The graphical presentation of riser diameter against section or modulus of casting seems to be the most favoured by workers in the field. In the light of results obtained in U.S.A., Europe and the U.K., Foseco have prepared a series of graphs relating diameter to section thickness of seven standard shapes and these are reproduced in Fig. 10a–g. Since these graphs can be used for all metals it should be noted that the four lines are:

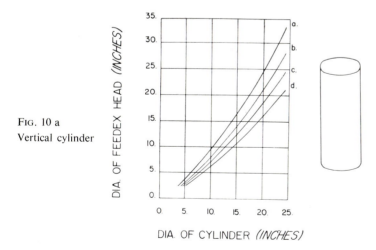

FIG. 10 a
Vertical cylinder

FIGS. 10a–g. "Feedex" head diameter for section thickness of varying casting shapes.

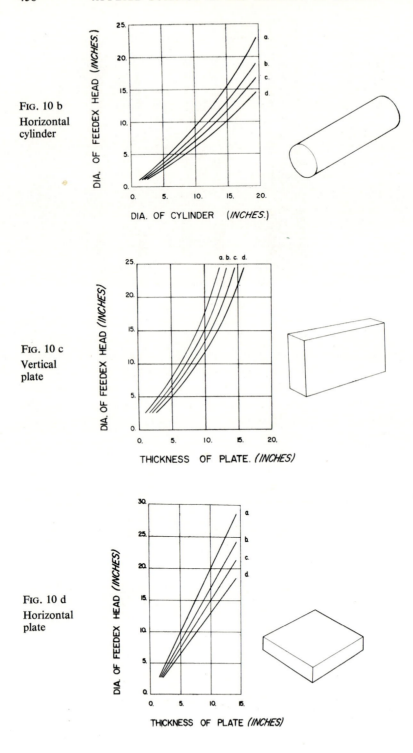

FIG. 10 b
Horizontal
cylinder

FIG. 10 c
Vertical
plate

FIG. 10 d
Horizontal
plate

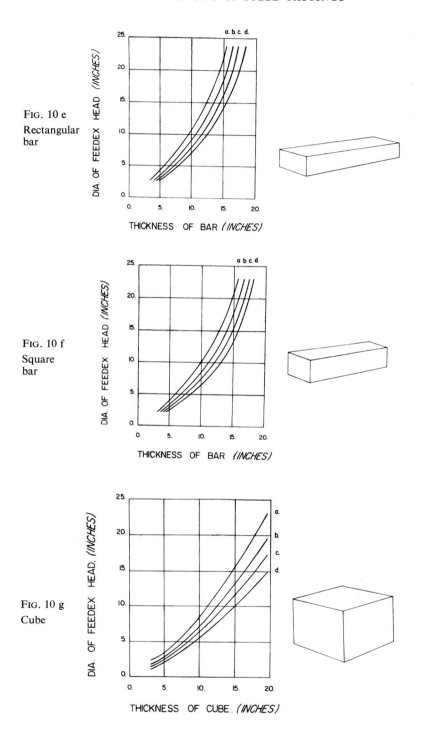

FIG. 10 e
Rectangular
bar

FIG. 10 f
Square
bar

FIG. 10 g
Cube

(a) Steel.
(b) Copper-based alloys and high duty irons.
(c) Light alloys.
(d) Grey iron.

Since it is not practical to represent every casting shape it is necessary to decide which shape is nearest to that of the casting and use the ruling section thickness to decide the riser diameter. Large diameter cylinders can be considered as either vertical plates or bars depending on the depth of sections. The transition from square bar to vertical plate is when the depth exceeds three times the section thickness and from a square bar to a rectangular bar when the width is greater than the section thickness but less than three times the thickness. A horizontal plate will have the two other dimensions greater than three times the section thickness.

With larger castings it will be necessary to consider them as separate portions governed by feed requirements and feeding distance between risers. Experience indicates that the feeding distance is equal to four times the diameter for a single feeder, i.e. three times the diameter between the outer edges of the riser, where multiple feeders are used. This feeding distance can be increased by the judicious use of chills midway between risers to cause steeper temperature gradients. From the graph nearest the casting section the minimum diameter of riser is ascertained and to ensure a sound feed the riser diameter should never be smaller than this, but can in many cases be exceeded.

For steel castings it is necessary to use 25% of the casting weight in risers to give adequate insurance against the varying factors of pouring temperatures, pouring speeds, alloying additions and mould materials, which can affect the cooling characteristics and thus the feed requirements. This will be the total riser requirement and where more than one riser is required will be shared according to the section demand. With the minimum diameter, decided from the graph, and the weight, we can easily calculate the height of the sleeve.

It may well be that this height will be more than one and a half times the diameter, a generally accepted ratio of height to diameter to give a reasonable volume/surface area modulus and in such cases the diameter should be increased so that the weight can be accommodated within the accepted ratio. The range of height to diameter ratio is from $\frac{1}{2}$:1 to $1\frac{1}{2}$:1 with 1:1 being preferable.

On multi-risered castings the calculations may result in differing heights, particularly where sections vary considerably and in this case all the risers should be brought to a common level equal to the tallest sleeve. However, where there is a discrepancy in pouring heights of a considerable distance, as, for instance, in stepped castings or large bossed wheels then one can resort to making the lower level sleeves into blind risers (equivalent to a Williams head). Where this practice is used venting becomes more critical and adequate precautions should be taken, not only to vent the outside of the sleeve, but to provide a "whistler" of $\frac{1}{4}$ in. diameter from the top of the sleeve itself, i.e. to allow easy escape of displaced air as the riser fills with steel. Since the exothermic delays the formation of skin in the riser there is no need for a cracker core, to keep the riser open to atmospheric pressure, without which the casting would most likely feed the riser.

The final dimension of the sleeve to be decided is the exothermic wall thickness and this can generally be taken as $\frac{1}{10}$ of the riser diameter plus $\frac{1}{8}$ in. with a minimum wall thickness of $\frac{1}{2}$ in. for practical manufacturing purposes, i.e. all sleeves under 5 in. diameter will have a $\frac{1}{2}$ in. wall thickness. The steps involved in risering a casting with exothermic sleeves can be summarized as:

1. Determine the geometric shape or shapes to be fed in the casting.
2. From the graphs determine the minimum diameter of risers required for each shape.
3. Determine the feed requirement (usually 25% of casting weight).
4. Determine the height of risers, knowing the diameter and weight required.
5. Ensure that the maximum feeding distance is not exceeded from any riser.
6. Determine the exothermic wall thickness from the riser diameter.

To illustrate this method consider the risering of the gear blank casting shown in Fig. 11a and b where the blank casting weight is 7400 lb. It is obvious that at least six risers will be required on the circumference to cover the "T"-junctions where the arms join the rim. This rim can be considered as a vertical plate of section $2\frac{1}{2}$ in. and Fig. 10c can be used to determine the riser diameter, but since the depth is considerably larger than the section some padding will be necessary to ensure complete soundness. Either metal padding or exothermic

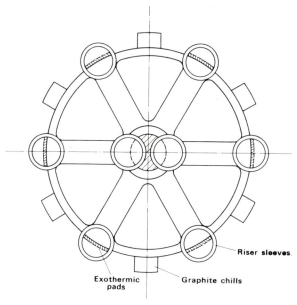

FIG. 11a. Gear blank—top view.

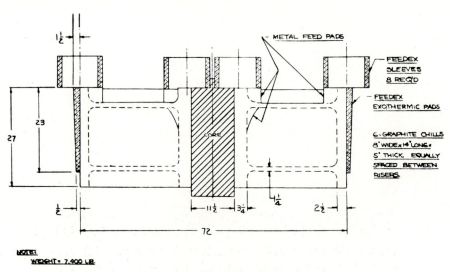

FIG. 11b. Gear blank—elevation.

padding can be used and both will increase or effectively increase the section to 5 in. and for this section thickness a minimum riser diameter of 8 in. will be required. The feed requirement of the casting as a whole is 1850 lb weight and the rim will require approximately 1250 lb of this to feed the wall and the arm junctions. From these details one can now calculate that each rim riser will need to contain 208 lb and so would be 15 in. × 8 in. diameter. When checking the feeding distance, however, one sees that the circumference of the rim is 226 in. and with six heads necessary to feed the "T"-junctions the feeding distance would be 38 in., i.e. $4\frac{1}{2}$ times the riser diameter. One is, therefore, justified in increasing the diameter to 9 in. and as an added safeguard chills are used midway between risers, low down on the rim wall. These chills require to be 5 in. thick (i.e. twice the section thickness) and will be made 8 in. wide by 15 in. long. Increasing the diameter of the risers means that one can reduce the height to 12 in. to give the required feed weight.

Now the centre boss has to be fed and from consideration of its section and weight, 500 lb of feed metal are required. The section, again a vertical plate, is $3\frac{1}{4}$ in. and would need a minimum of $6\frac{1}{2}$ in. diameter riser and two such risers will satisfy feeding distance. However, for the sake of uniformity in pouring height, since $6\frac{1}{2}$ in. diameter risers would need to be 27 in. high, 9 in. diameter risers were decided upon and would satisfy the feed demand. A full ring head could be used, using an 18 in. outside diameter and an $11\frac{1}{2}$ in. inside diameter to give the required $6\frac{1}{2}$ in. effective riser which at 12 in. deep would contain 600 lb of feed metal. However, this method would prove more costly than two separate heads since the sleeves would take far more exothermic and the cutting of the residual head would be more expensive. The final aspect of this exercise is that of padding and Griffiths, Hall and Neu[12] showed that with slight modification of the standard exothermic it is possible to obtain a skin which is in many ways equal, if not better, than that obtained with moulding material. One can, therefore, effectively replace metal padding with exothermic material and this practice reduces considerably the fettling required on the casting. Experience has shown that the thickness of the exothermic pads should be between 40% and 60% of the section thickness at the top, tapering down to a minimum of $\frac{1}{2}$ in. at the bottom and the depth is governed by the casting requirement, e.g. in this case the pads need to be taken below the lower arm junctions.

It will be seen that this gear blank will require exothermic pads $1\frac{1}{2}$ in. thick tapering to $\frac{1}{2}$ in. whose width will be equal to the contact distance of the riser on the outer rim. Figures 12 and 13 show the actual casting which was made to

FIG. 12. Gear blank at knock-out.

the above method and which proved to be perfectly sound when fully machined, and with a poured weight, including sprues and gates of 11,350 lb gave a yield of 65%. The reason exothermic feeding has been described in detail is because many advantages are possible in its use:

1. Increased yields, leading to higher productivity in the foundry.
2. Improved soundness and quality because of the more favourable temperature gradients which can be developed within the mould.
3. An increased weight range of castings possible for foundries with limited melting facilities.
4. A reduction of up to 75% in fettling costs can be made by the combined use of padding exothermics and sleeves.
5. Lower moulding costs by reducing cope depths due to reduced riser height requirements.

FIG. 13. Gear blank fettled.

FIG. 14. Pinion casting in steel.

Fig. 15. Pinion—same casting as Fig. 14 after heads have been removed.

INSULATING PREFABRICATED SLEEVES

Recent trends have indicated more interest in insulating sleeves.

They are essentially lower cost, self-heating insulating sleeves of low density. After casting, there is limited exothermic reaction with little fume, developing very good long term insulation properties. One version is manufactured in a fibrous insulating material ensuring a strong sleeve able to stand up to foundry handling much better than exothermic sleeves; they are very suitable for use with moulding machines, sand slingers, etc., where conventional exothermic sleeves are too brittle.

These insulators combine the exothermic properties of "Feedex" with very good insulation properties. The sensitivity of the exothermic reaction is retarded and heat is developed over a longer period. For smaller diameter sleeves, therefore, (under 5 in. diameter), performance will not be expected to be as good as that of exothermic sleeves.

In view of the unique insulation properties of these materials, it is essential to use a very good, highly insulating top cover on the riser head to maintain maximum efficiency. Figures 14 and 15 show the application of "Fabrex 10" to a mill pinion.

Figure 15 shows Pinion Casting after removal of heads. Note knock-off type of sleeves used, eliminating head cut off. Casting fed with four 12 in. high, 8 in. inside diameter "Fabrex" heads with 3 in. diameter breaker core.

REFERENCES

1. Foundry Services Ltd. and STRAUSS, K., and BODDEY, R. F., British Patent No. 627678.
2. RAVAULT, F., and BRADDY, D., The chemistry of the reactions which occur in aluminium-based exothermic materials. Foseco Research and Development Report No. 120, April 1958.
3. EMM, J. C., and HICKMAN, J., The nature and control of exothermic reactions. Foseco Research and Development Report No. 175, August 1963.
4. LOCKE, C., BRIGGS, C. W., and ASHBROOK, R. L., *Trans. AFS*, **62**, 589–599 (1954).
5. WLODAWER, R., The control of the soundness of steel castings. Procs. of Eighth Annual Conference of the B.S.C.R.A., Harrogate, 1962, 38–59.
6. ADAMS, C. M., and TAYLOR, H. F., *Trans. AFS*, **60**, 617–623 (1952).
7. GERTSMAN, S. L., *Trans. AFS*, **57**, 332–342 (1949, 1957).
8. CHVORINOV, N., *Giesserei*, **27**, 177–186, 201–208, 222–225 (1940). B.I.S.I. Translation No. 117.
9. WLODAWER, R., The control of the soundness of steel-castings. Proceedings of the Eighth Annual Conference of B.S.C.R.A., Harrogate, 1962, pp. 38–59.
10. BISHOP, H. F., TAYLOR, H. F., and POWELL, R. G., *Foundry*, **86**, June, 54–59 (1958).
11. SNELSON, D. H., *Brit. Foundrymen*, **51**, 486–501 (1958).
12. GRIFFITHS, M. H., HALL, C., and NEU, M. G., *Brit. Foundrymen*, **55**, 93–102 (1962).

CHAPTER 15

THE FEEDING OF IRON CASTINGS

J. L. FRANCIS and P. G. A. PARDOE

Superheat

Components of an infinite variety of form and shape can be produced by pouring molten metal into appropriate moulds. To ensure complete filling of the mould the pouring temperature of the metal must be sufficiently above its melting point so that solidification does not commence before the mould is full. This added increment of temperature is termed superheat.

Liquid Shrinkage

When molten metal cools, its density increases and its volume decreases. Molten cast iron obeys this law. A steady and uniform reduction of volume occurs as the cast iron cools from its pouring temperature down to the temperature at which solidification begins. F. G. Sefing[1] gives a value of $1 \cdot 1 \%$ of the volume for each $100°C$ of fall in temperature down to that of incipient solidification. Thus, an unnecessarily high pouring temperature increases the liquid shrinkage.

Solidification Shrinkage

During the phase change from liquid to solid a further diminution of volume ensues which is known as solidification shrinkage. The amount of this varies according to the chemical composition of the cast iron, especially the carbon and silicon contents. When the composition and the rate of cooling is such as to allow the precipitation of the carbon in the form of graphite, some of the solidification shrinkage is compensated. Carbon present in solution as iron carbide (Fe_3C) has a specific gravity of about $7 \cdot 66$. Graphite has a specific gravity of $2 \cdot 25$ and will, therefore, occupy nearly $3\frac{1}{2}$ times the space it filled when in the form of carbide.

Total Liquid Shrinkage

It follows that total liquid shrinkage is the summation of the volume decrease, resulting from the temperature drop from pouring temperature, to the onset of solidification and the shrinkage which takes place during the change from liquid to solid. For some compositions of grey cast iron this can be little or nothing when the expansion due to graphitization is taken into account. On the other hand, cast iron of all white structure containing no graphitic carbon has a total liquid shrinkage approaching that of carbon steel. Total liquid shrinkage for different kinds of cast iron, as given by Sefing,[1] is shown in Table 1.

TABLE 1. *Total Liquid Shrinkage Values for Different Kinds of Cast Iron*

Type of cast iron	Pouring temperature	Total liquid shrinkage
Soft grey T.C. 3·5–4·0%	1350°C	0·6%
Medium Carbon 3·0–3·25%	1450°C	4·2%
High Test T.C. 2·5–2·75%	1450°C	5·0%
Ni-Resist	1450°C	5·6%
White chill cast	1370°C	5·75%
White malleable	1510°C	7·25%

Solid Contraction

As castings cool, in the solid state, from red heat to atmospheric temperature their volume continues to diminish. This is referred to as solid contraction. In order to produce castings free from liquid shrinkage defects, the founder must supply feed metal capable of making good the volume decrease resulting from total liquid shrinkage only. Solid contraction is the concern of the pattern maker who dimensions the patterns according to the contraction rule appropriate to the metal or alloy from which the castings are to be made. The average allowance for grey iron castings is $\frac{1}{8}$ in. per foot and for white iron castings $\frac{1}{4}$ in. per foot. That is to say, all pattern dimensions are increased by this ratio. Machining allowance, where required, is extra.

Mould Dilation

This is another factor which influences the feed requirement of castings. Total liquid shrinkage can be calculated and so allowed for but, mould expansion is a variable of unknown quantity; it depends on mould rigidity. Green sand moulds are often not rigid enough to withstand the pressure created by head of metal and the expansion force caused by graphitization. The latter, especially, may result in enlargement of the mould space just at the onset of solidification when the extra feed metal, thus rendered necessary, is difficult to supply. Any expansion of the cores, where present, is a compensating factor which helps to cancel out mould expansion. Shrinkage defects caused by mould expansion can show up as widespread microporosity or as holes due to concentrated shrinkage.

Temperature Gradient

Due to the variation in sectional thickness of castings and the positioning of runners and ingates, the temperature attained by moulds and cores will not be uniform throughout. Hot spots are caused by designs with re-entrant angles. Feeding difficulties are also set up when heavy bosses connect with other parts of the casting by relatively thin walls. The latter solidify while metal remains liquid at the interior of the thicker sections. As the liquid interior solidifies and shrinks it will contain porosity unless liquid feed metal is available for it to draw on. Alternatively, it may be induced to solidify at similar speed to the thinner sections by the application of chills or denseners. Feeding is promoted by any agency which can induce or assist uniform temperature gradients between the part to be fed and the supply of feed metal.

Chills and Denseners

These are a means of creating temperature gradients in the required direction and are formed from materials such as iron, graphite and silicon carbide which can conduct heat away more rapidly than sand moulds or cores. Metal denseners create a hazard of their own because if they are colder than the mould, moisture may condense on them and cause steam holes in the castings, also if they become rusty or oxidized, the molten metal will react against them to produce unsoundness. Again, care and judgement are necessary when metal chills are applied to iron castings for, if they are too big, the area of the casting adjacent will contain iron carbide and become too hard to machine, or it may contain incipient cracks.

Directional and Progressive Solidification

This principle should be considered for all castings; for plate and bar type castings, poured through a feeding head, it is easy to achieve as shown in Fig. 1.

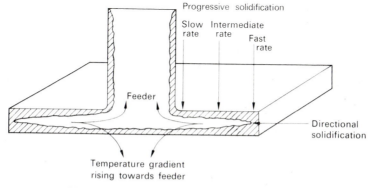

FIG. 1. Directional and progressive solidification.

The molten metal runs to the remotest part of the mould, loses temperature on route and solidifies. This process can be envisaged as an infinite number of small layers or increments of molten metal which follow a similar but increasingly shorter course as they approach the source of supply. There is thus set up a uniform temperature gradient between the first and last metal to enter the mould, in which the highest temperature is at the feeder end where any liquid shrinkage can be made good. At the far end of the casting, the sand surface acts as a mild chill, and helps the directional solidification due to what is termed the "end effect". For example, no "end effect" would be present in a continuous ring casting.

Unfortunately, the majority of castings are not simple shapes of uniform section. The presence of cores interferes with directional progressive solidification and smooth temperature gradients, so does the necessity for placing ingates in adverse positions, such as at the base of moulds when dross forming alloys must be poured free from turbulence. In practice there are so many variables and conditions to be considered that no universal formula can be evolved by means of which the exact feed volume could be calculated for all castings. However, much has been accomplished along such lines which, combined with experience, compromise and empiricism, is of great assistance in solving feeding problems.

Efficiency of Feeder Heads

It is at once apparent that any feeder head must not only contain a sufficient volume of feed metal but also should maintain it liquid until after the casting has become solidified. As we have seen, it is possible to ascertain by calculation, how much feed volume is required to compensate the total liquid shrinkage of any volume of cast iron. However, no feeding head is 100% efficient, so that it is not sufficient to provide only the theoretical volume of feed metal. Something greater is necessary and this is often attained by multiplying the theoretical feed value by a factor K. Sefing gives the maximum feeding efficiency of cylindrical feeding heads as 40%. Thus a theoretical feed requirement of 10 lb supplied by a circular feeding head would need to be $\frac{10}{40} \times 100 = 25$ lb.

In other words, the factor K in this case is 2·5. This value is considered optimistic for general practice. A more likely value is 25 to 30% for heads moulded in sand.

Chvorinov has shown that the time taken for a mass of liquid metal to solidify is a function of the ratio of its volume to surface area. That is to say, a given volume having the smallest possible surface area will remain liquid longest. The geometrical shape which fulfils this requirement is the sphere. Spherical feeder heads do not lend themselves to practical moulding and so cylindrical heads have become commonplace as the shape coming next to the sphere in prolonging the time taken for any molten mass of metal to solidify.

All calculations for deriving feed volumes are based on Chvorinov's rule relating volume to surface area which is expressed as follows:

$$\frac{\text{Volume of feeder}}{\text{Surface area of feeder}} \quad \begin{matrix}\text{must be equal to}\\ \text{or greater than}\end{matrix} \quad \frac{\text{Volume of casting}}{\text{Surface area of casting}}$$

In the majority of cases the finding of the volume and surface area of castings involves laborious calculation and most work in this field has concentrated on evolving less complicated procedures by means of which satisfactory results may be obtained with the expenditure of less time and effort. Bearing in mind the number of variables which can influence the feed requirement and the use of the safety factor K, a working approximation suffices.

Derivation of Feed Requirements

Many able workers have tackled this problem, especially for steel castings, and the results of their efforts are published in the relevant text books and technical journals. Account is taken of the fact that it is necessary to apply feeding only to the heavy sections of castings or to places of heat concentration. M. Jeancolas[2] has dealt with the subject concisely. Each component part of a casting which needs feeding is considered separately and provided with its own feeding head. The solidification time or cooling modulus of the feeder (Mf) is related to the cooling modulus of that part of the casting to be fed, as has been explained previously. This can be expressed in the form $Mf = KMc$, where K is the safety coefficient and Mc represents the cooling modulus of the part to be fed. When the value K has been derived for standard sand moulded feeder heads it can be reduced to $\frac{1}{2}$ to $\frac{2}{3}K$ where exothermic sleeves and covers, or feeding aids of like nature, are employed.

It is clear that the volume of feed Vf must equal the total liquid shrinkage, multiplied by the safety factor K. This latter is given as $K = 2$ for exothermic heads and $K = 6$ for standard sand moulded feeders. The formula for calculating the feed volume then becomes $Vf = K \times r \times Vc$, where r is the percentage total liquid shrinkage of the metal or alloy from which the casting is made. A number of values for r have been given in Table 1. The parts of castings which require feeding can, in most cases, be approximated to standard geometrical shapes, such as bars, plates and cylinders, the volumes of which are easily calculated.

Feeding Aids

A number of these have been developed over the years and some have become obsolete or discarded with the passage of time. One of these is rod feeding, where a steel rod of appropriate diameter is pumped up and down in a carrot-shaped feeding head. In the hands of a skilled operator this method is effective in keeping open the connecting neck or channel between the feeder and the casting. Fresh

Q

molten iron had to be added to the feeding head as the feeding operation proceeded. On large castings it was a hot, tiring and laborious operation, sometimes calling for a relay of men. There was also risk of penetrating too far into the mould with the feeding rod or breaking the rod off inside the casting. Little science entered the design of rod fed heads which were left to the discretion and experience of the moulder.

Later on it came to be realized that gating techniques could influence feeding and a study of these resulted in one or two methods that achieved success becoming known throughout the world.

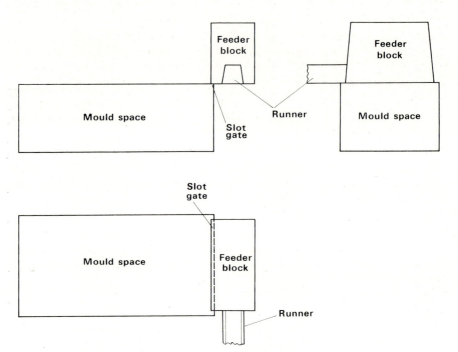

FIG. 2. Diagram of block-runner.

Pencil Gating

This gating system was developed by Ronceray and applied mainly to chunky castings with little or no coring. The gates came direct from the pouring head on to the casting and had a diameter similar to that of an ordinary pencil. The number of these gates depended upon the weight of the casting. By distributing the molten iron through a number of such small gates the hot spot created by a single gate was avoided. Furthermore, these small gates usually resulted in a longer pouring period before the mould filled. In fact the aim was to pour at such a speed that the bulk of the liquid shrinkage would be automatically fed as the casting proceeded. This is, in fact, the principle of progressive solidification upon which all recently developed continuous casting processes work. The hazards attending pencil gating are premature freezing of the metal before the mould is full and the tendency for the moulding sand to scab during the lengthened pouring period.

Block Running

Attributed to F. J. Connor, this is a combined gating and feeding technique which, again, is suited mainly to thick sectioned castings with minimum coring. The gate is in the form of a narrow slot formed from the overlap of the block into the mould space, as indicated in Fig. 2. The width of the overlap does not exceed $\frac{1}{16}$ in. Requisite gate area for large castings is obtained by using more than a single block. The downgate and the ingate, entering the base of the block, must each be 25% greater in cross-sectional area than the slot. It is intended that the block should fill as quickly as possible and remain full during the pouring of the mould. In this way it forms a preheated feeding head capable of supplying the required feed metal. L. W. Bolton[3] and W. D. Ford give the following dimensional details of block or lip runners (Table 2).

TABLE 2.

Casting weight lb	Weight of runner lb	Ingate in.	Block runner in.
108	10·0	$\frac{1}{16} \times 3$	$4\frac{1}{2} \times 3 \times 3$
15	7·5†	$\frac{1}{16} \times 2\frac{7}{8}$	$2\frac{7}{8} \times 3\frac{1}{4} \times 3\frac{3}{8}$†
11	7·0†	$\frac{1}{16} \times 2\frac{1}{4}$	$2\frac{1}{4} \times 3\frac{3}{4} \times 3\frac{3}{8}$†
20	7·0†	$\frac{1}{16} \times 3$	$3 \times 3 \times 3\frac{1}{4}$†
110	6·5‡	$\frac{1}{16} \times 3\frac{1}{2}$	$3\frac{1}{2} \times 2\frac{1}{8} \times 3\frac{1}{4}$‡
135	30·0	$\frac{1}{16} \times 4$	$4 \times 5\frac{1}{4} \times 5\frac{1}{4}$

† This runner serves two castings.
‡ These figures apply to each of two feeders.

With this system of running and feeding the work involved in trimming the castings is greatly reduced.

Side Feeding

A system of gating and feeding by a side runner-feeder technique was developed and described by T. Barlow[4] and H. C. Winte in 1945. Since then it has been tested and proved satisfactory in many different foundries and applications. Although designed for grey iron castings it has application in principle to all castings. All the necessary details are given in Fig. 3. The following points are important and must be observed in order to achieve maximum results:

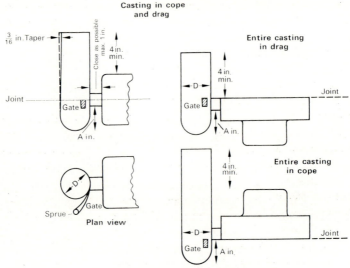

Casting in cope and drag

Entire casting in drag

Plan view

Entire casting in cope

Principal dimensions of shrinkbobs

Dia. of shrinkbob 'D' inches	Depth of drag bob below neck 'A' inches	Dimensions of gate into shrinkbob inches	
2	$2\frac{1}{4}$	$\frac{3}{8}$, $\frac{1}{4}$ 1 in.	**Light castings** $\frac{1}{2}$ to $\frac{3}{4}$ in. square or $\frac{3}{16}$ to $\frac{7}{8}$ in. dia.
$2\frac{1}{2}$	$2\frac{1}{4}$		
3	$2\frac{3}{4}$	$\frac{1}{2}$, $\frac{3}{8}$ 1 in.	**Medium castings** 1 to $1\frac{1}{4}$ in. square or $1\frac{3}{16}$ to $1\frac{7}{16}$ in. dia.
$3\frac{1}{2}$	$2\frac{3}{4}$		
4	$3\frac{1}{4}$		
5	$3\frac{3}{4}$	$\frac{3}{4}$, $\frac{1}{2}$ 1 in.	**Heavy castings** $1\frac{1}{2}$ to 2 in. square or $1\frac{3}{4}$ to $2\frac{1}{4}$ in. dia.
6	$4\frac{1}{4}$		
8	$5\frac{1}{4}$		

FIG. 3. Effective shrinkbob risers for grey iron castings.

1. The exact dimensions of the connecting neck are determined by the weight of the casting.
2. The length of neck should never be greater than 1 in. in order to avoid preliminary solidification and prevention of full feeding.
3. A short neck ensures thorough heating of the surrounding sand sufficient to permit adequate flow of metal until the final solidification of the casting. This is true in spite of the relatively small cross-section of the neck recommended in this design.
4. The short, small neck reduces the work necessary in trimming.
5. The shrinkbob feeder is divided into two parts known as the cope-bob and the drag bob, both of which must be proportioned correctly.
6. The diameter of the shrink-bob is determined by the size of the casting and the type of iron being poured. Larger shrinkbobs are required for low-carbon high-strength grey cast irons than for the softer higher carbon irons.

THE FEEDING OF IRON CASTINGS

7. The cope section of the shrinkbob feeder must be at least 4 in. above the highest point of the casting so as to give sufficient feed pressure for correct feeding.

8. A most important feature of the drag section is that the depth should extend below the connecting neck sufficiently to ensure the maintenance of molten metal in that part of the feeder directly opposite the neck.

9. If the drag-bob is too shallow, or if it is omitted, the last portion of the feeder to solidify will be fairly high up in the cope section and will be separated from the connecting neck by solid iron.

10. The drag section is also necessary to provide sufficient heating up of the sand surrounding the neck to maintain the metal liquid during the entire period of feeding.

11. The gate to the feeder must provide a choke in order to obtain the necessary swirling effect and cleansing of the metal. It should enter the feeder at a tangent.

12. Only three sizes of gate are used for castings from 10 lb to 2 or 3 tons in weight, but multiple applications can be used where necessary.

13. The down runner should have a cross-sectional area from 1·4 to 1·6 times that of the ingate at its smallest part.

Light castings include those of the smallest size likely to require feeding. Medium castings are those weighing from 50 to 400 lb, and heavy castings are those the weight of which is measured in tons rather than pounds.

Atmospheric Feeders

After it has been pointed out, the influence of atmospheric pressure as a feeding aid becomes very obvious. Its application is in the form known as the Williams feeder illustrated by Fig. 4.

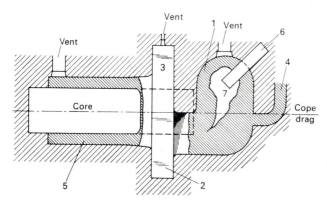

FIG. 4. Williams feeder.

The principle on which atmospheric feeding relies is that shrinkage cavities which form in the interior of castings are at a lower pressure than atmospheric. This is true provided that the solidified skin which forms on the outside of the poured metal is continuous and strong enough to withstand the pressure of the atmosphere without collapsing or puncturing. Then, if a supply of liquid metal can be maintained from the feeder head to the place where the shrinkage is taking place, the air pressure being in excess of that at the shrinkage area, will force the feed metal into the shrinkage void. By making use of atmospheric pressure in this way, feed metal can be forced to areas well above the level of the feeder head. Thus, it is of great advantage in the case of blind feeders as shown in Fig. 4.

Casting metals and alloys solidify by one or other of two main systems, known as skin forming and mass solidification respectively. These are explained fully in the section on the feeding of non-ferrous castings. The skin formers provide the strong continuous outside skin which is essential to atmospheric feeding but, the mass feeders do not. Thus in the latter case, the weak and discontinuous skin collapses under the pressure difference between the atmosphere and the shrinkage area, thereby allowing a short circuiting of the pressure from the feeder which is reduced to the effectiveness of gravity forces only.

Broadly speaking all unalloyed metals and those alloys which solidify over a narrow range of temperature, solidify according to the skin forming system and are thus amenable to atmospheric pressure feeding. Examples are steel, aluminium bronze and aluminium–12% silicon eutectic alloy. On the other hand, gun-metals and aluminium–4% copper, are alloys of wide freezing range and tend to solidify as mass feeders. Cast iron has such a multiplicity of grades and compositions that it embraces both methods of solidification. The high strength cast irons tend towards skin formation, while the lower and softer grades are mass feeders, or show intermediate tendencies. Therefore, attempts to use atmospheric feeders on iron castings are successful only in special cases.

In the diagram (Fig. 4) the blind atmospheric feeder is designated 1. It is connected to the casting, 5, by a connecting neck joining the flange, 2. The ingate into the blind feeder, 1, is shown at 4.

As soon as the metal enters the mould, a solid skin forms and envelopes the casting at all surfaces, including that of the blind feeder, 1. Due to liquid shrinkage of the casting, 5, and flange, 2 and 3, a shrinkage void, 7, forms in the feeder, as it supplies liquid metal to the casting. The sand core, 6, is fixed in the mould so as to penetrate the solid skin surrounding the feed and rest in contact with the liquid metal at its centre. This core transmits air pressure to the molten feed metal forcing it into shrinkage voids which are likely to form at the hot spots formed by the radius joining the flange to the body of the casting. The fact that feed metal is transferred in this way is proved by the formation of the cavity, 7, which remains to be seen in the feeder.

Apart from the necessity to provide the core, 6, to transmit air pressure, it is also essential that the cross-sectional area of the ingate, 4, is less than that of the connecting neck attaching the blind feeder to the flange. If these areas are equal or reversed, then the ingate will not solidify in time to prevent the atmospheric pressure acting in the reverse direction and forcing the liquid metal back through the running system.

Exothermic Feeding Aids

Another big step forward in the march of casting feeding progress was the introduction of heat generating aids. In the first place these took the form of surface cover fluxes for delaying the freezing over of the feed and thereby allowing atmospheric pressure to function throughout the feeding period. As in the case of blind feeders, once the surface becomes solid, the pressure of the air is cut off and gravity alone acts to drive the feed metal to its required destination. Most exothermic anti-piping fluxes make use of the Thermit principle in which finely divided aluminium and iron oxide react to form alumina and iron with the generation of heat. It is also termed the Goldschmidt reaction and is represented by the following equation:

$$2\,Al + Fe_2O_3 = Al_2O_3 + 2\,Fe$$

The oxidation of aluminium by means of iron oxide results in the evolution of 200 Cal of heat, and of molten iron to the extent of about 50% of the weight of thermit mixture used. When aluminium is oxidized directly to alumina by oxygen gas the evolution of heat is about doubled:

$$2\,Al + 3\,O = Al_2O_3 + 400\,Cal$$

Anti-piping flux mixtures for cast iron are compounded so that little or no molten iron is formed, some of the oxygen being obtained from nitrate and from the atmosphere once the reaction is under way. Other components are added to the mixture so as to produce a good heat insulating residue on completion of the reaction, with the object of reducing heat loss by radiation and convection.†

Mouldable Exothermic Material

A logical development of the exothermic surface cover is the heat generating sleeve. Heat escapes from feeding heads not only from their surface but is also conducted away by the sand walls which surround them. Maximum feeding efficiency is achieved by containing the molten feed metal in a material which abstracts no heat and by stopping radiation and convection losses from its surface with an insulating or heat generating cover.

The manufacture of anti-piping surface fluxes which react consistently and leave satisfactory end residues, is not easy. It is even more difficult when bonding materials are added to the exothermic mixtures for the purpose of

---

† For a detailed description of the developments of exothermic feeding compounds, see Chapter 14, p. 443.

moulding sleeves and other required shapes. For reliability in use they must be controlled for speed of reaction, rate of burning and heat development. They must not ignite at the temperature used for drying and when dried must be strong enough for handling and transportation. All these problems have been satisfactorily overcome in the production of "Feedex" mouldable exothermic feeding aids, a number of which are illustrated in Fig. 5.

FIG. 5. Variety of exothermic sleeves.

Heavy sections of cast iron subjected to slow cooling are apt to develop coarse graphite and an open structure of large grain size. This is very liable to occur at areas directly under feeding heads. More especially so if exothermic sleeves are used since they are applied deliberately to delay cooling of the feed metal. Also they generate high temperatures of the order of 1800°C. So as to minimize this disadvantage the base of the exothermic sleeve is formed from core sand which functions as a protective buffer between the feeder sleeve and the casting. Sleeves so fitted can be seen illustrated in Fig. 6.

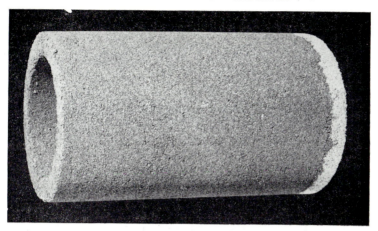

FIG. 6. Exothermic feeder sleeve fitted with sand buffer.

Washburn Breaker Cores

The sand buffer can be modified to serve a dual purpose in that it may form a breaker core which greatly facilitates the removal of the feeder head in the trimming shop. In many cases it can be knocked off by a hammer blow. Recommended sizes for the connecting neck diameter of breaker cores have been prepared, for steel castings, by the Technical Research Committee of the Steel Founders' Society of America and they are as shown in Table 3, together with other dimensions indicated in Fig. 7. Although designed for steel castings they form a good guide to the practice required in the case of iron castings.

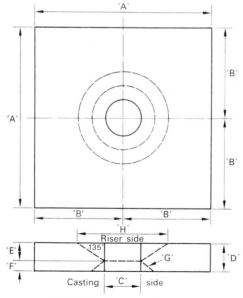

FIG. 7. Breaker core diagram.

Q*

TABLE 3. *Dimensions for Breaker Cores (in.)*
(*see Fig. 7*)

Feeder diam.	Dim A	Dim B	Dim C	Dim D	Dim E	Dim F & G	Dim H
2	3	$1\frac{1}{2}$	$\frac{3}{4}$	$\frac{5}{16}$	$\frac{3}{16}$	$\frac{1}{8}$	$1\frac{1}{8}$
3	4	2	$\frac{7}{8}$	$\frac{3}{8}$	$\frac{7}{32}$	$\frac{5}{32}$	$1\frac{5}{16}$
4	5	$2\frac{1}{2}$	$1\frac{1}{8}$	$\frac{3}{8}$	$\frac{7}{32}$	$\frac{5}{32}$	$1\frac{9}{16}$
5	6	3	$1\frac{1}{2}$	$\frac{1}{2}$	$\frac{5}{16}$	$\frac{3}{16}$	$2\frac{1}{8}$
6	7	$3\frac{1}{2}$	$1\frac{5}{8}$	$\frac{1}{2}$	$\frac{5}{16}$	$\frac{3}{16}$	$2\frac{1}{4}$
7	8	4	$1\frac{7}{8}$	$\frac{5}{8}$	$\frac{3}{8}$	$\frac{1}{4}$	$2\frac{5}{8}$
8	9	$4\frac{1}{2}$	$2\frac{1}{4}$	$\frac{3}{4}$	$\frac{1}{2}$	$\frac{1}{4}$	$3\frac{1}{4}$
9	10	5	$2\frac{1}{2}$	$\frac{7}{8}$	$\frac{5}{8}$	$\frac{1}{4}$	$3\frac{3}{4}$
10	11	$5\frac{1}{2}$	$2\frac{7}{8}$	1	$\frac{3}{4}$	$\frac{1}{4}$	$4\frac{3}{8}$
12	13	$6\frac{1}{2}$	$3\frac{3}{8}$	$1\frac{1}{8}$	$\frac{7}{8}$	$\frac{1}{4}$	$5\frac{1}{8}$

Wall Thickness of Exothermic Sleeves

In the small sizes the wall thickness must be sufficient to allow for the mechanical strength necessary for handling and transportation. While this is also true for the large diameters it must at the same time be sufficient to generate the greater amount of heat required for the much larger volume of metal contained in the feeder heads. Recommendations for wall thickness relative to internal diameter of the sleeve are:

TABLE 4.

Internal diameter in.	Wall thickness in.
below 3	$\frac{1}{2}$
4–6	$\frac{3}{4}$
7–9	1
10–12	$1\frac{1}{4}$
13–15	$1\frac{1}{2}$
16–18	$1\frac{3}{4}$
19–21	2
22–24	$2\frac{1}{4}$
25–27	$2\frac{1}{2}$
28–30	$2\frac{3}{4}$

Venting Exothermic Sleeves

In discussing the thermit reaction it was shown that there must be an adequate supply of oxygen in order to achieve maximum generation of heat. It was mentioned that some of the oxygen required is obtained from the air. Thus, it is necessary to provide vents in the walls of the sleeve so as to allow free access of air. These vents can be seen in the illustration, Fig. 5. They should extend to within $\frac{1}{2}$ in. of the bottom of the sleeve and also serve to allow free escape of the gases of combustion. Similar venting should be carried out in the backing sand immediately surrounding the sleeve.

Height-to-diameter Ratio

It is important to use a sleeve of suitable height-to-diameter ratio. For iron castings a height equal to one and a half times the internal diameter of the sleeve is recommended. Feeding heads contained in exothermic sleeves with their surface covered by exothermic flux powder experience little or no heat loss during the initial feeding period. Thus the feeder heads do not adopt the usual inverted cone-shaped sink common to sand lined heads. They sink uniformly with a flat or slightly concave surface and this explains their increased feeding efficiency. More than 50% of the traditional head is not used in actual feeding but merely supplies heat capacity.

Padding

Mention was made in an earlier paragraph of casting design where thick sections and bosses connect with other parts of the casting by thinner and faster cooling walls which obstruct feed supply to the areas of delayed solidification. Use of chills and denseners were referred to in this connection. An alternative way of solving the problem is by padding. This involves the addition of extra metal to the thinner connections so as to delay their cooling and allow feed metal to flow through.

Padding is a practice frequently employed in producing steel castings, but is not so prevalent in iron foundry practice. However, it is made use of for malleable castings which are poured from white cast irons having a high liquid. shrinkage. This extra metal allowance, which forms part of the pattern, is not required by the customer and must be ground off or machined away. Therefore, it reduces casting yield and involves extra time and cost in the trimming shop. A malleable iron casting provided with metal padding is shown at "A" in Fig. 8.

Another way of overcoming this feeding difficulty is by the application of exothermic padding such as depicted at "B" in Fig. 8. The pads can be per-formed and dried so as to fit into prints left by the pattern in exactly the same manner as sand cores. When contacted by the molten iron they react to produce sufficient heat to keep the section liquid long enough to allow the passage of feed metal from the feeder head. Subsequently, the casting is knocked out, the pad is left behind and no extra trimming or grinding is required. Also the casting yield is higher.

(a) Conventional Metal Padding

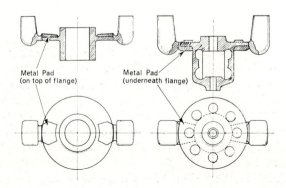

(b) FEEDEX Padding

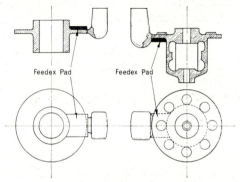

FIG. 8. Malleable iron casting with: (a) conventional metal padding; (b) exothermic padding.

Production of Exothermic Feeder Sleeves

Sleeves of any dimensions can be produced from standard core boxes by hand ramming or by core blowing machines. Many sizes in common widespread demand, can be purchased ready for use. In fact, almost any shape and dimension can be supplied in the ready-to-use form, provided that they are required in sufficient quantity.

The position has been further simplified by the development of the product in a flexible form. This is illustrated by Figs. 9 and 10. The latter illustration shows clearly how the flexible form of the material can be used to produce sleeves of greatly varying diameters. Height, too, can be adjusted as required by superimposing sections one on top of the other.

It is most important that all the feed metal is contained in the exothermic sleeve and none of it in sand. For example, if the pouring bush is higher than the calculated height of the sleeve, then the latter must be increased so that the two are level. Under no circumstances should a part of the sleeve be made in sand.

FIG. 9. Exothermic sleeve in flexible form.

FIG. 10. Method of use of flexible exothermic sleeve.

Insulating Sleeves

Experience has shown that in many cases a non-heat generating insulating material can perform satisfactorily as feeder sleeves. They act in a similar way to enable reduction of feeder head volume and prolong the period during which the feed metal remains liquid. Also there will not be the same tendency towards the creation of a coarse grained structure immediately below the feeding head. Insulating sleeves are particularly applicable to grey iron castings as well as those of aluminium and copper-base alloys.

In the early stages of development heat insulating sleeves were made from foamed plaster of Paris. They function very well provided that pouring temperatures remain below about 1200°C. Above this temperature, plaster sleeves break down with the liberation of sulphur dioxide. Insulating materials based on vermiculite and fibrous substances such as asbestos do not suffer this disadvantage.

A point worth emphasizing in favour of exothermic sleeves is their help in overcoming adverse temperature gradients. This applies especially in cases where for reasons of expediency it becomes necessary to gate castings at the base of the mould or in other positions which set up temperature gradients acting against directional solidification. In these circumstances the temperature of the metal, when it arrives at the feeder heads, is lower than that in the vicinity of the gates. However, on entering the exothermic lined head, the heat generation increases its temperature and assists in restoring the desired condition.

Balanced Feeding

No matter whether feeding aids are used or not, the feeding system must always be balanced where more than one feeding head is used on a casting. An example can be taken from the work of S. T. Jazwinski[5] and his collaborators.

Figure 11 depicts a crankshaft fed by means of two whirl-gate feeders and a gravity feeding head. Molten metal passes down the gate into the lower parts of "B" and "C", through the mould and into head "A". When the mould is

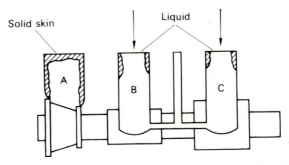

FIG. 11. Unbalanced feeding system owing to solidification at surface of head "A".

completely full there will be relatively cold metal in head "A", surrounded by the cold sand of the mould since in this example neither exothermic nor insulating sleeves are used. In the actual casting the metal will be hotter and it will be surrounded by sand which has been heated up by the flow of the metal. Still hotter metal will be located in heads "B" and "C" surrounded by sand which has been considerably pre-heated by the passage of all the metal, which enters the mould, passing through. In these circumstances head "A" begins to solidify at the surface before heads "B" and "C". No surface liquidizing fluxes are present on any of the feeding heads. Immediately the surface of head "A" freezes, the atmospheric pressure upsets the balance of the whole feeding system.

Atmospheric pressure is shut off from head "A" by its solidified surface but is acting fully on the still liquid surfaces of heads "B" and "C". Thus, no further feeding takes place from head "A", in fact, the opposite occurs. Liquid shrinkage in head "A", after the solidification of its surface is compensated by metal from the casting because of the pressure transmitted by heads "B" and "C". Unless heads "B" and "C" are sufficiently large and efficient to feed the whole of the casting plus head "A", shrinkage cavities will form in the casting while head "A" will be sound.

The size and extent of the cavities within the casting will be governed by the difference in time between the solidification of the surface of head "A" and that of heads "B" and "C". That is to say the greater the time period, the larger the cavities are likely to be.

Immediately the surfaces of heads "B" and "C" solidify, the feeding system becomes balanced again, but acting under the force of gravity only. By this time, however, head "A" will most likely have become completely solidified and consequently all the metal used in making head "A" free from shrinkage defects must be compensated by feed metal from heads "B" and "C". If the latter are not able to fulfil this demand shrinkage cavities will be formed where the heads join the casting.

REFERENCES

1. SEFING, F. G., *Proc. Inst. Brit. Foundrymen*, **38**, A142–A154 (1944–5).
2. JEANCOLAS, M., *Foundry Trade Journal*, **120**, Feb. 24, 255–264 (1966).
3. BOLTON, L. W., and FORD, W. D., *Foundry Trade Journal*, **88**, May 25, 551–558 (1950).
4. BARLOW, T., and WINTE, H. C., *Foundry*, **93**, July, 44–49 (1965).
5. JAZWINSKI, S. T., WELLS, E. D., and FINCH, S. L., *Proc. Inst. British Foundrymen*, **38**, A155–A174 (1944–5).

CHAPTER 16

THE FEEDING OF COPPER BASE CASTINGS

T. A. BURNS

Of all the alloy groupings the copper base group exhibits the widest range of properties and characteristics. In terms of composition they are normally divided into two basic groups—those containing not more than 5% zinc (bronzes, gun metals, etc.), and those with above 5% of the element (brasses and high tensile brasses, nickel silvers, etc.). When considering feeding, however, another division, which is independent of composition, is necessary on the basis of solidification characteristics and whether the freezing range is nominally short or long. Table 1, p. 488, shows the two principal groups for the freezing ranges of the commonly used alloys.

GENERAL CONSIDERATIONS FOR SAND CASTINGS

Some incidental factors, in addition to mode of solidification must be borne in mind when considering solidification of a given casting. In a sand mould the rate of heat extraction from the cast metal is governed by the properties of the sand and, from a flat plate, will be more or less constant after the initial chilling effect of the damp surface. Few castings are geometrically simple, however, and re-entrant and external arcs and angles will affect the rate, the former slowing it down and the latter increasing it. Additionally, casting shape, conduction through thick and thin sections, rate of mould fill, position of ingates and feeders, etc., will all have important effects. A strong effect can also be exerted by chills which are quite commonly used to initiate a preferred solidification pattern. There are, therefore, very many variables that can affect the manner and speed of solidification of copper base castings and hence, casting quality and feed requirements. The complexity of the field is reflected in the vast amount of literature that has been published and to which reference can be made, dealing in detail with every aspect mentioned above.

Certain general principles can, however, be extracted and these will apply irrespective of the casting or the alloy used.

1. Running System

The method of introducing the metal into the mould cavity will obviously have a pronounced effect on the temperature of the metal in the mould and a significant, if not critical, effect on features such as local hot spots due to over-heating of the sand, final temperature gradients, dross/oxide inclusions, amount

TABLE 1. *Freezing Ranges of the Principal*
Sand Cast Copper Alloys

Alloy	BS.1400	Approxi- mate freezing range deg C
Group A		
Copper	HCC1	Nil
	HCC2	Nil
Aluminium-bronze	AB1	10
	AB2	20
Copper-manganese- aluminium	CMA1	45
	CMA2	40
Die-casting brass	DCB1	5
	DCB2	5
	DCB3	10
High-tensile brass	HTB1	10
	HTB2	15
	HTB3	20
Sand-casting brass	SCB1	70
	SCB2	40
	SCB3	25
	SCB4	10
	SCB5	60
Group B		
Phosphor-bronze	PB1	175
	PB2	185
	PB3	180
	PB4	175
	LPB1	140
Leaded-bronze	LB1	170
	LB2	175
	LB3	175
	LB4	120
	LB5	120
Gun metal	G1	175
	G2	170
	G3	140
Leaded-gun metal	LG1	150
	LG2	160
	LG3	170
	LG4	160

of top feeding required, etc. The importance of a correctly sized and situated running system must never be underestimated since it can often exert an over-riding effect on the soundness or otherwise of the casting.

Single ingate systems should have the downsprue tapered to ensure complete filling during the pour, should be provided with a sump to take the thrust of the first metal in and should be mildly choked, i.e. between 1:1 and 1:1·2. (Ingate cross-sectional area equal to or only slightly larger than the sprue cross-sectional area at its narrowest point.)[2]

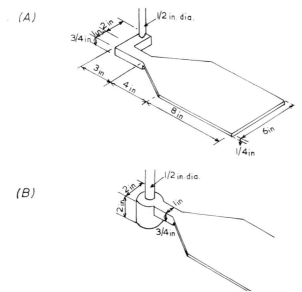

FIG. 1. (a) Right-angled runner gating ratio 1:4:1·3. (b) Well base, gating ratio 1:4:1·3.

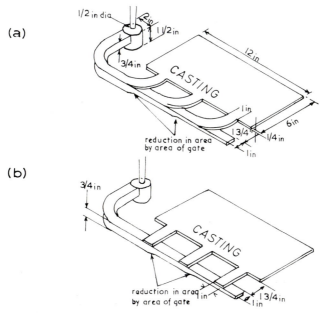

FIG. 2. Running and gating systems with gating ratio of 1:4:4. (a) Design of Eastwood, *et al.* (b) Modification of (a).

Such systems are generally inadequate for aluminium bronze and other dross forming alloys where special traps or skimmers are necessary to prevent dross entering the casting.

In horizontal multi-gate systems mild choking is again recommended in terms of entry and exit cross-sectional areas. The cross-sectional area of the runner bar itself is not so critical and it may be sized to act as a dross trap, to reduce turbulence in the metal and to act as a feeder to the casting. Ratios between 1:2:1 (simple systems) and 1:4:4 (tapered sprue, large runner possibly stepped at gates) have been successfully used.[3]

Slight uphill running is also often extremely useful, smoothing out metal flow in the system, distributing metal evenly and avoiding squirting into the casting. More elaborate arrangements may again have to be made for the dross forming alloys, although with care Eastwoods system in particular is often very satisfactory.

2. Pouring Temperature

The pouring temperature for copper and nickel base castings must be related to composition and melt quality. The most suitable pouring temperature for a particular casting will be that at which solidification first commences, plus the amount of superheat required as determined by practical trials and which will normally be markedly affected by the running system used and the type of mould material. With the long freezing range alloys in particular, casting soundness is dependent on short solidification periods or steep thermal gradients and some compromise may be necessary. Thin section castings may need a relatively high pouring temperature, the amount of superheat being drastically reduced as the section increases. With gun metal (BS.1400 LG2. 85/5/5/5) for instance, the superheat for a thin walled valve could be 200°C reducing to 60°C or thereabouts for a thick section casting.

The short freezing range alloys are not so critically dependent on pouring temperature to achieve casting soundness provided the running/risering system is suitable but even here it is likely that only a relatively narrow range will produce good castings. A small change in pouring temperature can cause a chain reaction of thermal alterations in the metal and the mould, some of which will be disadvantageous to casting soundness.

3. Use of Chills

The use of chills to initiate and enforce a preferred mode of solidification has been referred to. Because of the rapid rate of heat extraction, steep temperature gradients are obtained and solidification is generally directional even with the long freezing range alloys. They are therefore widely used. Chills may be made from steel, cast iron, copper or nickel base alloy or preformed from bonded silicon carbide, graphite or iron shot. Bronze or nickel base chill material is preferred, since surface condition is extremely important to prevent blowing and these materials do not rust or absorb moisture. They cannot, however, be recovered magnetically from the knocked-out sand.

THE FEEDING OF SHORT FREEZING RANGE ALLOYS

Assuming that the metal has been introduced into the mould cavity in a satisfactory manner (this group includes the principal dross forming alloys such as aluminium bronze, high tensile brass, chromium bronze, etc.) and recognizing the importance of the factors previously discussed, the next stage is to consider manner of freezing and the location of risers. Because of the strong directionality of solidification in these alloys, which has been taken into account when designing the running system, it is possible that uniform section castings of, say, less than $\frac{1}{2}$ in. thick may need feed metal only in those regions which are last to solidify, i.e. in the region of the ingates. This can often be provided by a shrinkbob at the ingate thus dispensing with separate risering and reducing fettling.

Thicker or uneven section castings will, however, require risering and it is accepted that these alloys need heavy feeding, 20–25% feed often being allowed.[4]

TABLE 2. *Liquid and Solid Contraction of Copper Alloys*
(after Pilling and Kihlgren)[4]

Alloy and composition	Weight of liquid at freezing point	Total volu-metric shrinkage	Freezing shrinkage	Solid shrinkage	
				Volume	Linear
%	lb/in³ g/cc	%	%	%	%
Aluminium bronze (90 Cu, 10 Al)	0·241 6·67	11·2	4·1	7·1	2·37
Copper (Deoxidized)	0·285 7·89	10·7	3·8	6·9	2·32
Bearing bronze (80 Cu, 10 Sn, 10 Pb)	0·293 8·11	11·2	7·3	3·9	1·29
Gunmetal (85 Cu, 5 Sn, 5 Pb, 5 Zn)	0·287 7·94	10·6	6·3	4·3	1·44
Brass (70 Cu, 27 Zn, 2 Pb, 1 Sn)	0·271 7·50	12·4	6·4	6·0	1·98
High tensile brass (56·7 Cu, 40 Zn, 1·3 Fe, 0·5 Sn, 1·0 Al, 0·5 Mn)	0·261 7·22	11·5	4·6	6·9	2·32
Nickel-silver (65 Cu, 20 Ni, 15 Zn	0·279 7·73	12·1	5·5	6·6	2·19

It will be observed from Table 2, however, that there is no real evidence for this assumption since the total volumetric shrinkage is no greater and, is often less than, other copper base alloys.

Again variable section castings can often be theoretically sectioned and in conjunction with the proposed running method probable direction(s) of solidification worked out. Feeding heads must be located on the last parts of the casting to solidify and be of sufficient mass to ensure that full feed requirements are satisfied. The importance of feeder shape and size are discussed in the chapter on Feeding of Aluminium and reference should be made to it as the basic requirements are very similar. With short freezing range alloys, however, a further technique which can sometimes be used is that of padding. To ensure directionality of solidification it has been described how freezing should begin at parts of the casting remote from the feeders and proceed evenly back to sections which can be externally fed. In order to maintain this overall pattern it is sometimes permissible or even desirable that an intervening thin section should be temporarily thickened, the surplus metal being removed during subsequent fettling. Costs and wastage can thus become excessive and the sounder as-cast skin is removed.

Recently, however, exothermic materials possessing thermal characteristics close to those of the solidifying metal have become available and pads made from these materials can be substituted. The burnt-out pads knock out with the sand and fettling of the padded area is minimal. From Figs. 3 and 4 it will be appreciated that directionality from the web/rim intersection (a hot spot) to the feeder has been preserved by the pad.

EXAMPLES OF THE APPLICATION OF FEEDEX PADDING.

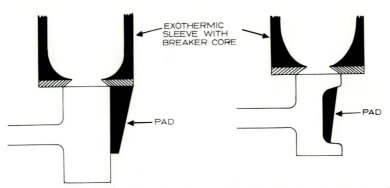

FIG. 3. Examples of application of exothermic padding: a gear blank.

FIG. 4. As Fig. 3—a double-flanged runner wheel.

These diagrams also serve to illustrate that exothermic sleeves can be advantageously used to line riser cavities and their use can usually allow a marked reduction in head size and therefore an increase in yield. The input of heat into the metal is extremely useful in restoring a favourable temperature gradient in the casting/feeder complex. The metal entering a feeder head may well be the first that has entered the mould and therefore the coolest at the completion of pouring.

More information on exothermic heads and insulating sleeves will be found in Chapter 13 on the theory and practice of feeding castings.

The Feeding of Long Freezing Range Alloys

For reasons previously discussed the long freezing range alloys are less amenable to both self and external feeding. The difference between these and the short freezing range group is well illustrated diagrammatically in Fig. 5.

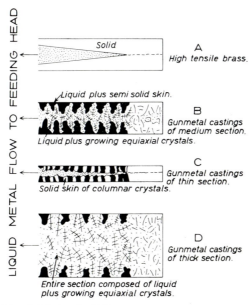

FIG. 5. Mode of solidification of gun metal castings.

The application of feeding heads can only provide movement of feed metal into the casting while the metal is still fully liquid or, during the early part of the freezing range, while the solid/liquid pasty mass retains some mobility. Massive heading may not, therefore, ensure sound castings—it can in fact have the opposite effect and increase porosity in the casting due to its effect on heat distribution in the mould. The most important factor is to provide sufficient metal to allow mass feeding to proceed as far as casting design will permit, assisted by some atmospheric and metallostatic pressure from the riser itself.

FIG. 6. Casting in high tensile brass (HTB2) showing bottom running system and exothermically sleeved risers to improve yield and reduce fettling. A column shoe for Coventry Cathedral, subsequently polished all over.

(By courtesy of Highton & Son Ltd., Dalston, London).

Thin Section (below $\frac{1}{2}$ in.) Sand Castings

Provided that the metal is introduced into the mould cavity with minimum turbulence despite a fairly rapid rate of pouring and that it has the requisite degree of superheat, then as indicated in Fig. 5 (c) the rate of solidification should be fast enough to promote a degree of directional solidification. The tendency will be for a surface layer of columnar crystals to form because of the initial steep thermal gradient, pushing any equi-axial growth or porosity towards the centre of the section. Such centre line porosity will not interfere with light machining or pressure testing of the casting and if the casting has been correctly designed and is of generally even section very many items can be satisfactorily made this way using no feeding other than, perhaps, a small amount from the runner.

Many thousands of small valves, cocks, pipe fittings, etc., are made using these principles but it must be stressed that success is conditional upon obtaining the correct balance between pouring temperature and pouring speed (running and gating) so that the mould material is not overheated. If overheating does occur then a mixed crystal or coarse equi-axial growth will take place giving dispersed instead of centre line shrinkage porosity and leaving the casting susceptible to leaking on pressure testing particularly from areas subsequently machined.

It has also been proposed (British Non-Ferrous Metals Research Association) that volume contraction during solidification in thin section castings that cannot be adequately fed can be partially offset by promoting a mild metal/mould reaction. Phosphorus is normally used as a deoxidant before pouring and any excess beyond that to ensure complete deoxidation, can react with moisture from a green sand mould. The resulting porosity, depending on the phosphorus content, can be in the form of fine, discrete, dispersed pin holes which in total will take up a fair volume offsetting normal contraction and which will not provide a route for leaking on pressure testing. Obviously the amount of metal/mould reaction will be dependent on the amount of residual excess phosphorus and on the rate of solidification (section thickness). Residual phosphorus contents of the order of 0·03–0·08% are suggested, the higher figure being used where the solidification rate is high, i.e. thin sections.

Apart from some runner bar feed to satisfy heated areas immediately adjacent to ingates it will be noticed that in general, thin sectioned castings in long freezing range alloys can be cast satisfactorily without conventional risering provided certain requirements are met.

Thicker Section (above ½ in.) Sand Castings

Reference to Fig. 5 (b) and (d) indicates that in the thicker sectioned castings, due to slower solidification rates, the grain structure is almost completely equi-axial and solidification of the surface skin is never greatly in advance of that at the centre of the casting section. The relatively large amount of heat introduced with the metal will obliterate any chilling effect of the sand face and the casting will solidify slowly. The pasty mode of solidification will ensure that supply of feed metal from feeding heads will be cut off at an early stage and such castings will, therefore, tend to be affected by a dispersed, interdendritic type of unfed shrinkage porosity, often in a marked layer formation. Fracturing of an affected area will show gross discolouration ranging from the dull, matt silky appearance of sound metal through yellow, grey and brown in the unsound area.

In these heavier section castings the only practical ways of ensuring an acceptable level of soundness in the casting are to pay particular attention to heat distribution in the mould by means of properly designed multiple gating and risering, close control of pouring temperature and careful location of correctly sized chills. With most commercial castings careful mould design in respect of relationship of ingates, chills and feeders can ensure some semblance of directionality during solidification and should always be attempted. With regard to feeding heads, alloys in this group normally exhibit between 5 and 7·5% volume contraction during solidification although this figure will vary to some extent by mould deformation and gas content of the metal (metal degassed prior to pouring may show an increase in liquid shrinkage if extra feed metal is not provided). For reasons already discussed it is unlikely that all contraction in the casting can be fed from the feeders and this might suggest that the feeders themselves can be relatively small or few in number. It should be remembered, however, that some liquid contraction must be compensated for in addition to possible feed in the pasty range and that metallostatic and atmospheric pressures can be utilized to keep the casting "pressurized". For these reasons normal heading involving around 20% of casting weight, as described for the short freezing range alloys, should be calculated and applied. It should be mentioned here that the technique of padding as discussed for short freezing range alloys does not apply to long freezing ranges as far as experience and theoretical considerations indicate at the moment.

EFFECT OF COMPOSITION

Where heavy section castings are subjected to pressure testing some benefit is often obtained from the presence of 2–3% of lead. The function of lead (which is insoluble in copper) is partly to increase the supply of low melting point liquid during the later stages of solidification but also to seal any continuous channelling formed by interconnected shrinkage porosity. Nickel (0·5%–1%) has also been shown to be of assistance in promoting soundness in thick sections

TABLE 3. *Effect of Composition on the Freezing Range of Copper-base Alloys*

Composition, %							Start of freezing °C	End of freezing °C	Freezing range °C
Cu	Sn	Zn	Pb	Ni	Al	Others			
90·0	10·0	—	—	—	—	—	1010	835	175
92·5	7·5	—	—	—	—	—	1030	880	150
95·0	5·0	—	—	—	—	—	1050	915	135
97·5	2·5	—	—	—	—	—	1070	990	80
91·9	8·0	—	—	—	—	—	1022	879	143
90·9	7·9	—	—	1·0	—	—	1027	888	139
90·2	7·8	—	—	2·1	—	—	1032	893	139
88·7	8·2	—	—	3·0	—	—	1038	896	142
79·0	9·3	0·3	10·2[1]	0·4	—	0·01 P	929	762	167
88·0	5·7	4·4	1·4[2]	0·6	—	0·01 P	988	826	162
85·0	4·4	4·6	5·1[3]	0·7	—	0·01 P	1010	853	157
75·8	2·7	14·8	5·6[4]	0·7	—	—	953	831	122
63·8	3·8	7·1	4·2[5]	20·2	—	—	1143	1109	34
95·0	—	5·0	—	—	—	—	1065	1060	5
90·0	—	10·0	—	—	—	—	1040	1030	10
57·8	0·3	39·3	0·2	0·1	1·1	1·0 Fe 0·5 Mn	880	862	18
90·0	—	—	—	—	10·0	—	1040	1035	5

[1] Lead phase freezes at 314°C. [2] Freezing of lead phase could not be detected. [3] Lead phase freezes at 316°C. [4] Lead phase freezes at 319°C. [5] Lead phase freezes at 309°C.

by giving a more uniform distribution of lead (very useful with the high lead bearing alloys) and shrinkage porosity and improving feeding characteristics generally.

Table 3 illustrates the effect of some variations of alloy composition on the freezing range and it should be noted, as discussed above, that where lead is present its freezing point has not been taken into account. Otherwise the freezing range would have been of the order of 700°C.[5]

At this juncture it is relevant to draw attention to the effect of contamination of the melt by aluminium or silicon—the latter specially in relation to the high lead content bronzes. Both these metals, if present in small amounts—say between 0·01% and 0·2%, will, because of their affinity for oxygen, form discontinuous oxide films that can easily become entrained in the molten metal. Their density is sufficiently close to that of the metal that they do not separate into the dross or slag easily. If this contaminated metal is then cast, these thin but extremely strong films will have a direct blocking effect on mass feeding. This is quite apart from the blemishes that will show on the casting surface and the route provided for leaking on subsequent pressure testing. In these respects aluminium particularly, as the more common, must be rigorously excluded from any alloy in which it is not a deliberate constituent and a level of 0·01% should be regarded as an absolute maximum for sand castings. Apparent fluidity of a contaminated alloy is also, for very obvious reasons, severely reduced.

Silicon will act similarly and a further danger here concerns the high lead content (say above 10%) bronzes. If present, silicon can alter the nature of the oxide film at the metal/mould interface, a very fluid lead silicate is formed and penetration of the mould by the metal and also a strong metal/mould reaction are distinct possibilities. With heavy section castings having an extended liquid period in the mould, as little as 0·005% silicon is sufficient to promote metal/mould reaction in leaded alloys.

With these exceptions the majority of the long freezing range copper base alloys in common use are not unduly susceptible to contamination in so far as the general feed requirements are concerned.

TOP POURING

For some heavy (and occasionally relatively light) section castings of regular geometric shape, e.g. large bushes or half bearings cast on end, and where the alloy is not of a dross forming nature, top pouring via pencil gating is often the best method. For reasons already discussed, every endeavour should be made to ensure some directionality of solidification even with the long freezing range group and conventional bottom running with top risering will usually provide a filled mould with the desirable temperature gradients inverted, i.e. cold metal in the risers and the hotter metal just inside the ingates. Particularly with bottom run large bushes, etc., any turbulence in the lengthy downsprue—a distinct probability—will inject air into the casting which may be unable to escape through the solidifying metal toward the top of the mould. In these cases top pouring has definite advantages since temperature gradients can be steep and maintained throughout the pour—in fact the casting can be largely self-feeding if the system has been properly designed and the correct pouring temperature is used. For bushes, etc., the pattern length should be extended by about 1 in. to allow machining clearance of any oxides/dross that might form in the mould cavity. A full circle runner trough should be located about 1 in. above the mould (this is usually most conveniently made in silicate bonded sand) with pencil ingates at intervals connecting the trough into the mould cavity. Depending on the bush diameter between 4 and 12 pencil gates will be necessary and these should be sized by calculation to give a rate of mould fill of approximately $1\frac{1}{2}$ in. per second. Pouring temperatures should be on the high side relative to the casting section involved.

REFERENCES

1. MACKEN, P. J., *Foundry Trade Journal*, **119**, 361–366, table 1 (1965).
2. GLICK, W. W., JACKSON, R. S., and RUDDLE, R. W., *The British Foundryman*, **52**, 91–107 (1959).
3. Ibid., p. 95.
4. PILLING, N. B., and KIHLGREN, T. E., *Trans. of the AFS*, **40**, 201–216 (1932).
5. HUDSON, F., and HALL, J. R., *The British Foundryman*, **56**, 261–277 (1963).

CHAPTER 17

THE FEEDING OF LIGHT ALLOY CASTINGS

T. A. BURNS

The light alloys normally include the whole range of aluminium base alloys together with the magnesium base range. Neglecting the fact that special precautions with respect to exposure to atmosphere have to be taken with the magnesium base range both can, for the purposes of considering feeding, be conveniently dealt with together.

Although the magnesium range of alloys normally require pouring temperatures of the order of 100°C or more above those for aluminium both groups can be considered as falling within a medium pouring temperature range and the requirements for aluminium, which is the more common group and will be discussed in detail, will also apply to magnesium.

FREEZING RANGE

Both short and long freezing range alloys occur within the group. The majority of commercial castings are made from the aluminium/silicon/copper alloys when the freezing ranges are variable but are, in general, long.

Only pure aluminium, which is little used except for slush casting and some pressure diecastings, together with the eutectic aluminium/silicon alloys possess a short freezing range. Theoretically the extent of the freezing range can be an important consideration since it will affect the way a casting solidifies. It may simplify some problems but enhance others.

When a pure metal or short freezing range alloy is cast against a mould there is a strong initial chilling effect from the mould that will give a thin layer of fine equi-axially orientated crystals but thereafter solidification will proceed against the thermal gradient (heat is being lost at a more or less constant rate to and through the mould material) by means of growth of columnar crystals. In other words a freezing front will advance in fairly regular fashion from the mould wall and virtually parallel to it. Unless adequate feeding arrangements are made for an alloy freezing in this manner gross unfed shrinkage defects are likely to result at the centre of the section or at hot spots. On the other hand a long freezing range alloy, after the initial skin chilling effect, will form random crystal nucleii dispersed throughout the body of the casting, these "seeds" then growing as the mass slowly cools, at a rate proportional to the rate of heat extraction, until they touch each other and form a coherent network or skeleton, further growth being restricted by availability of liquid feed metal.

$$D = k\sqrt{t} - c$$

where D = thickness of solidified metal,

   k = a constant depending on casting size and mould conductivity,

   c = a function of degree of superheat.

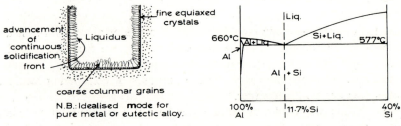

FIG. 1. Solidification by skin-formation (narrow freezing range alloys).

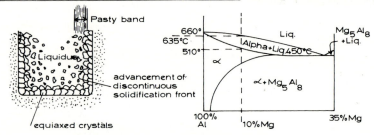

FIG. 2. Solidification of wide freezing range alloy.

At any intermediate stage, therefore, these alloys will consist of a mushy mass of solidified crystals surrounded by still liquid metal. This type of solidification pattern is most likely to give unfed shrinkage defects in the form of small dispersed interdendritic cavitation with, however, some tendency again to concentrate at heat centres. The two forms of solidification are neatly summed up by the descriptions "skin formers" and "pasty solidifiers".

SOLIDIFICATION PATTERN

Bearing the foregoing in mind it will be appreciated that it should only be necessary to ensure that metal in those parts of the mould remote from the feeders should begin to solidify first and that solidification should then proceed continuously back to the riser to ensure completely sound castings. This will apply equally to long and short freezing range alloys provided the thermal gradient is sufficiently steep. The term "directional solidification" is used to describe this pattern and provided certain requirements are met then a sound casting must result. These requirements are principally that the desired axis of solidification must proceed at a faster rate than a lateral advance from side walls otherwise bridging will occur to hinder the flow of feeding metal. It also assumes that in a long freezing range alloy the "pasty" band between fully solidified and fully liquid metal is relatively narrow, which condition is seldom approached in a practical casting due to interference from casting shape, moulding, running and feeding factors, etc. It is probable that under practical conditions of casting light alloys and particularly where the rate of solidification is relatively slow, e.g. in sand and gravity (permanent mould) die castings almost the whole mass of solidifying metal consists of a mixture of solid and liquid irrespective of the classification of the alloy in terms of freezing range.

This is because of the necessity for bottom gating and consequent temperature inversion together with the effect of turbulence. For the purposes of running and risering a casting therefore it is not usual to take undue account of the characteristics of the particular alloy but to apply certain general considerations covering the light alloy group as a whole. Nevertheless, these will be aimed at satisfying the needs of good general directional solidification and may be summarized as follows:

1. Correct gating and feeding.
2. Correct pouring temperature.
3. Use of chills.
4. Nucleation.
5. Correct mould dressing (mainly gravity diecasting).

1. Correct Gating and Feeding

(a) Sand Castings

Because of their ready oxide-forming properties the light alloys must be cast via bottom gates to eliminate or control oxide formation within the mould cavity and possible subsequent entrainment within the casting proper. Since also, to take advantage of gravity and for convenience, risers are situated on the top of a casting it is apparent that the first metal into the mould, and therefore the coolest, is most likely to be displaced into the feeder heads. An inverted undesirable temperature gradient could therefore be set up that would quite likely persist throughout solidification unless the risers were particularly massive. Adverse temperature differences of the order of 20°C between the top and bottom of an uphill run sand cast test bar 1 in. diameter × 11 in. long in aluminium and magnesium alloys have been observed.

With most light alloy sand castings moulding and/or coring considerations will dictate which way up the casting will be made and it follows that sites for ingates and risers will probably be restricted to "obvious" areas to conform to the general scheme of bottom gating and top feeding. Some correction of the consequent unfavourable temperature gradient can often be obtained by pouring the casting through the running system until the metal just enters the base of the risers and then topping up with hotter metal from the crucible direct into the risers. This method is often useful with medium or even some heavy section castings but is unlikely to succeed with light castings since the rate of solidification is generally too rapid.

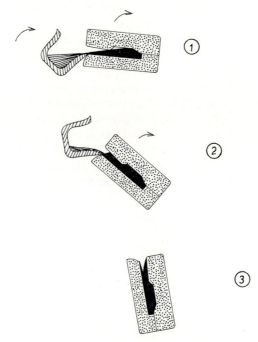

FIG. 3. Semi-Durville casting process.

Some complicated castings may even require a form of semi-Durville type running system in which a pronounced tilt is given to the mould at the commencement of pouring, the mould being returned to horizontal as pouring proceeds. In addition to its principal aim of reducing turbulence the system can also assist feeding in that the first metal in is relatively localized and return to horizontal can, depending of course on casting shape, ensure that this cool metal is displaced gently to a remote mould region leaving the hotter follow-up metal in a more favourable position from the temperature gradient point of view. If this mould tilting method is used, the ingate/downsprue should always be at the lowest point in the tilted mould. The method can also be used in conjunction with subsequent topping up of risers as discussed previously.

It has sometimes been proposed that a secondary gate or bridge from the downsprue into the base of the riser, if conveniently located, can ensure hot metal into the riser and so provide available feed potential. The principal objections to this system are essentially that the flow of molten metal down the sprue, once established, is difficult to change and the bridge seldom performs as it is supposed to, the metal continuing to enter the casting via the normal gate. Additionally the bridge is idle during the early part of the pour, interferes with the geometry of the running system and usually only provides an ideal configuration for the aspiration of air into the metal stream by virtue of a venturi effect. Such bridging may also introduce mechanical stressing of the casting due to differing rates of solidification, so much so that some casting may hot tear or "spring" during fettling. The objections are considered to outweigh the advantages by a wide margin and any proposed use of the method should be carefully considered as a more acceptable and more efficient system can most probably be substituted.

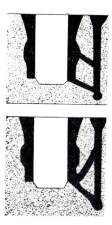

FIG. 4. Use of branch gate or bridge.

So far running and risering have been considered only as two distinct systems performing separate functions but this need not necessarily always be so. In addition to introducing metal in as quiet and clean a condition as possible to the mould cavity the running system, suitably sized and laid out, can also perform a useful function as a source of feed metal. To prevent pressure build up and avoid squirting into the mould, also to exercise a skimming effect, running systems for light alloys are often sized around 1:2:1 or thereabouts in terms of ratios of effective cross-sectional areas of downsprue/runner bar/ ingate(s). Control of the metal velocity and volume depends only on the downsprue and ingate(s) cross-section and it is possible to increase cross-sectional area and mass of the runner bar without detriment to the mechanics of the system as a means of controlling metal flow. The law of continuity states that in a system of filled channels the flow rate is the same at all points and where the channel cross-section is smaller the velocity will be correspondingly higher whereas where the cross-section is larger the velocity will have to be correspondingly lower to maintain the system filled. A suitably enlarged runner bar

R

can therefore function as a reservoir of liquid feed metal and since the running system as a whole must pass the entire volume of metal to make the casting it will pick up sufficient superheat to enable it to function effectively as a feeder. The weak point is of course the ingate area since despite the preheat due to the passage of metal it may still freeze off prematurely. The system as outlined is often extremely useful and further variations are possible extending to the use of shrinkbobs or even side risers or full box height risers located on the running system rather than on the casting proper. Very seldom, however, can the method be used as the sole means of feeding a light alloy casting and it is usually used supplementary to conventional risering. It should be remembered that by definition the light alloys have a low specific gravity and the displacement factor with respect to mould air is therefore low. Top risers provide an escape route for displaced air and so help prevent air locking. This is a further consideration that should be borne in mind when methoding a casting with respect to siting of risers and it should not be confused with mis-running due to low pouring temperatures.

The difficulties, both practical and theoretical, of ensuring favourable thermal gradients in a solidifying light alloy are, therefore, well recognized bearing in mind that the liquid-to-solid volume contraction, which will require feeding, is around 7% theoretical but, as has been indicated, about 3 times this figure is required practically, it is understandable that relatively heavy risers, etc., must be used. The average aluminium alloy sand casting will carry virtually its own weight of risers and running system, i.e. casting yield will be around 50% assuming an all green sand mould without additional feeding aids (which will be discussed later).

Riser Shape

The function of a riser is to retain a supply of available feed metal for as long as there is a demand from the casting. It is, therefore, important that it should lose its heat at as low a rate as possible and thermodynamic considerations indicate a sphere as having the lowest surface/volume ratio and hence the ability to retain heat longest. Except possibly in the case of disappearing" expanded polystyrene or polyurethane premoulded forms a spheroidal riser is a practical impossibility and the next thermally most efficient shape—a cylinder— should be used. Again, however, efficiency will decrease as the height : diameter ratio increases and as far as possible this should be kept fairly close to 3:1, i.e. a 2 in. diameter riser will be 6 in. high. The area of contact with the casting is also important since this is the route for feed metal and necking down must be restricted, to prevent premature freezing, only to allow access for fettling tools.

Sometimes the casting shape cannot be conveniently fed from cylindrical risers and kidney shapes or wedges must be used. The isotherms in an open riser, assuming steady-state conditions, indicate that the heat centre, i.e. the last part to solidify, will be at or slightly above the geometric centre due to rapid heat loss from the surface to atmosphere. The geometric centre of a

V section wedge will be nearer the open end of the V and thus relatively distant from the casting. Such a riser is likely to bridge across in the early stages of solidification and its full potential will not be realized. A much better section from this point of view is a chunky U that brings the heat centre down much nearer the casting.

Heat loss to atmosphere from an open riser can be an important factor affecting temperature gradients in a cast mould; obviously its effect will be greater the larger the surface area of the riser. The major proportion of a casting/ riser complex will be in contact with the sand and while the mould will have a strong initial chilling effect—a cold damp surface—once the immediate surface layers have heated up and moisture has been driven deeper into the sand, further temperature loss will be at a slow steady rate. On the other hand both radiation and convection will continuously assist rapid loss from a riser surface. Full riser potential will, therefore, be better obtained if a layer of insulating, or better still, a suitably sensitized exothermic material, is applied immediately on completion of the pour. An insulating layer will equate heat loss all round the riser surface but some exothermic materials, if correctly chosen, release sufficient heat both to have a powerful self-heating effect and also increase riser temperature. Burnt out exothermic materials should subsequently provide good insulation and in both cases, exothermics and insulators, further heat loss to atmosphere is severely restricted and the riser is that much more effective.

Sleeved Risers—Insulating and Exothermic Sleeves

Pursuing the above idea a stage further there would obviously be considerable advantage in lining the whole riser with heat conserving materials. While heat losses from the riser to the sand of the mould are not excessive any improvement would be welcome since it would increase efficiency and yield and also provide a safety factor to compensate for other minor variations that might affect casting soundness.

Materials having better insulating values than green sand and which have been used for fabricating riser sleeves are numerous and include resin bonded sand, asbestos paper and board, plaster of Paris, oil bonded sand, etc. For a long period the most popular insulating sleeve material was an aerated or foamed form of plaster of Paris. A particular refractory type of plaster is mixed with water in the usual way and a small quantity of surface active agent then added before transferring to a high speed mixer whose principal purpose is to entrap thousands of small air bubbles in the slurry. These bubbles are stable and are trapped in the solidified plaster thus further improving the insulating properties. The external surfaces of such sleeves are continuous and show no evidence of the internal honeycomb effect. After proper drying the best insulating characteristics of such sleeves are markedly superior to green sand. Recently even further improvements have been obtained from the use of a bonded fibre type of sleeve where in addition to improved heat conservation other advantages include resistance to damping back from the green sand mould in which they are rammed and a toughness and resilience notably absent in the extremely brittle plaster sleeves. Better technical performance, better handling characteristics and a faster production rate (plaster sleeve production is extremely slow due to mixing and pouring requirements and core box limitations) have meant these felt-like fibre sleeves have practically superseded all other forms of insulation and use has been extended to launders, furnace plugs, semi-continuous cast die heads, etc., where their particular properties have proved attractive.

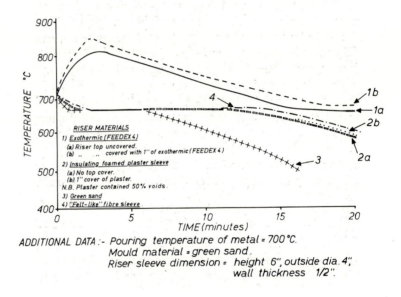

FIG. 5. Cooling curves of commercially pure aluminium showing the effect of riser material on solidification of the metal.

A further gain in thermal efficiency of a riser, as can be seen from the cooling curve diagram, can be obtained from an exothermic sleeve. A range of exothermic materials is available to suit all commonly cast metals and riser diameters and from this range can be selected a type suitable for the light alloys. Generally speaking, since they need to be rapidly sensitive to medium range metal temperatures the sleeve sensitivity will fall into the high sensitivity grouping, i.e. they will fire quickly when the molten metal touches them, burn hot and provide good subsequent insulation. Internal temperatures of a burning high sensitivity exothermic material can go as high as 1700–1800°C so even allowing for an approximation that some of the heat will escape to the mould material, a considerable input to the riser itself is possible. This dramatic restoration of favourable temperature gradient often means that appreciable economy in head size, and therefore improvement in yield, is possible.

In both cases, of course, i.e. insulating and exothermic sleeved risers, the full effect must be maintained by application of a layer, of approximately the same thickness as the riser wall, of a similar material on to the riser/air interface. In the case of insulating sleeves it is permissible and advantageous to apply an exothermic carbonaceous type powder as the anti-piping compound or hot topping material.

Fig. 6. Casting in "Alufont 3". A railway carriage bearing plate with sand heads and exothermic sleeved risers showing the considerable saving of metal.

Riser Size

As previously discussed a continuous struggle exists in light alloy founding between the necessity for bottom running and the high feed requirements of a casting against a usually unfavourable temperature gradient.

For plain, unsleeved risers in green sand this means that the effective diameter (in the case of cylindrical risers) must be appreciably greater than the section it is intended to feed.

The theoretical aspects of riser dimensioning in respect of a casting/riser system in thermal equilibrium are dealt with elsewhere. Almost no fundamental work on heat flow or even thermal characteristics of the light alloys at or near the solidification point is available to put green sand riser dimensioning on a scientific basis and as a result the subject has been treated essentially to an empirical approach. In practice this usually means that a large safety factor is allowed and any subsequent adjustments reluctantly made, if the casting is successful, on a "hit or miss" basis. This view is understandable in the light of the complexity of the problem where the interplay of such factors as casting shape, pouring temperature, metal/sand mass, absence or presence of coring and metal/core relationship, mould material, composition of the alloy and its freezing range, etc., can all affect results to varying degrees.

Most castings can be visually sectioned into regular geometric shapes and the heat centres (thicker sections that will need feeding or chilling) will suggest themselves. Arrangements can then be made to supply feed metal to these areas from suitably placed and sized risers or from a modified running system as previously discussed. To isolated bosses and thicker sections it is usual to allow 100% feed metal, i.e. the riser will be the same mass as the visually isolated section to be fed, and riser/casting contact area must be carefully considered to prevent premature freezing and bridging over.

Surface/volume ratios and probable temperature gradients indicate that the riser should be somewhat thicker in section than the section to be fed and factors of 1·3 to 1·5 times the casting thickness are usual. Similar calculations must be applied to any blind risering (side risers that do not reach the top of the mould—unusual in light alloy practice) or runner bar feeding.

The general feed requirements of a given casting can, therefore, be built up into a composite picture and the optimum running and risering systems to satisfy these demands worked out.

(b) Gravity Die (Permanent Mould) Castings

Although the overall rate of solidification in a metal mould is very much faster than in sand, similar arrangements in respect of ensuring a preferred solidification pattern must be made to obtain good castings consistently. Because of the speed of solidification, temperature gradients are steep and feed requirements short and sharp. It is, therefore, usual to apply very generous risering (especially runner/risering) to gravity die castings, the lower yield being accepted in view of the increased production rate possible with the process combined with the quicker return of clean metal for remelting.

Similar considerations with regard to bottom running, risering, etc., will apply as described for sand castings and a proposed casting may be visually assessed in the same way to determine location of gate(s), riser(s), etc. Here, however, more regard must be paid to die construction since it may not be possible to incorporate all the desired features in a rigid mould. Gravity dies must obviously be kept as simple as possible with the minimum of coring, loose pieces, draw backs, etc. Die operating temperature is obviously an important factor in promoting or opposing a good solidification pattern and running and risering is often overgenerous in terms of actual requirements in order to maintain die temperature alone. For further details on these points and on some of the expedients and coating arts commonly used to produce acceptable castings see under the appropriate chapter.

2. Correct Pouring Temperature

To an important extent the pouring temperature of an alloy can have an effect on grain size, shape and distribution in the solidifying casting and this factor can sometimes be used to advantage, e.g. with some copper base alloys in ensuring a columnar pressure resistant skin. With aluminium alloys, however, because the solidification rate, whatever the method of moulding, is quite rapid, the effect is not so useful and in general only a pouring temperature just sufficient to ensure a completely run casting is employed. Long solidification periods in the mould must be avoided because of possible gas pick up, high volume contraction and the normal pasty method of crystallization. The correct pouring temperature must therefore be decided upon—and adhered to—in the light of casting section, alloy used and running method employed. High pouring temperatures—say above 760°C/780°C—and consequent longer time at high temperature in the furnace must be avoided from the points of view of enhanced possibility of gas absorption (see chapter on Aluminium) and grain coarsening due to aggregation of nucleii. Where a large surface area-thin section casting is involved it is preferred that, rather than increase the pouring temperature to run the thin sections, the running/ingating system be enlarged and the casting made more quickly using multiple ingates if necessary.

3. Use of Chills

The desirability of directional, controlled solidification has been stressed in earlier sections. Chills in the form of metallic inserts in the mould (sand castings) obviously remove heat from the casting at a more rapid rate, at least during the early part of solidification, than the relatively insulating sand and they can, therefore, often be used to advantage to propagate a preferred gradient. Localized surface sinks at isolated bosses and section junctions can often be overcome with suitably sized and located chills, since they will have the effect of evening out small section variations. Chills can be of steel, cast iron, copper or even aluminium. They should be at least as thick as the section requiring chilling and massive individual contact areas with the casting should be avoided: instead use multiple smaller or well drilled chills to allow escape of gas. Chills should always be kept in good condition and coated with a suitable dressing to reduce damping back and to decrease the effect of any hard initial chilling that could leave a shut on the casting. Blowing from chills is usually the result of sand moisture condensation on the chill face so they should be preheated and the moulds closed and cast without delay. A proper dressing will also prevent welding—especially important in the case of copper or aluminium chills. Running/ingating systems should be arranged so that incoming metal does not continuously flow over any chills thus strongly heating them and destroying the effect required.

4. Nucleation

The chapter on Aluminium (Chapter 7) has described how aluminium and its alloys can be grain refined by means of the addition, in salts or metallic hardener form, of suitable foreign nucleii of similar lattice structure. It follows that during solidification, particularly with the wide freezing range alloys, better mass feeding will be obtained if the solids content of the pasty mass is in a finely divided, i.e. nucleated, form.

Mechanical bridging will be less likely to occur and effective mass movement and hence better feeding, more likely to persist to a later stage in the solidification process.

Provided, therefore, that the methoding of a particular casting is correct, nucleation of the metal can provide some assistance in the feeding process. If mould design and layout is unsatisfactory, however, the reverse may apply and grain refinement will succeed only in collecting a dispersed shrinkage distribution into a gross cavitation.

5. Correct Mould Dressing

In a sand mould and similarly in a pressure die casting, the nature and applied thickness of any dressing can have no significant effect on the mode or speed of solidification. On a gravity die, on the other hand, the die dressing exists as a layer of (usually) insulating material between two masses of metal. Its characteristics and thickness can therefore play an important part in governing heat flow from the cast metal to the die. It follows that the type of die dressing and the way it is applied can have a controlling effect on the mechanism and direction of solidification and therefore on the quality of the casting. These aspects are discussed in greater detail in Chapter 12 on gravity die casting.

INDEX

R*

Metal
  penetration, after binder burn-out 361,
    366, 367
  powder interface, coating ingot moulds,
    for 93, 94
Metallostatic pressure, coating to withstand
  363
Metals, die casting for 389
Methanol, carrier for coatings, properties of
  363
Methyl celluloses and additive 378
Methylene chloride, carrier for coatings,
  properties of 364
Mica, coatings, in 362
Microporosity in long freezing range alloys
  432
Mineral blacking, coatings, in 370
Mineral oils, unsaturated 353
Misch metal
  high lead bearing bronze, in 232
  introduction of cerium, for 195
Miscibility gap in high lead bearing bronze
  232
Mixer
  batch type, bonding sands 341
  continuous, bonding sands 341
  desulphurization, in 5
  mill, bonding sands 340
  pig iron 4
Mode of freezing
  long freezing range alloys 427
  short freezing range alloys 418
Modification of eutectic aluminium silicon
  alloys 259, 281
  with strontium 283
Moisture in ingot moulds 91
Molasses
  binder for coatings 366
  cement bonded sand, effect on 349
Molochite in coatings 369
Molybdenum
  additions to cast iron 181
  P–C–Fe, quaternary eutectic 181
Monel metal 223
Montmorillonite 375
Mould
  additives, steel ingots 94, 111
  atmosphere, S.G. iron 204
  coatings, steel ingots 111
  condition of, steel ingots 110, 111
  and core paint, inhibiting 218
  design, steel ingots 112
  dilation, influence on feeding castings 468
  dressings for steel ingots 88, 91, 93
  expansion, steel ingots 89
  extender, continuous casting of steel 152
  flux, continuous casting of steel 143
  ingot design 12
  inoculation for iron castings 192
  temperature, steel ingots 112

Mouldable exothermic feeder material 477
Mouldable exothermics 450, 454
Moulding
  cement 322
  high pressure 320
  skin dried 321
Moulding and core sands (see also Core
  and moulding sands) 313
  chromite 315
  compression strength comparisons 319
  definition of 313
  olivine 315
  processes 317, 318
  properties required 313, 314
  zircon silicate 315
Moulds
  block, continuous casting of steel 140
  cooling
    primary, continuous casting of steel
     140
    plates, continuous casting of steel 152
    water, chemical "balance", continuous
     casting of steel 140
    water corrosion, continuous casting of
     steel 140
    water corrosion inhibitor, continuous
     casting of steel 140
  "H" section, continuous casting of steel
    152
  influence of, on steel quality 110, 111,
    112, 113
  lubrication 143
  multiple section B.I.S.R.A. continuous
    casting of steel 152
  plate, continuous casting of steel 140
  protection, continuous casting of steel
    140
  reciprocation, continuous casting of steel
    140
  steel ingots, for 89
  solidification rate in continuous casting
    of steel 140
  tube, continuous casting of steel 140

NaCl–KCl equilibrium diagram 253
NaCl–KCl–NaF equilibrium diagram 254,
  260
NaCl–NaF equilibrium diagram 260
Nailing feeder heads, of 53–4
Natural bonded sand, typical mixture, grey
  iron 320
Natural resin as binder for coatings 367
"Navac" vacuum melted sodium for
  modification of aluminium silicon
  alloys 261
Nickel
  additions to cupola charges 178
  additions to ladle 181
  aluminium bronze, in 221